Oliver Wendell Holmes Jr., Pragmatism and Neuroscience

Jay Schulkin

Oliver Wendell Holmes Jr., Pragmatism and Neuroscience

palgrave
macmillan

Jay Schulkin
Department of Neuroscience
Georgetown University
Washington, DC, USA

ISBN 978-3-030-23099-9   ISBN 978-3-030-23100-2   (eBook)
https://doi.org/10.1007/978-3-030-23100-2

Cover illustration: Architectural Images/Stockimo/Alamy Stock Photo

This Palgrave Macmillan imprint is published by the registered company Springer Nature Switzerland AG
The registered company address is: Gewerbestrasse 11, 6330 Cham, Switzerland

*For Tom Grey, noted scholar of Holmes and of pragmatism,
and valued friend.*

*Danielle Schulkin with all hopes for the future in the law and for all that
matters; and her patient and thoughtful help with this manuscript.*

*And for my original teacher of the pragmatic tradition
and friend for nearly half a century, Robert Neville.*

Preface

I thank my colleagues and friends. In particular, I want to thank Brian Butler, Andrew Hartzell, Mark Johnson, Fred Kellogg, Nora Peck, and David Weissman.

The lifeblood of the mind resides in you, in the interactions of meaning. Many thanks to Tom Grey and Hank Greely at the Stanford Law School where I found a patch of intellectual affinity and friendship.

Once again, I thank Tibor Solymosi for his help on the manuscript and more importantly for his philosophical friendship.

Washington, USA   Jay Schulkin

Contents

List of Tables

1

Introduction

We are at a point in time so replete with neuroscientific breakthroughs that one might say we live in an age of neuroscience. This book seeks to provide perspective in the way neuroscience is considered by placing it in the context of the law and more specifically, a setting in which pragmatism, the law and Oliver Wendell Holmes Jr. interface.

Oliver Wendell Holmes Jr., a well-known Supreme Court Jurist, understood, espoused and breathed the intellectual oxygen of his age. He embraced the continuous flow of science into the larger body politic and cultural evolution, including the law. Although today Holmes is considered to be part of the American school of philosophical pragmatism along with well-known American pragmatists C. S. Peirce and William James, his relationship to pragmatism is not often clear. He revered the new sciences of biology, anthropology, and statistical inference, and he repeatedly commented on the integration of these sciences with judicial reasoning.

Even though Holmes's link to pragmatism continues to be a subject for debate among many pragmatists, I suggest that he is very much linked. And that a fruitful link is from the "Common Law" of Holmes to the "Common Faith" of Dewey for a naturalized epistemic engine

© The Author(s) 2019
J. Schulkin, *Oliver Wendell Holmes Jr., Pragmatism and Neuroscience*,
https://doi.org/10.1007/978-3-030-23100-2_1

ready to engage and incorporate a neuroscience perspective within the law and the larger culture. The importance of neuroscience and the limitations of the law is grounded in this pragmatic excursion. The pragmatic view of the law is an evolving one, not frozen in geometrical first principles, something that Holmes well understood and contributed and warned against, but an experimentalist logic of discovery and argument and correction. While the path is indirect, high ideals for what counts in human life is something Dewey provided insight into. Neuroscience anchored to the law needs to capture this pragmatist sensibility. Holmes naturalistic orientation to the law, while historic is an important road in this conversation.

But Why Holmes? Who Was He?

He was born in 1841 and died in 1935. During his lifetime, he witnessed an explosion in science and law and a changing relationship between science and culture. For a period of time, Holmes was actually at the heart of the law as an important contributor to what became the most original part of American legal reasoning, i.e., legal pragmatism (e.g., Grey 1989; Posner 1992). "Laws greatest pragmatist remains Holmes", one noted legal scholar would assert (Posner 1995, p. 13).

After the war, Holmes married Fanny Dixwell, the granddaughter of an esteemed Massachusetts lawyer who was part of the moderate wing of religion and the larger sense of epistemic pursuits, religious tolerance and cultural evolution. While they had no children, the couple were together until the end of her life (she died in 1929, and he in 1935). Holmes's correspondence revealed that while his life was with others, he would say of his wife that "for sixty years she made life poetry for me" (letter to Frederick and Lady Pollock, May 24, 1929). Oliver Wendell Holmes Jr. led a very long and rich life.

Nonetheless, Holmes, known for his flirtatiousness, carried on with the Lady Castletown over a thirty-year period, to whom he expressed a wide range of emotions (1896–1929, White 1993; Baker 1991). Indeed, in one of his letters to Lady Castletown (August 19, 1897), he would write, "putting to death the inadequate to get the world to limit procreation…" He would say, "existing society is founded on the death of men." Eugenics, unwisely, was real for Holmes.

Holmes was curious and intellectual, but he was also elitist, with little real sympathy for the downtrodden and the diverse minorities swelling the American population (Hollinger 2005). One modern American historian commented that Holmes was "cynical-anti-democratic" (Kloppenberg 2010, p. 19).

Importantly, Oliver Wendell Holmes Sr., a physician, Dean of the medical school at Harvard, and a writer, penned a biography of Ralph Waldo Emerson that displayed a life of thought, scholarship, and spirituality. He also wrote poetry and studied medicine. Holmes's world was quite rich. In a letter to Morris Cohen (August 5, 1919), the philosopher, Holmes Jr. wrote, "my father was brought up scientifically – i.e. he studied medicine in France – and I was not. Yet there was with him, as with the rest of his generation, a certain softness of attitude toward the interstitial miracle – the phenomenon without phenomenal antecedents that I do not feel." Two sentences later Holmes wrote, "*The Origin of the Species* I think came out while I was in college." That work made a great difference to the way in which Holmes viewed the world; he understood the continuity of science with the larger body politic, and more specifically, with the law.

Understanding Holmes means placing him in the larger culture of ideas, the larger culture in which he breathed intellectual oxygen. Pragmatism is a recurrent theme in placing Holmes. But it is not so transparent. His philosophical roots are largely in the Anglo-American traditions of which pragmatism and positivisim are two features.

Holmes Jr. grew up with respect for biology. Holmes Sr., the physician, theorized about human health and discerned something about crib death. Holmes Jr. sat early in a context of conversation. Moreover, Holmes Sr. participated in the invention and use of instruments in medicine and science (Holmes Sr., see Gibian 2001). The fluidity of biology and the larger sense of inquiry perhaps ran through his intellectual veins, a key metaphor in the expansion and continuity of our sense of being in the world (Johnson 2007; Winter 2001). Certainly, both Holmes Sr. and Jr. participated in the wider intellectual milieu. Holmes's father was a well-known literary physician (Gibian 2001).

Holmes Jr. was born to explore all fields of human inquiry, but he would be less interested in grounding something on subjective reports and more about grounding law and life in things outside the subject. The new world of statistical inference that his colleague C. S. Peirce understood would be the future grounding of human judgment.

Holmes grew up with a poetic flair for language. He was a literate person who became an editor of the Harvard paper and, during the war, a daily witness to human slaughter. His rejection of the traditional religions was not unusual in his circles, especially among war veterans. His major intellectual influence on the biological sciences was Chauncey Wright, a staunch defender of Darwin in this newly freed colony (Wright 1877). In his article on "The Evolution of Self-Consciousness," Wright called attention to the evolution of the courts, judges, legislation and decision-making as well as to the practical consequences of our laws.

The New England mindset that Holmes inherited was part wilderness and part salvation. It placed a particular emphasis on piety, stoicism, plain talk, and a sense of nature (Miller 1939/1982) as well as on an experimental sensibility. This is not surprising since experimentalism is at the heart of our cultural scientific ascent and hypothesis formation underlies both the instruments we build and the ideas that we have (Peirce 1868, 1877/1992, 1878a, b).

What had begun to emerge in the colonies was a transition from common to statutory law and a larger sense of instrumentalism in the transactions that dominate human interaction. Commerce and property were key legal features in a growing economy with an entrepreneurial spirit. Contract and commercial law also blossomed through this period (Horwitz 1992). For Holmes, it was naturalism, influenced by utilitarianism, positivism and eventually a variant of pragmatism that led him to reject the seventeenth century New England ethos (cf. Grey 1989; Kelley 1989–1990). Indeed, when pragmatism was later criticized for its lack of depth, an argument was made that the depth might be found in the social policy and progressivism of John Dewey (Niebuhr 1932/1960; Miller 1939).

While he goes off to a war that spares his life, which marked him for a lifetime, but perhaps not for the worst. He always seems detached to many people, but James wrote to him in 1868 that "my friendship for you is more a sort of physical relish for your wit and wisdom." James said of Holmes "this must lead to Chief Justice, United States Supreme Court" of his desire, and capability. But James and Holmes also said unflattering things about one another (Alschuler 2000; White 2000), as Holmes would say about Peirce or Dewey, the primary faces of pragmatism.

Holmes gobbled up the Harvard intellectual milieu and the naturalistic sentiments of Emerson, and was inspired by the life of John Marshall, a federalist with democracy a route for the competition of ideas. He was no fan of big government. But Holmes never idealized the vulnerability of our condition, our institutions, and the situation that we are historically thrown into and then have to survive. Holmes did not glorify democracy.

From Marshall to Taney was an amazing step downward; Marshall was Holmes's hero, Taney was an apologist for slavery, in so far as he is tied to the supreme judgment about slavery, freedom and citizenship in America, and for sustaining slavery through the law. Roger Taney is best remembered for the decision with regard to Dred Scott (1856), a freed slave, whom Taney refused to recognize as a full person, and certainly not with rights even though now living in Wisconsin, a free state. Ironically, Taney himself understood that slavery was not a good thing for the American political soul. Appointed to the court by Andrew Jackson, he was a Jeffersonian in orientation and came from a slave holding family (Faigman 2004).

The Dred Scott case was an embarrassment for the Supreme Court even at the time. Taney was as a court leader writing and defending a conception of personhood that more resembled the Nuremberg laws than the American tradition. But, then, our original sin in the formation of our republic is the constitutional position on slavery.

The appraisal and reappraisal of Holmes has continued for 75 years (Howe 1951; White 1971; Grant 2016). As we will continue to see, in the minds of his critics and admirers, there are many different Oliver Wendell Holmes Jr.s.—one version being the authoritarian aristocrat (White 1993). James categorized thinking as either "tough" or "tender minded"; Holmes, not a legal formalist like some of his colleagues (e.g., Langdell 1880; see Grey 1983; LaPiana 1994; Kelley 2002) was undoubtedly "tough minded" in the Jamesian sense. Holmes has alternatively been viewed as a "cultural hero" (Hollinger 1992, 1996) or a contemptuous atheist whose placement in a powerful position could result in dire consequences (Alschuler 2000).

Like many of the founding intellectuals vital to the United States' formation, Holmes was an avowed naturalist and science oriented. But he was not, by any means, a consistent biological or social determinist and his mystical sense could be tied to some Emersonian and later Deweyan notion of nature. Holmes' adumbrated a sized down sense of human rationality and human decision making.

Demythologized Rationality: Pragmatist Sensibilities and Neuroscience

Rationality demythologized to human size is a pragmatist goal, for endless situational contexts. What counts is accountability, rational choice within the bounds of reasons, good reasons, sane choices and the sense of responsibility. These are, of course, normative goals in a participatory democracy. Thus, the sense of responsibility, of holding people accountable, on some accounts, will dissolve into an "illusion" of free will (e.g., Greene and Cohen 2004; Wegner 2002). Though, to be sure, we are endlessly in deception about our choices. My view is that the freedom/non-freedom issue is better couched in terms of perceived capability, which is in part discerned by the tools of modern neuroscience and human competence and performance testing. Indeed, our freedom evolves as our choices do, as we expand our capabilities (Dewey 1925; Dennett 2003; Schulkin 2007). This is the general theme of our evolution butilt into pragmatist sensibilities.

We hold people accountable not because there is an illusion of choice, or that all things are reduced to predictable brain events, or that brain events or genes determine all action. Rather, all action is probabilistic, and choice is considered within options, within capability, including neural capability (Garland and Glimcher 2006). We have added the language of neuroscience into the context of choice. And we are prepared to recognize the many ways in which self-deception is a feature of a compromised brain (Hirstein et al. 2005, 2018). Freedom for a pragmatist is thus not a lofty property; it is hard earned, tied to innovation, cultural expression, choice and capabilities. It is utterly demythologized. It is hard earned and easily lost. Lost by tyranny; which is why democracy is emphasized. And lost by neural function; by aging and dementia, by developmental deficiencies; both genetic and epigenetic, by cultural deprivation and economic woes, of which there are many.

Freedom, some sense of it in our shared understanding of accountability and choice is part of understanding human action. It is bounded by context and experience (Neville 1974). Neuroscience highlights this by bringing further considerations of relevance. For instance, the development of habits is a cornerstone of classical pragmatism (Schulkin 2007), as is considering diverse regions of the brain (Waldbauer and Gazzaniga 2001; Gazzaniga 2008). Indeed, one region of the brain, the basal ganglia, underlies the organization of action and is also tied to discernment of probabilistic events (Schulkin 2015).

Neuroscience does not replace philosophical conundrums but is rather a part of the conversation. In the naturalization of the knowing process, neuroscience is part of understanding something about human knowing, human judging, etc.—another place in which to house discoveries and insights from neuroscience (Buckholtz and Faigman 2014).

Neuroscience will contribute to an understanding of self-control, our sense of choice, effort and freedom and the neural systems that orchestrate control. Neuroscience, like law, is a part of our cultural expression. Each is nature and culture folded into one another. And they are both tied to our sociality.

In this text, I am interested specifically in two cultures: philosophy and the law as they interact with neuroscience, and neuroscience as a subset of biology. One goal of the book is to suggest that some of the facts of neuroscience fit easily into discussions of human experience and the law. What we don't want, however, is to oversell the neuroscience. "Not so fast," as Judge Rakoff (2016) reminded us recently in the *New York Review of Books*, while considering the impact of neuroscience on the court. He cautions us about the rush of neuroscience into the court-room. Judge Rakoff also admits that "neuroscience is mindboggling."

A meeting of law and neuroscience is unlikely to prove persuasive in the courtroom any time soon. But, as neuroscience knowledge becomes more reliable and more easily accepted by both the larger legislative public and in the larger public through which neuroscience filters into epistemic and judicial reliability, the two will ultimately find themselves in front of a judge. A pragmatist view of neuroscience will aid and underlie these events.

The perspective in this book is that human problem solving and of the law is tied to a naturalistic, realistic and an anthropological under-standing of the human condition. The situated character of legal rea-soning, given its complexity, like reasoning in neuroscience, can be notoriously fallible. Legal and scientific reasoning is to be understood within a broader context in order to emphasize both the continuity and the porous relationship between the two.

My own version of pragmatism—and this is a book within the prag-matist tradition and about pragmatism—which I refer to throughout, is close to C. S. Peirce's with regard to problem-solving and inquiry, and to John Dewey's with regard to issues that matter to the human condi-tion. As one reviewer noted, John Dewey, not Oliver Wendel Holmes Jr., is the real hero of the book. Indeed, he is. But Holmes is endlessly interesting.

I am both a pragmatist and a neuroscientist. The book reflects, as have others I have written, both these interests. But here is the con-sideration of one pragmatist, often not noted as such: Oliver Wendell Holmes Jr. presented in the context of diverse considerations of the law, reasoning, decision making, and ethics. The reader will see that I am much closer to John Dewey, the well-known pragmatist, than I am to Holmes. Both, nevertheless, were in search for external measures to cod-ify human behavior. In this regard, the brain is one epistemic anchor.

But Holmesian sensibility, or reasoning, yields a cautionary note about neuroscience, and still an appreciation of neuroscience. Neuroscience is both important and has limits and not to be oversold in the context of the law. The pragmatism, associated with John Dewey is the larger context in which the science is incorporated in the larger sense of our lives. It is the larger sense that Holmes alludes to and that Dewey captures in the philosophy of pragmatism, pragmatism ripe with human meaning and possibilities. The theme of the book like others of mine is couched within pragmatism and neuroscience. In this text, it is shifted towards considerations of the law, through the lens of Holmes, but more tied to Dewey.

Hardly a day goes by when there is not some sort of focus in the press about neuroscience and the study of the brain. Why so much neuroscience? One reason for this focus on the brain is a fleeting desire to "know thyself." The Socratic belief that self-knowledge brings fulfillment is thought to entail having knowledge of our brains. But self-knowledge is itself splinted. We are not really one thing. Even the brain is not one thing—let alone more complicated concepts about us. Knowledge is spread across the panoply of human activities. Understanding the brain is an important one amongst others.

To understand the brain is to place understanding as a part of biology. Doing so requires something of our evolutionary pasts and our propensities as a species. Our evolutionary understanding of species, variability, and adaptation to niches and diversity (along with good-enough adaptation, as nothing has perfectionistic reasoning) is in itself the stuff of biological design. The brain is no different.

This view on adaptation and instrumental innovation were rooted in Dewey's *Experience and Nature* (1925), especially in the metaphors and hard-felt realities of life, both the precarious and the stable. In *Experience and Nature*, these core metaphors are intended to capture human experience across the great array of human expression and to house a sense of nature embedded in our cultural milieu. Dewey unabashedly embraced an evolutionary point and demythologized human reasoning. It was in biology that Dewey was able to identify the key features of our condition.

Neuroscience is an advancing field that is just starting to make noise within the larger culture. But the link between self-knowledge and neuroscience is still barely audible. In the end, what emerges from studies in neuroscience may produce little that resembles self-knowledge.

Yet it will be continuous with understanding ourselves (Schulkin 2015). A pragmatist view of the law and of neuroscience underlies the epistemic transitions and embodiment into a culture of inquiry.

What is it that we want to know when we pursue knowledge of ourselves? We might begin with an enormous list of things but will most likely focus on the things that we think matter to our current or imagined lives. One question often posed is "what kind of life is worth living?" At the beginning of any philosophical inquiry, this basic question is most relevant to what we mean by self-knowledge. Dewey's intellectual imprint lines the pages of this book.

A task Dewey made repeatedly was to break diverse dualisms, such as mind and body, sense and cognition, action and thought, science and art, and politics and law. Thought runs throughout it all. The issue is what are the ends or goals that run through thought, the means for achieving them, and the fluidity and appropriateness of the tools to meet goals.

I think Holmes would be quite intrigued by the neural sciences as brain events and capturing them with external measures as an objective measure. Holmes would be eager to see the continuity of neuroscience with the practice of jurisprudence along with the body of knowledge that is evolving in the indefinite formation of law. Dewey and Holmes both emphasized external events and brain activation as one form of an external event. Whether the emphasis is helpful or informative is another matter. Dewey, however, would celebrate the expansion of human participation, of a conception of law and ethics that expanded human participation in governance and self-direction. Holmes would be silent or pessimistic about such endeavors. Here the conception of neurolaw or neuroethics would be very different.

But this book highlights a cultural evolutionary trend; pragmatism, of which Holmes is one and legal reasoning being a subset of demythologized sense of reasoning in general. This is not controversial, and the links are suggested throughout the present text—the continued blend of our sense of ourselves within the broader context of capturing our experiences is a pragmatist theme.

Here the emphasis is on Holmes as an endlessly interesting figure as many investigators that are cited in this book have noted. The book makes no apologies in bringing a discussion of Holmes, pragmatism and neuroscience in a common context. It is not an expected one. But experts will want more. Those who know about Holmes will find common knowledge, those versed in pragmatism will see familiar themes, and those familiar with neuroscience and the law will also see some common themes.

It is Holmes the outstanding flawed figure, major figure in law, who crawls back from a war that scarred him and enriched him that understood the continuity of science, statistical inference and the broadening arena of science into the law; neuroscience. But for those looking for a neuroscience, they should look elsewhere. I do tie neuroscience in nearly every chapter in the book, and tie it to core themes rich in the pragmatist tradition of broad human inquiry and discovery and human well being. This is the pragmatism found in the works of John Dewey.

References

Alschuler, A. W. (2000). *Law Without Values: The Life, Work and Legacy of Justice Holmes*. Chicago: University of Chicago Press.

Baker, L. (1991). *The Justice from Beacon Hill: The Life and Times of Oliver Wendell Holmes Jr*. New York: Harper & Row.

Buckholtz, J. W., & Faigman, D. L. (2014). Promises, Promises for Neuroscience and Law. *Current Biology, 24,* 861–867.

Dennett, D. C. (2003). *Freedom Evolves*. New York: Penguin Books.

Dewey, J. (1925/1989). *Experience and Nature*. LaSalle, IL: Open Court Press.

Faigman, D. L. (2004). *Laboratory of Justice*. New York: Time Books.

Garland, B., & Glimcher, P. W. (2006). Cognitive Neuroscience and the Law. *Current Opinions in Neurobiology, 16,* 130–134.

Gazzaniga, M. S. (2008). The Law and Neuroscience. *Neuron, 60,* 412–415.

Gibian, P. (2001). *Oliver Wendell Holmes and the Culture of Conversation*. Cambridge: Cambridge University Press.

Grant, S. M. (2016). *Oliver Wendell Holmes Jr. Civil War Soldier, Supreme Court Justice*. New York and London: Routledge.

Greene, J. D., & Cohen, J. (2004). For the Law, Neuroscience Changes Nothing and Everything. *Philosophical Transactions of the Royal Society of London. Series B, 359,* 1775–1785.

Grey, T. C. (1983). Langdell's Orthodoxy. *University of Pittsburgh Law Review, 45*, 1–53.

Grey, T. C. (1989). Holmes and Legal Pragmatism. *Stanford Law Review, 41*, 787–870.

Hirstein, W., Poland, J., & Radden, J. (2005). *Brain Fiction: Deception and the Riddle of Confabulation*. Cambridge: MIT Press.

Hirstein, W., Stifford, K. L., & Fagan, T. K. (2018). *Responsible Brains: Neuroscience, Law and Human Culpability*. Cambridge: MIT Press.

Hollinger, D. A. (1992). The Tough Minded Justice Holmes, Jewish Intellectuals and the Making of an American Icon. In R.W. Gordon (Ed.), *The Legacy of O. W. Holmes*. Palo Alto: Stanford University Press.

Hollinger, D. A. (1996). *Science, Jews and Secular Culture*. Princeton: Princeton University Press.

Hollinger, D. A. (2005). The One Drop Rule and the One Hate Rule. *Deadalus, 18*(Winter), 28.

Horwitz, M. J. (1992). The Place of Holmes in American Legal Thought. In R. Gordon (Ed.), *Legacy of Oliver Wendell Holmes Jr*. Palo Alto: Stanford University Press.

Howe, M. D. (1951). The Positivism of Mr. Justice Holmes. *Harvard Law Review, 64*, 529–546.

Johnson, M. (2007). Mind, Metaphor, Law. *Mercer Law Review, 58*, 845–868.

Kelley, P. J. (1989–1990). Was Holmes a Pragmatist? Reflections on a New Twist to an Old Argument. *Southern Illinois University Law Review, 14*, 427–467.

Kelley, P. J. (2002). Holmes Langdell and Formalism. *Ration Juris, 1*, 26–51.

Kloppenberg, J. T. (2010). James's Pragmatism and American Culture, 1907–2007. In J. Stuhr (Ed.), *One Hundred Years of Pragmatism* (pp. 7–41). Bloomington: Indiana University Press.

Langdell, C. C. (1880). *Summary of the Law of Contracts* (2nd ed.). Boston: Little, Brown.

LaPiana, W. P. (1994). *Logic and Experience*. Oxford: Oxford University Press.

Miller, P. (1939/1982). *The New England Mind of the Seventeenth Century*. Cambridge: Harvard University Press.

Neville, R. C. (1974). *The Cosmology of Freedom*. New Haven: Yale University Press.

Niebuhr, R. (1932/1960). *Moral Man and Immoral Society*. New York: Charles Scribner.

Peirce, C. S. (1868). Questions Concerning Certain Faculties Claimed for Man. *Journal of Speculative Philosophy, 2*, 103–114.

Peirce, C. S. (1877/1992). The Fixation of Belief. In C. Kloesel & N. Houser (Eds.), *The Essential Peirce: Selected Philosophical Writings Volume 1* (p. 115). Bloomington: Indiana University Press.

Peirce, C. S. (1878a). Deduction, Induction and Hypothesis. *Popular Science Monthly, 13*, 470–482.

Peirce, C. S. (1878b). Doctrine of Chances. *Popular Scientific Monthly, 12*, 604–615.

Posner, R. A. (1992). *The Essential Holmes*. Chicago: University of Chicago Press.

Posner, R. A. (1995). *Overcoming Law*. Cambridge: Harvard University Press.

Rakoff, J. S. (2016, May). Neuroscience and the Law: Don't Rush In. *The New York Review of Books*.

Schulkin, J. (2007). *Effort: A Behavioral Neuroscience Perspective on the Will*. Mahway: Erlbaum Press.

Schulkin, J. (2015). *Pragmatism and the Search for Coherence in Neuroscience*. London: Palgrave Macmillan.

Waldbauer, J. R., & Gazzaniga, M. S. (2001). The Divergence of Neuroscience and the Law. *American Bar Association, 41*, 357–364.

Wegner, D. M. (2002). *The Illusion of Conscious Will*. Cambridge: Harvard University Press.

White, G. E. (1971). The Rise and Fall of Justice Holmes. *University of Chicago Law Review, 39*, 51–77.

White, G. E. (1993). *Justice Oliver Wendell Holmes: Law and the Inner Self*. Oxford: Oxford University Press.

White, G. E. (2000). *Oliver Wendell Holmes Sage of the Supreme Court*. Oxford: Oxford University Press.

Winter, S. L. (2001). *A Clearing in the Forest: Law, Life and Mind*. Chicago: University of Chicago Press.

Wright, C. (1877/1971). *Philosophical Discussions*. New York: Burt Franklym.

2

Holmes's Critical Experience in War: Trauma and the Brain

Holmes famously chose experience over logic in adjudication and the law. But he most likely also understood that forms of logic in inquiry or problem-solving run through experience. In fact, logic runs through experience like oxygenated coherence. Holmes never lost sight of the active sense of mind and how it *runs through* the many faces of pragmatism.

From the start of this book, I have suggested that ***one view held by pragmatists was that the deepening of human experience was an important goal in life.*** **But Holmes was glaringly flawed and limited.** He was not a friend to those who needed one or a champion of civil liberties. As Mencken (1930) noted, Holmes was a great and learned person, but he was no friend of the Bill of Rights: "to call Holmes a liberal is to make the word meaningless."

He *exhibited* a rapture of prose highlighting expectations, enhancing the causal efficacy and responsiveness that runs through our experiences, heightening them in a rhetoric of poetic grace. Within a language of the law that was vibrating and evolving, Holmes articulated a common law expanding in a participatory democracy. There was no one framework for Holmes except the aliveness of experience and the prevailing

© The Author(s) 2019
J. Schulkin, *Oliver Wendell Holmes Jr., Pragmatism and Neuroscience*,
https://doi.org/10.1007/978-3-030-23100-2_2

sense that the metaphors used were grounded in nature and the verse of strength and expectation.

Holmes's war experience may be the key to understanding him properly and no doubt he would agree. Even his tombstone at Arlington highlights his earlier experience as a soldier.

We now know that trauma of the kind experienced by Holmes during the war between the States changes the shape of the brain. Those who are lucky enough to persevere, despite the neural scars of war, still have to deal with long-term changes in their neural function.

This chapter is focused on Holmes's war experience. It is about his personal trauma and the impact—or probable impact—this primary experience had on his brain. Holmes's experience of war would pervade the rest of his life. And so it is considered within the context of the larger body politic of his time—the lens through which he understood the world. It is also considered as it is linked to pragmatism and survival and social contact and good enough problem solving.

Holmes in War and Peace

There is nothing like war to bring out the worst and the best in people. Holmes survived one of the worst wars in history. In many ways, the American Civil War was equal to World War I in its impact both during and after the battles. In each of these wars, advanced military technology ran up against antiquated military strategy, resulting in a death toll that was staggering.

Like Dewey and Bertrand Russell, Holmes lived into his mid-90s and because his very public position enabled him to maintain many rich correspondences, we have been able to learn something about Holmes's views when he was off the bench. From his correspondence, we know that Holmes's experience, like that of most survivors of the Civil War's horrors, would remain with him for the rest of his life. Speaking at a "Fraternity of Arms" meeting ("A Soldier's Faith", 1897), Holmes expressed a stoic sensibility in the way he thought about war while enjoying the calm of peace. Conveying his solidarity with the soldiers' experience, he told an audience of veterans, "As I look into your eyes I feel, as I always do, that a great trial in your youth made you different – made all of us different from what we could have been without it. It made us feel the brotherhood of man." Although Holmes addressed the duty and heroism involved, he also stated that "war, when you are at it, is horrible and dull."

Holmes enlisted in the army in 1861 and remained in service through almost the full span of the conflict, which ended in 1864. Other luminaries in his intellectual group, James and Peirce for instance, did not join or fight. Henry Abbott, a close friend whom Holmes would never forget, died in battle. Though not a fan of emancipation (Baker 1991; Wells 2015; Mendenhall 2015), Abbott nevertheless gave his life in a war where both sides claimed God's righteousness. Holmes wrote this poem in Abbott's memory (October 17, 1864):

> *"THE BRIGHT AND PARTICULAR STAR"*
> *HENRY LIVERMORE ABBOTT 1842-1864*
>
> *"He steered unquestioning nor turning back,*
> *Into the darkness and the unknown sea;*
> *He vanished into the starless night, and we*
> *Saw but the shining of his luminous wake*
> *Thou sawest light, but ah, our sky seemed black,*
> *And all too hard the inscrutable degree,*
> *Yet noble heart, full soon we follow thee,*
> *Lit by the deeds that flamed along thy track…"*

Though his life was molded by the hardships of war and degradation of the human experience he had witnessed, Holmes was, at least publicly, mute about his experiences much of the time. A well-known quote from Holmes about the experience of war states that it was "incommunicable," and indeed Holmes seldom talked about the war directly. What he communicated in "Soldier's Faith" was the faith that underlies duty, with little knowledge or understanding.

After the battle of Ball's Bluff (October 1861), Henry Abbott had written Holmes's father with news of his friend, Holmes Jr.: "Lt. Holmes shot through the breast, will recover." Holmes's father was not an avid abolitionist before the war but during the war, became a convert. His son, at the beginning an abolitionist, become less so with the toil of war (see Touster 1982; Wells 2015). Though notions of death and dying are not abstract, dying for the rights of others *is*. Doubting that "the butchers' bill" was a very meaningful phrase, Holmes ultimately became beleaguered and then beaten down. War quickly wipes away innocence, replacing it only with indelible scars. As one can see in Holmes's letters to his parents, a poetical sensibility tears flesh from a moral soul.

Holmes was repulsed by the evils of slavery. And duty was one of his fundamental values; it was something Holmes understood and believed in and a value that would prevail throughout his life. He had fought in a just war that mattered to a country that was on the line and in battles that would define the trajectory of his country. But it was a war in which death prevailed to a tragic degree. That Holmes had kept his battle-stained wartime clothing was only discovered after his death. Still hanging in a closet were Holmes's soldier's uniforms with a note that read "these uniforms were worn by me in the Civil War and the stains upon them are my blood" (Baker 1991). The closet also contained two musket balls that had been removed from his body during the war.

Drew Faust, a noted historian, frequently refers to Holmes in her book on the Civil War, *The Republic of Suffering* (2008). She references the abject pain and suffering that occurred—the slaughter, dying and disease and the ultimate impact on families. Many have written about the decay in war and the cannon of death. Faust asserts that soldiers like Holmes were "better prepared to die than to kill" (2008, p. 6).

Holmes was only one of these many soldiers who were badly wounded and still returned to battle. At Ball's Bluff, at the onset of the war, Holmes's sergeant squeezed the bullet from his chest (Adams 2014). Faust notes that when Holmes was brought to the hospital severely wounded and in excruciating pain, he called out "a deathbed recantation" to be sent home to his parents (Faust 2008, p. 14).

Holmes was also wounded at Antietam, which remains the deadliest one-day battle in all American military history.

A note still hanging on a wall in the Harvard Law Library reads "I am Capt. O.W. Holmes 20th Mass V., son of Oliver Wendell Holmes MD Boston. I wrote the above when I was lying in a little house in the field of Antietam which was a while within the enemy's lines, as I thought I was faint and so would be unable to tell who I was."

Oliver Wendell Holmes Sr. went to find his son in Antietam to see if he would live. This was the act of a worried father who was very much part of the New England elite—many of whose sons had avoided the war. A physician and a member of the the American Philosophical Society, founded by Ben Franklin and devoted to scientific investigation, Holmes Sr. was devoted to the amelioration of suffering. As a poet, he was close to Emerson, sharing with him a sensitivity toward the human condition. He was also the grandson, on his mother's side, of Charles Jackson, Associate Justice of the Supreme Court of Massachusetts. But none of these things mattered at Antietam. There he was just another father worried about a son who was suffering and very much alone.

Many have detailed the useless pain and suffering in a war that ultimately kept the Union together. Holmes modestly noted that after the war "the world was never right again." He spent the rest of his life within the culture of American jurisprudence. Advancing from his role as lawyer extraordinaire with a literary legal flair to the highest court in Massachusetts and finally to the highest court in the US, he remained a product of his time, class, education, and military experience. But we cannot know just when or how the impact of the war influenced his judicial decision making.

War and Memory

In a letter to Sir Frederick Pollock in 1895 (October 21), Holmes would write, "Thirty-four years ago today was my first battle. I was shot through the breast at Ball's Bluff and it always seems something of a university to me." Later in this letter (and in others) he says that he "expected to die." His parents always remained close to his exploits during the war; he may not have been close to his father, but he surely was loved by him (Baker 1991).

Sir Frederick Pollock would be one among a list of individuals with whom Holmes would have an extended correspondence. In his letters, Holmes wrote as a public intellectual while privately revealing his personal experience. So the war came up; it was a highlight of his experience he could never forget.

In this same letter, Holmes later wrote: "I happened to have a bottle of Laudanum in my pocket and resolved if the anguish became unbearable to do the needful. A doctor (I suppose) removed the bottle and in the morning I resolved to live" (Pollock Holmes letters 1874–1932).

At Antietam, Holmes was left discarded near the corpse pile, awash in his own blood and still bleeding. As he lay in the ruins of the West Woods section of the battlefield (McPherson 1988), Holmes lost hope and later wrote that he thought "…the arm is tired" and "the south may have achieved their independence" (McPherson 1988, p. 590; Holmes 1946). Like Joshua Chamberlain, a prominent military figure from Maine, Capt. Holmes was wounded in battle not only at Antietam but several times, and still came back to the Army after each recovery. Like James, Peirce, or others from the Cambridge elite, surely, he could have found his way out of the ordeal but chose not to.

Holmes was wounded in battle not only at Antietam but several times in campaigns that included Ball's Bluff, Antietam, and Chancellorsville. By 1864, he had finally had had enough (Adams 2014) and when his initial three year-enlistment contract had expired, he resigned from military service. Toward the end, Holmes believed that the South would very likely win the war. As duty was always a key feature for Holmes, years later he would commemorate the soldiers of the south and their cause from the point of view of the soldier doing his duty.

After Cold Harbor, Holmes commented that the troops had begun to lose it and had "gone crazy" (McPherson 1988; Holmes 1946, Civil War Diaries). Before there was a term for post-traumatic stress disorder (PTSD), there was still PTSD. Soldiers suffered from multiple horrors—slaughter, pain, misery and the utter devolution of function and meaning. Where war ushers in the worst of humankind, however, it also calls up the best—loyalty, courage and the gratefulness for survival. Holmes was loyal, civic minded and stoic. He fought in a war whose cause he was not vocal about nor was he diehard about its meaning. The pain of it all within a commitment to a federal government emboldened with efficacy and power (Donald 1947/1984; Burt 2013).

Trauma, Fear and the Brain

Let us return again to trauma and the brain. Holmes was altered by the war, as were all others who fought in it (see his Civil War Letters and Diary, 61–64). In his Memorial Day Speech in 1884, Holmes would recall a doctor remarking on the boy in the next bed who had just passed, "he was a beautiful boy."

Critical events are remembered in part by the sculptured induction of neurotransmitters or neuropeptides in the brain. One neurotransmitter tied to memory in general and memory of trauma in particular, is norepinephrine and the neuropeptide, corticotrophin releasing hormone (CRH) (McGaugh 2003). Cortisol facilitates the expression and induction of norepinephrine or CRH. One result is a remembered event. Holmes remembered his war experiences to the end of his life.

Cortisol released during trauma acts in regions of the brain, including that of the amygdala, to release adrenalin neurotransmitters (McGaugh 2003). The release of these neurotransmitters highlights and sculpts the account of the traumatic events in neural architecture. The adrenalin neurotransmitter, strategically expressed in diverse forebrain regions, creates a reverberating circuit of neural meaning that captures vital human experience.

And yet, neurons in the brain decay. Brain regions are rendered vulnerable by wear and tear over time (Sapolsky 1992; McEwen 2007). Holmes would go on to live over another 60 years after the war despite the toll that the war had taken. The genetic hardware that underlay his physical and mental capability were outstanding.

We know that war and other traumatic events produce changes in gene expression (epigenetics). The genes that underlie CRH expression, as noted earlier, are anchored to unfamiliar, dangerous events and memory. As I've indicated, we know that this neuropeptide is altered by traumatic events of the kind that Holmes survived. Long-term changes in CRH expression occur in regions of the brain tied to fear, pain, anxiety, and suicide (Charney 2013; Schulkin 2017).

A true understanding of Holmes requires understanding something about his war experiences (Rogat 1964; Grey 1992; White 1993). Holmes understood fear. His war experiences left palpable battle scars and enabled him to express his solidarity with others.

We live with endless imperfections trying to cope, (under siege and evolving in a culture) or not with a nature that is continuous, but perhaps more plastic than Machiavelli could see. Holmes understood something about the endless flaws and conceits of the human condition—something Machiavelli made no mythology about. Both men seemed dour with fatalistic expectations.

The debates about religion were settled for Holmes. None were acceptable to him in the forms in which they existed at the time. To Holmes, the world was one adaptation, harsh with an increasing population weeded out by selection and disease. Hobbes and later Malthus or Ricardo were not so far off at one level. A lot of events reveal the hardships, the brutish and the shortness of our lives. Selection principles of diseases and wars were construed as population control as both resulted in less mouths to feed.

These events were just a continuous function with the creation of laws and regulation of safety (LeDoux 2015). Information molecules in the brain are tied to wanting and then securing safety, fear reduction or satisfaction (Schulkin 2017). These same aims run through the veins of legislation. The fact that neuropeptides such as CRH, for instance, exist, is a gripping fact about our central nervous system. They run across an information highway of a chemical nature that mobilizes the protective features of our viability. Our nervous system is keen on detecting differences and danger; it is designed to promote safety.

These systems are meant to be regulated: turned on and off. Unrelenting fear leads to something we have come to call by another name: PTSD. This condition is not well understood, but we know roughly what it means. Consider the trauma of trench warfare survivors in World War I or of young Syrian children whose normality has become war and fear. These were things well understood by Holmes (Grey 1992; White 1993). We also know something about the brain regions tied to such trauma. But they are regions of the brain tied to many behavioral expressions (e.g., prefrontal or cingulate cortex, hippocampus, amygdala) (Charney 2013).

Regions, such as the amygdala, are responsive to events that are unfamiliar. Something unfamiliar is potentially dangerous. It has long been known that the amygdala is involved in fear responses to unfamiliar events and in giving value to diverse events (LeDoux 2015; Rosen and Schulkin 1998; Phelps et al. 2004). Excessive activation of the amygdala has been linked to racial features in the US (Olson and Phelps 2007). Recent neuroimaging evidence revealed that brain systems involved with social aversion, including the insula, lateral frontal cortex, and amygdala, are recruited by mechanisms of racial bias and social exclusion. Both in the brains of the agent and the target of prejudice, this recruitment is an important factor in the development of negative beliefs towards out-groups, unlocking social aversion mechanisms associated with emotions such as contempt, anger, and disgust (Rozin 1998).

Indeed, the neurobiological systems are tied to attention to unfamiliar events or the anticipation of satisfying our wants. Neurobiological systems include information molecules such as CRH, a neuropeptide produced in regions outside the traditional hypothalamic pituitary adrenal (HPA) axis (e.g., amygdala). The amygdala is tied to survival systems, to essential, motivational systems that underlie fear (Rosen and Schulkin 1998; LeDoux 2015). Holmes's brain was no doubt changed by his being shot several times, in multiple battles and within the sight and smell of fright. He had been surrounded by the sounds of pain and then, finally, the silence.

In this context, the adrenal steroid increases the gene expression that underlies CRH expression; it is the existentialist nightmare. One result is the angst of excessive fear or uncertainty—as we all know a very unpleasant experience. It has consequences from alertness about danger, something that the amygdala for instance is heavily involved with, to depletion of resources and a burgeoning sense of hopelessness. *These are metabolically expensive events* (Sapolsky 1992).

This peptide is produced in the peripheral nervous system as well as in the central. Peptides are strings of amino acids. CRH is one kind of peptide usually associated with the hypothalamus and in this case the HPA axis. Cortisol, a steroid, is usually associated with stress or coping and coding for adversity (McEwen 2007).

While cortisol is associated with adversity, it is not simply that; most of it is tied to energy expenditure. Cortisol is fairly characterized as the steroid hormone tied to energy balance. Fear is metabolically expensive. It wears and tears on tissue: brain regions are vulnerable to deterioration, bone is demineralized, immunity is compromised, etc. (Sapolsky 1992; Schulkin 2017).

Moreover, this particular information molecule is tied to the gut. When we speak of "gut reactions" to events embedded in decisions about what is safe and what is not, we are wiser than we know. The same information molecules are expressed in both the brain and in the peripheral regions with a continuance of contact via these information molecules that are one synapse from each other. Additionally, there are direct projections in the gut from regions like the amygdala (Swanson 2000). These systems are tied to problem solving and surviving and underlie the organization of action and of getting the feel for things in a fluid, uncertain and precarious world. Holmes's precarious world lay in the vivid memory of war—a war in which slaughter was common and survival understood.

Wrong-Doing and Fear

Our species' evolutionary past is tied to small groups in which there are coalitions of support: we really need each other and we know it. Through his wartime experiences, Holmes understood that group formation is vital and necessary for viability. He never forgot that what mattered was to stand up with others. There is something eternal to our species and history that grounds us in ties to one another. Though the violence and worry may vary, we will always have plenty of worry and violence is omnipresent. Despite a sense of detachment, Holmes never forgot this.

Danger and *want*, states that are rich in anxiety, pervade the lives of many, regardless of class or ethnicity. Hobbes keyed into this. Human nature is tied to these existential facts. Existence, as conveyed by Sartre and other existentialists, is, in literary and philosophical terms, a big part of life, at any time in our evolutionary and cultural lives. Angst, worry, choice, and its consequences are the stuff of our lives. Or perhaps as Dewey (1925/1989) later suggested, we are always coping with the precarious; our biology is in tune with this capability.

Violence is the plague of human existence. Law, supported by evolving forms of behavioral sustenance, has and perhaps will continue to have, an impact. Finding ways to limit human violence is the greater human goal, reminiscent of what we have wanted for some time. The "bad man" theory of the law is one approach to violence, an expression of Holmes for what some do in following the law.

Before Holmes, Ricardo, Malthus, or Darwin, there was Thomas Hobbes, a seventeenth century realist, who wrote about the difficulties of life and of adapting to circumstances. He also wrote about the endless lure of the sensation (the allure of pleasure, the avoidance of pain), and about how fear permeates the human condition. Writing amid wars both within and abroad, Hobbes continually reminded us of these difficulties.

Life, at the time, was hard and only excess seemed to eliminate strife. But Hobbes was part of the larger social context that was moving away from the dominance of the Greeks and Romans and the larger culture of the Church. Experimentalism had begun to emerge with the work of Harvey, who detailed the circulation of the blood. What mattered most to Hobbes were clarity of principles and the search for definition; he was certainly not an experimentalist.

Of course, fear is a dominant emotion at any time, but this was especially true during Hobbes' era and in his experience. Fear is also metabolically expensive and Holmes understood this from his war experiences. Fear churns the body when we are in coping mode. We have come to understand that when there is little relief, the chronic non-relief results in diverse effects on neural function—the tearing down of neural tissue and breakdown of chemical constitution. Cortisol, for instance, is elevated partly from fear; chronic, unrelenting release from the adrenal gland can result in adrenal pathology and the breakdown of neural decision making (McEwen 2007). Adaptation, however, is only one end of fear; the other is chronic anxiety. One is adaptive. The other not: "continual fear, and danger of violent death; and the life of man, solitary, poor, nasty, brutish and short" (p. 106, Hobbes in the Leviathan). These were daily staples of the common experience for Hobbes.

Hobbes recounted some folklore saying that we are born in fear. As he reflected on his mother's pregnancy, he noted that "she brought forth twins – myself and fear," while his mother was viewing the ocean on the coast of England wondering where the Spanish war ships might land. Perhaps Hobbes came prepared, with good reason, to see our existence as a war against all and with a propensity for excessive fear.

We know now that there are genetic differences in the experience of fear. Our vulnerabilities to excessive fear are emboldened by social facts that exacerbate this state of being, a natural condition lacking in satisfaction or much ability to predict what's ahead (LeDoux 2015). Predictability with no safeguard is an essential cephalic state. There is much at stake in predicting. These events start early.

Law, War, and Morality: Social Considerations of US Civil War

Holmes fought in a war whose goal was first to preserve the Union and second free the slaves. He was committed to both but remained distant from the African American plight over the course of his life.

Holmes displayed an independent mind and a strong will, while acknowledging the adjudication of ideas in the competitive atmosphere of a democratic public where ideas were fought and tested. His stoic sensibility revealed a clear, though sometimes inconsistent, sense of morality. He wrote with reasonable and balanced logic, little sense of hate, a philosophical disposition, and a tendency to moderation and in the pursuit of excellence and living well. He was unrelenting at the Law Bar as law embodies beliefs that have triumphed in the battle of ideas (Law and the Court). One key event was the granting of freedom to African Americans and, following that, the right to vote for men. Although Holmes understood and fought for these rights, he was never outspoken and perhaps even slightly ambivalent about them. He only acknowledged that there were continued issues in the south with making these rights a reality.

While Holmes's voice was mostly mute when it came to African Americans, he had independently decided on what he thought was right. He didn't believe that what was legislated was necessarily the same thing. Holmes had quit school and fought for what was right. He had suffered greatly but mostly in silence. Although Holmes's voice might not have been a forceful one in the struggle for equality (McPherson 1975), he did his duty and lived a moral life.

Years later, as a Supreme Court Justice of the United States, Holmes would lament the injustice of having an all-white jury in a county where 2/3 were black (in Moore vs Dempsey 1923). But this did not mean that he was an advocate for racial justice like other motivated justices and lawyers of his time. W. E. B. Du Bois, a fellow Harvard graduate, New Englander and eminent scholar, would lament this fact (see Lewis 2000).

Du Bois would eventually become a moral arm of African American injustices, cataloging the list of harms done. Laski (1925), Holmes's socialist colleague in the UK, commented that Du Bois's book, *Dark Waters*, was "a very brilliant but hateful book" and perhaps "what the white southerner would write if he turned negro" (Lewis 2000).

In Holmes's day, mixing of the races, or miscegenation, was a major social issue. Even if someone had a distant relative who was black but looked white, it meant that they were considered black by society. Having African American blood as anything more than one-sixteenth of your racial make-up was enough to be viewed as "tainted" (Hollinger 2005), just as having a Jewish great-grandparent made one Jewish under the Nuremberg laws. White supremacy, an original American ethical sin, was commonplace in Holmes's day. Anti-miscegenation laws were observed until 1967 in some states (e.g., Virginia) until struck down by the Supreme Court in *Love v. Virginia* (see Posner 2003). But the blood issue reeks of the biological and science—a science of exclusion.

Holmes fought to keep the union together and for emancipation. Along with defending the Union on the bloody battlefields, he stood up for and protected abolitionists prior to the war's outbreak. But unlike his Harvard classmate and friend Norwood Hallowell, who fought and shed blood in many of the same battles as Holmes and went on to lead African American troops (Miller 2005; Wells 2015), he was no radical with regard to any position. At the time of Hallowell's death in 1914, Holmes would remember him as a "generously gallant spirit" and "oldest friend."

Holmes was not a vocal abolitionist to the same extent as his friend Hallowell; but, in contrast to his friend Henry Abbott, he was committed to the ideals of the war—at least until becoming fatigued and demoralized by the blood and decay of battle.

Very much a creature of his time, Holmes, the judge, did not champion the rights of African Americans. In voting against the interests of African Americans and denying them voting access (see *Giles v. Harris*, Alabama. *Nixon vs Herndon*, 1927, Texas), he may have lamented the (in)justice or (im)morality of this decision but not the legality. So, while Holmes may have recognized the unfairness of the decision, he was not willing to intercede in a state matter. Holmes remained very conscious of his role as a judge and Supreme Court Justice, which created an immeasurable distance between him and the plights of those with less.

Holmes understood and largely did not challenge the greater body politic of white racial and class dominance—who won rules and whose laws reflected their dominance. Although nominally given basic rights through Amendments to the constitution after the Civil War, African Americans were simply not on an equal footing. In later Supreme Court decisions, all talk about "separate but equal" would be judged as bad faith, morally corrupt and empirically blind. Yet miscegenation was a law in many states that also included other ethnic groups, notably American Indians (Hollinger 2005). Our history is the expansion of rights much contested but building to broaden the body politic by legal acceptance and democratic participation (Foner 2006/2012). Holmes did not choose this path.

Holmes seemed to choose a middle course, more like that of Lincoln than of radical Republicans. With regard to the rights of African Americans, Lincoln had come to a place that was different from the one where he started. He could not be called an abolitionist; the preservation of the union was foremost in his mind. Holmes may have been closer to Lincoln than to Horace Greely, Thaedeus Stevens or others of his own background (e.g., Shaw). But his thinking continued to evolve as he suffered through a war to preserve the union—a union that itself evolved into an intent to expand democracy, human rights, and human dignity (Foner 2006/2012; McPherson 1991, 2008). But this was not Holmes; it was Lincoln.

The moral anchor, critics say, is not present or present enough in Holmes (Kelley 1989, 2000). Holmes didn't defend the rights of African Americans, Asians and laborers in ways he might have been expected to. But there is no mythology. Holmes was a mere mortal— flawed, detached and imperfect. He was also an excellent expositor of the law, with a delicate pen to prose eloquently of matters of the law within the larger ruminations about culture and consequences (Rogat 1964).

A Common Faith, the title of a book by Dewey, is about *our* faith and the cultural features that matter to having a meaningful life—as Dewey understood it. Holmes's book, *The Common Law*, captures the cumulative features of the law as Holmes understood it. Both books are tied to the pragmatism that the two seemed to embrace as well as to the advances in our cultural evolution. For both men, natural law is replaced with cultural trial and error. But, unlike Dewey, to Holmes this might have seemed frail and perhaps hopeless.

Dewey's "common faith" smacks of something for the "common people" like those who populated the poems of Whitman as the celebration of a growing democracy of inclusion and aesthetics that is common in human life (Butler 2010; Reynolds 1995). Walt Whitman was a witness to the Civil War while Holmes fought it. Both saw the suffering and the wanton destruction of human life.

Conclusion

Holmes was never the same after the war and, like most other survivors, had difficulty talking about it (Menand 2001, 2002). He was also resilient, and this can be both seen and felt in the talks he gave to his fellow soldiers. Sounding like Spinoza, he urged his audience, in an 1884 Memorial Day address, to "think of life" and "not death." Holmes's good friend, Abbott, had died at the battle of Fredericksburg and that memory caused him endless sadness.

An imagined war fills the landscapes of a many a historical page, autobiography, novel or work of art. Whether it takes place in the western front of the first World War or the eastern front of the second (Fussel 1990; Grey 1989), what exists isn't just fear and trembling in an existential moment, as described by Kierkegaard, but a very real fear of war. Fear can, of course, turn to violence, and violence often goes hand in hand with ruthless aggression in the fluidity of the market place of motivational expressions generated by the brain.

The faces of war are endless as is the deafening effect on humanity and eternal bad stream or recurrent—*caused by the all too recurrent frequency with which wars occur.* On the biological and cultural side, it is not hard to understand, nor are the post-traumatic related events. Holmes may have been less vulnerable than others. Yet, as he lay wounded, thinking that he was dying, he had, in fact, come to a different kind of end—one that he may have desired because he had finally had enough of the senseless destruction.

Indeed, "imagining" and "doing" recruit activity in many of the same brain regions. So-called mirror neurons act so that the idea of imagining people will be reflected in visual spaces if it is about scenes and objects. The visual regions of the brain underlie not just visual experience but visual imagination (Kosslyn 1986). The imagining of events is tied to action because our minds are always active within our bodies.

Holmes witnessed hell on earth in his war experience. We have a sense now for how the brain is impacted by traumatic experiences in vulnerable people. Even when these survivors emerge relatively unscathed from a physical standpoint, their brains are still permanently marked by their individual experiences of war.

What Holmes learned in the war he carried with him for the rest of his life. They were embodied experience metaphors packed with a punch for life: the real flavor of blood and guts, charge and fear, terror and survival, pain and groans of agony.

Holmes was a complicated figure. On the one hand, revered, heroic, aristocratic; on the other, reviled and detached. This is just a sampling of the wide variety of words used to describe him. Holmes himself was inconsistent. A naturalist, atheist, and an elitist, he was also a fatalist and pessimistic about the human condition. Holmes was literary and also tough-minded. It is in his inconsistencies that he represents something in all of us and probably why he continues to be intriguing.

Because Holmes's brain was transformed by his experience of war, reference throughout the text is made to fear and trauma. There is no one brain signature of fear or PTSD that would be conclusive in a court of law. Holmes's brain changed as a result of the destruction he had witnessed during the Civil War. In spite of this, he still knew what he wanted, writing, in July 1861, "If I survive this war I expect to study law as my profession, or at least as a starting point."

Holmes would often be thought of as someone who lived a solitary life, but he does not strike me as solitary. His was a life enriched by intellectual expression and a high level of social capability. Holmes's solidarity was with the soldiers from both sides. He honored the duty of bloody battle in war, saturated as it was with decay, smell, and remembrance.

Though Holmes was deeply impacted by this disastrous war, he had also emerged from battle very much alive, filled with ambition, a stoic and a naturalist. No doubt these were tendencies that had already existed when he was a young man within the wider Cambridge sphere of affluence, nuance, and desire for knowledge. But the impact of his war experiences served to solidify them.

Discussions in the Metaphysical Club, of which Holmes was a member (as will be discussed in the next chapter) also had a significant impact. A wide range of subjects were explored: the wonders of the day, breakthroughs in thought and the naturalization of science with other forms of human practice like the law and literature. There were a bulk rock of narratives about biology; stories of persevering and of linking a stoic sense of balance with ongoing dreams of possibilities served to remind Holmes of what is not.

The war experience always seemed fresh in Holmes's mind. He saw the glimmers of the possible amidst the reality of the discrepancy.

Holmes (1881) was committed to the science of biology and to some sensibility to anthropological views about the human condition, e.g., righting wrongs, liability for harm and the regulation of harm. But he was limited and not particularly wise in his expression regarding the plights of others. Within him, though, is an Emersonian sense of being historically bound and swept up in the plight of the age in which he lived—a sense, perhaps, of what Emerson called "the over-soul," heroic moments that Nietzsche portrayed in prose about struggle and strife.

Like Nietzsche, Holmes provokes one to think (Posner 1992; Luban 1992) rather than to think about something in particular. As he stated in "The Soldier's Faith," his experience in the war helped him to understand that "joy of life is living… overcoming obstacles." Though a catalyst for thought, Holmes is a figure of striking mystery whose experience of the horrors of war left a permanent mark on his brain (Grey 1992; Rogat 1964). Grey (2000, p. 144) suggested that "Holmes had in fact shaped his own view of forward-looking practical judgment as a young officer on the battlefields of the civil war".

But social contact is critical and perhaps Holmes understood that social contact is basic mental health adaptation. This is apparent in his letters, in his walks with colleagues, and in his remembrance of his fallen colleagues. He remembered them to the end of his long life. Holmes's world was rich with social contact through discourse with others, his "pen pals." Holmes deep meaning is in his social contact—an important adaptation, a fundamental feature in the human condition (Jaspers 1913/1997).

His cosmic view was Emersonian, and for poetic moments on nature and union of the self. Perhaps as in: "His self and the sun were one and his poems, although making of his self, were no less makings of the sun" (Wallace Stevens).

Holmes's rejection of religious belief, his suspicion of James on this account (see Pollock Holmes letters 1942/2015) and his rejection of natural law, or an etiology of meaning that underlies the human condition that is expressed in law, was not a fall from a moral sense of the world (cf. Kellogg 2007; Grey 2014; Howe 1963). There is just no inherent plan, in reason, inherent in history. Holmes is far from the rationalism about morality and law, or necessary historical epochs so dominant in the view of the law that emanated from Germany. But morality, Holmes over and over asserted, is inherited in our institutions. In the law, it is like breathing. It is just frail, and brute power predominates the lower bastions of our human expression.

There is no question that Holmes was an instrumentalist. Ideas were set in the context of instrumental use. Ideas were not reified. They were contextualized in use. But there was more.

Holmes certainly understood what Dewey emphasized in his great book, *Experience and Nature*, namely the sense of the precarious and the stable. Human orientation and investigations stabilize transactions with others and the world, in an effort to transform the precarious into the stable or the viable.

Holmes and Peirce definitely saw James as soft when it came to such subjects. But was he 'soft' or did these two individuals look inward at all? After all, it takes courage to look inward. Neither Holmes nor Peirce valued introspection (see Brendt 1993), and certainly not a basis for a sound epistemology. Both Peirce and Holmes were looking for steady anchors to external events to build reliable anchors for epistemic discovery and solidification in a community of inquirers.

James would write to Holmes on May 15th 1868 that "I am tending strongly to an empiricist view of life." Neuroscience, one should note, is one such external anchor. Holmes, in a letter to James, would write (May 24, 1896), "I long ago made up my mind that all that one needed was a belief in the significance of the universe. And it has come to me lately that even that might be too ambiguous." What was significant was pragmatism (July 24, 1907; Perry 1935), but in a Holmesian way he tells James "that the only promising activity is to make my universe coherent and livable, not to babel about the universe" (March 24, 1867; Perry 1935).

Holmes and James became friends after Holmes came back from the war and entered law school at Harvard. Their exchanges were about materialism and the meaning of life (Perry 1935). Holmes shows signs

of vulnerability and insecurity when, in a letter to James, he writes, "…
writing is so unnatural to me that I have never before dared tried to
write to you unless in connection with a subject" (December 12, 1867).

Indeed, despite his massive physical appearance, Holmes could
be quite vulnerable. It comes through in his letters to his colleagues.
Writing to a young colleague in 1924, he would say "the friendship
of a young man like you is dear to me" (3/1924, Letters to Dr. Wu).
Holmes's engagement with others, from an Irish Priest in Cork County,
Ireland (see Holmes-Sheehan Correspondence) to a decades-long correspondence in the very first part of the twentieth century with a young
man named Einstein, revealed ways in which Holmes could make himself vulnerable and available to others (Holmes-Einstein Letters).

But while he moved away from James—indeed James would die
almost 30 years before Holmes—he would reveal himself in his letters. James continued the quest inward. Holmes, on the contrary, was
always looking for something external, some way to anchor and provide reference points, looking to fix reference for coherence and meaning (Schulkin 2015). One palpable fact was the trauma of war, as he
would put it to his young friend Dr. Wu and sculptured in the memory
of war and a sense of Malthusian clearing of human population through
war and other forms of human wanton behaviors and the exposure
to viruses, famine, and climate change. For Holmes, nature and culture fused at every turn, across the array of experiences and historical
contexts. But then so is inquiry. He would write to Dr. Wu (June 16,
1923), "philosophy wisely understood is the greatest interest there is."

Indeed, Holmes understood that foraging for coherence in law was
the good judgment that evolves over time and experience. In other
words, foraging for coherence in neuroscience similarly evolves over
time and experience.

We come to realize that the allure of choice is a feature that varies
with context, situation, temperament and longer-term vision. Excessive
consumption as in addictive behaviors is indeed a feature of the brain;
but then all human experiences weave across the synaptic and other
neural features that underlies a changing brain (McEwen et al. 2014).

Holmes's fatalism and perhaps his "extreme detachment"—a consequence of the war (Grey 1992; Rogat 1964; White 1993)—are

sometimes features of PTSD (Charney 2013). His experience was ultimately tied to the horrors of war. A war some 50 years later summed up so elegantly by a Canadian soldier, John McCrae, in May 1915: "we are the dead. Short days ago, we lived, felt dawn, saw sunset glow, loved and were loved, and now we lie in Flanders Fields."

Holmes never forgot the slaughter he had survived and that "in our youth our hearts were touched with fire" (March 1884). Or, as Wallace Stevens wrote in *Death of a Soldier*, "Death is absolute and without memorial, as in season of autumn, when the winds stop." Herein lies much of what Holmes was ultimately tied to: the duty, solidarity and memory of a soldier and war within an historic epoch.

References

Adams, M. C. (2014). *Living: The Dark Side of the Civil War*. Baltimore: Johns Hopkins University Press.

Baker, L. (1991). *The Justice from Beacon Hill: The Life and Times of Oliver Wendell Holmes Jr*. New York: Harper & Row.

Brent, J. (1993). *Charles Sanders Peirce*. Bloomington: Indiana University Press.

Burt, J. (2013). *Lincoln's Tragic Pragmatism*. Cambridge: Harvard University Press.

Butler, B. E. (2010). Democracy and Law: Situating Law Within John Dewey's Democratic Vision. *Ethics and Politics, 12*(1), 256–280.

Charney, D. (2013). *Neurobiology of Mental Illness*. Oxford: Oxford University Press.

Dewey, J. (1925/1989). *Experience and Nature*. LaSalle, IL: Open Court Press.

Donald, D. (1947/1984). *Lincoln Reconsidered*. New York: Vintage Books.

Faust, D. G. (2008). *The Republic of Suffering*. New York: Knopf.

Foner, E. (2006/2012). *Give Me Liberty*. New York: Seagull.

Fussell, P. (1990). *Wartime*. Oxford: Oxford University Press.

Grey, T. C. (1989). Holmes and Legal Pragmatism. *Stanford Law Review, 41*, 787–870.

Grey, T. C. (1992). Holmes, Pragmatism and Democracy. *Oregon Law, 71*, 521–542.

Grey, T. C. (2000). Holmes on the Logic of the Law. In S. J. Burton (Ed.), *The Path of the Law and Its Influence: The Legacy of Oliver Wendell Holmes Jr*. Cambridge: Cambridge University Press.

Grey, T. C. (2014). *Formalism and Pragmatism in American Law*. Boston: Brill Press.

Hollinger, D. A. (2005). The One Drop Rule and the One Hate Rule. *Deadalus, 18*(Winter), 28.

Holmes, O. W., Jr. (1881/1952). *The Common Law*. New York: Dover.

Holmes, O. W., Jr. (1897). *The Fraternity of Arms*. Remarks at a meeting of the 20th Regimental Association.

Holmes, O. W., Jr. (1946). *Touched with Fire: Civil War Letters and Diary 1861–1864*. Cambridge: Harvard University Press.

Holmes, O. W., Jr. & Pollock, F. (1942/2015). *Letters, Volumes 1 and 2* (M. DeWolfe Howe, Ed.). Cambridge: Cambridge University Press.

Holmes, O. W., Sr. (1864). *Soundings from the Atlantic*. Boston: Ticknor and Fields.

Howe, M. D. (1957/1963). *Justice Oliver Wendell Holmes: Volumes 1 and Volumes 2: The Proving Years*. Cambridge: Harvard University Press.

Jaspers, K. (1913/1997). *General Psychopathology* (Vols. I & II, J. Hoenig & M. W. Hamilton, Trans.) Baltimore: The Johns Hopkins University Press.

Kelley, P. J. (1989–1990). Was Holmes a Pragmatist? Reflections on a New Twist to an Old Argument. *Southern Illinois University Law Review, 14*, 427–467.

Kelley, P. J. (2000). Critical Analysis of Holmes's Theory of Contract. *Notre Dame Law Review, 75*, 1681–1773.

Kellogg, F. R. (2007). *Oliver Wendell Holmes: The Legal Theory as Judicial Restraint*. Cambridge: Cambridge University Press.

Kosslyn, S. K. (1986). *Image and Mind*. Cambridge: Harvard University Press.

Laski, H. J. (1925). *Socialism and Freedom*. Fabian Society.

LeDoux, J. (2015). *Anxious: Using the Brain to Understand and Treat Fear and Anxiety*. New York: Viking Press.

Lewis, D. L. (2000). *W. E. B. Dubois: The Fight for Equality and the American Century*. New York: Holt.

Luban, D. (1992). Justice Holmes and Judicial Virtue. *Nomos, 34*, 235–264.

McEwen, B. S. (2007). Physiology and Neurobiology of Stress and Adaptation: Central Role of the Brain. *Physiological Reviews, 87*, 873–904.

McEwen, B. S., Gray, J. D., & Nasca, C. (2014). Recognizing Resilience: Learning from the Effects of Stress on the Brain. *Neurobiology of Stress, 1*, 1–11.

McGaugh, J. L. (2003). *Memory and Emotion: The Making of Lasting Memories*. New York: Columbia University Press.

McPherson, J. M. (1975). *The Abolitionist Legacy*. Princeton: Princeton University Press.

McPherson, J. M. (1988). *Battle Cry of Freedom*. New York: Ballantine Books.

McPherson, J. M. (1991). *Abraham Lincoln and the Second American Revolution*. Oxford: Oxford University Press.

McPherson, J. M. (2008). *Tried by War: Abraham Lincoln as Commander in Chief*. Princeton: Princeton University Press.

Menand, L. (2001). *The Metaphysical Club*. New York: Farrar, Straus and Giroux.

Menand, L. (2002). *American Studies*. New York: Farrar, Straus and Giroux.

Mencken, H. L. (1930). Review of the Dissenting Opinions of Mr. Justice Holmes. *The American Mercury*, 122–124.

Mendenhall, A. P. (2015). *Oliver Wendell Holmes Jr., Pragmatism and the Jurisprudence of Agon*. Lewisburg: Bucknell University Press.

Miller, R. F. (2005). *Harvard's Civil War*. New Hampshire: University of New England Press.

Olson, A., & Phelps, E. A. (2007). Social Learning of Fear. *Nature Neuroscience, 9,* 1095–1101.

Perry, R. B. (1935). *The Thought and Character of William James* (Vol. 1 & 2). Boston: Little, Brown.

Phelps, E. A., Delgado, M. R., Nearing, K. I., & LeDoux, J. E. (2004). Extinction Learning in Humans: Role of the Amygdala and vmPFC. *Neuron, 43,* 897–905.

Posner, R. A. (1992). *The Essential Holmes*. Chicago: University of Chicago Press.

Posner, R. A. (2003). *Law, Pragmatism and Democracy*. Cambridge: Harvard University Press.

Reynolds, D. S. (1995). *Walt Whitman's America: A Cultural Biography*. New York: Knopf.

Rogat, Y. (1964). The Judge as Spectator. *University of Chicago Law Review, 31,* 213–256.

Rosen, J. B., & Schulkin, J. (1998). From Normal Fear to Pathological Anxiety. *Psychological Review, 105,* 325–350.

Rozin, P. (1998). Evolution and Development of Brains and Cultures: Some Basic Principles and Interactions. In M. S. Gazzaniga & J. S. Altman (Eds.), *Brain and Mind: Evolutionary Perspectives*. Strasbourg: Human Frontiers Science Program.

Sapolsky, R. M. (1992). *Stress: The Aging Brain and the Mechanisms of Neuron Death*. Cambridge: MIT Press.

Schulkin, J. (2015). *Pragmatism and the Search for Coherence in Neuroscience.* London: Palgrave Macmillan.

Schulkin, J. (2017). *The CRF Signal: Uncovering and Information Molecule.* Oxford: Oxford University Press.

Swanson, L. W. (2000). What Is the Brain? *Trends Neuroscience, 23,* 519–527.

Touster, S. (1982). Holmes a Hundred Years Ago: The Common Law and Legal Theory. *Hofstra Law Review, 10,* 673–708.

Wells, C. P. (2015). Oliver Wendell Holmes Jr. and the American Civil War. *Journal of Supreme Court History, 40,* 282–313.

White, G. E. (1993). *Justice Oliver Wendell Holmes: Law and the Inner Self.* Oxford: Oxford University Press.

3

Experience, Prediction, Surviving

Our cultural evolution in part rests on an evolution of understanding the very basics of life and the heights of human achievement as part of human experience. *Art as Experience*, a book by John Dewey about the evolution of craft to fine art, music and dance, addresses not just language and articulation, but also probes experience.

Experience and the brand of pragmatism that circulated during this time-period revolved around an active sense of exploration and discovery, invention and tool use, coherence and adaptation. Experience is the recurrent thread of the pragmatist, but an emphasis on the active sense of experience (Bernstein 2010; Janack 2012; Siegfried 1996; Schulkin 1991).

We find, in *Experience and Nature*, Dewey's capturing our sense of experience in adapting, and coping with the precarious and seeking the stable or the viable. Despite the ambiguities rightfully noted by many critics of Dewey, in particular, the sense of experience is critical in learning, knowing, and coming to understand others. It is also critical to understanding in a context of sharing experiences in a participatory democracy, but it is not in contrast with mechanism or subjectivity, as some might suggest (Pappas 2008; Neville 1974).

© The Author(s) 2019
J. Schulkin, *Oliver Wendell Holmes Jr., Pragmatism and Neuroscience*,
https://doi.org/10.1007/978-3-030-23100-2_3

In this chapter and the next, I explore the philosophical forms of pragmatism, to which Holmes is linked, set in a context of appreciating probabilistic inferences and good enough explanations in adapting to events and in our social evolution. Pragmatists demythologize human reasoning but emphasizes predictive coherence in human decision making. This is a feature of how we understand the brain (Sterling and Laughlin 2015; Schulkin 2015) and what figures in the development of statistical inference across human cultural activities (Gigerenzer 2000).

Varieties of Pragmatism: James, Holmes, and Peirce

The many meanings of human experience are nicely depicted by James in his *Varieties of Religious Experience*. This is an elegant book and perhaps James at his best, as he describes diverse forms of human experience. James runs a naturalistic thread throughout the text and includes depictions of the brain.

Pragmatists like James, Dewey, or Peirce emphasized an active mind. For James, influenced by Darwin, attention, memory, associate mechanisms, and the stream of consciousness were active states of mind; they were receptive, impositional. Mind was a piece of biology, psychobiology—part of nature's design of the brain in the context of behavioral coherence.

James was not an experimentalist, but he was a phenomenal phenomenologist. He was able to provide vivid descriptions of human experience (James 1890). James's treatise on psychology was a text Holmes quibbled with, but I doubt he had the patience or willingness to really study it. Holmes tended to see James as "soft," as did Peirce. But what prevails in James is the utterly active mind, and the contrast between the empiricism or the rationalism of the preceding centuries which couched science and law on a path. His *Principles of Psychology*, a book that lays out an evolutionary depiction of the brain, is still somewhat modern in terms of levels of function and distributed broad-based neural networks that underlie diverse capabilities in space and time.

James was on the solitary side. His psychology reflects not so much the essential nature of social contact with others, but the solitary self in a group, which becomes the social self. Holmes as well was a solitary person, in matter and in social presentation, which was perhaps facilitated by battle and loss. James was endlessly psychological, beyond what Holmes or Peirce were capable of, or wanted to be.

Holmes and Peirce eschewed the psychological turn, the endless abyss of the inward view. Both were more inclined to look to social context. Both Peirce and Holmes looked to statistical outcomes in a community of people struggling to make coherence (see Wells-Hantzis 1988), to link science and social context. Peirce, a logician and experimentalist, went to great lengths to show that "the theory of probabilities is simply the science of logic quantitatively treated" (1878a, b). The new tools in statistics were beginning to shape the epistemic contexts of discovery.

For James, psychology, though under the guise of biology and neurology, was understood within available science. But James also had another side, that certainly Holmes rejected, which was his mystical parapsychic side. Holmes did not understand and indeed rejected these non-scientific orientations that James harbored. But James was rich in the sense of experience and our vulnerabilities.

James was anchored to biology, or psychobiology, at his core. One of his great books, the *Principles of Psychology* (1890) came out nearly ten years after Holmes's *Common Law* (1881). Both are true literary achievements. James was anchored to capturing experience and understanding the brain. Both men were naturalists; Holmes perhaps more so. Both received great acclaim for these treatises, which came to dominate a landscape of inquiry, both psychological and legal.

Holmes's elevation was rapid. He was elevated to the status of Henry Maine, well known for the breadth and depth of the law—see his *Ancient Law* (1861), a noteworthy treatise describing customs, practices and codes of behavior. While Holmes was not a great admirer of Maine (Howe 1963), one reviewer, AV Dicey, wrote that "Mr. Holmes's book is the most original work of legal speculation which has appeared in English since the publication of Sir Henry Maines's *Ancient Law*" (Dicey 1882). One important aspect of Holmes's book is the expansion of legal concepts in an evolving culture, and the other is the separation of moral blame from what is accountable "outward conduct" (Dicey 1882). Holmes would emphasize that standards are external, and the body politic is a continuous variable with the expression of law.

James emphasized the individual, though he alluded to the "social self." It was left for John Dewey and certainly George Herbert Mead to emphasize the social side in which individuals exist. After all the political milieu that Holmes alluded to is social in nature where ideas are worked in competition. It is deciding together (Moreno 1995) where rationality reaches a pinnacle of meaning.

James worried about a "block universe," i.e., a social stifling. The individual is at the heart of a reformation dear to James and Holmes, yet the breakdown of a catholic unification, the one City of God (Augustine), or the one relation, or set of historical inevitabilities. Indeed, being alone is as fundamental as being with others. What James understood and captured is the individual experience; what Dewey understood less elegantly in literary capabilities is the social, being with and sharing experiences with others (McDermott 2007).

Indeed, James (1907) was willing to go to places where Holmes was probably uncomfortable, namely religious experiences. And in the correspondence of Holmes and James right after the war, there is an affectionate back and forth between the two of them (see James-Holmes correspondence). In a letter to James on December 15, 1867, Holmes talks about reading Kant and other intellectual items and gets quite personal, addressing James as "Oh Bill, my beloved" and writes in warm personal terms. In other letters during this period, Holmes writes much like Emerson and Thoreau about the glorious nature of robins, squirrels, and blackbirds. For Holmes, nature was alive, intellectual expansion as well.

But the personal, with which James would expand, Holmes did not. Some sort of falling out might have occurred, or a dilution of respect from Holmes with regard to William James (Grey, November 20, 2015, personal communication). Something had changed (Grey 1989). William would write to his brother Henry in 1876 about Holmes's "cold-blooded conscious egotism and conceit …all the noble qualities are poisoned."

Holmes never left the discussion of ideas, however arrogant and self-involved and inflated he was at times. He was humbled and reminded by his ordeal in the war where he thought himself superior to be reminded in which he found himself "inferior to men that I might have looked upon had not no experience taught me to look up" (letter to Laski, December 15, 1926; Posner 1992).

Self-revelation became more a mystery over time with Holmes and his letters. Moreover, James's inquiry is broadly into the human condition. James pondered the evidence of saints, and sinners; he labored over what mattered in a life of meaning and the importance of the moral life (Wells 2000) and well-being. Both men were philosophical. But James kept looking to understand the individual, to grasp the real phenomena of human experience in a context of empiricism, at least in principle within a background of biology and neurology, and what has become neuroscience.

Holmes in a letter to Lewis Einstein (September 27, 1909) commented upon James, writing, "he believes in miracles if you will turn down the lights…I think skepticism should be humble and be content with saying the universe has consciousness, significance…."

The Club and the Pragmatists: Holmes, Peirce, Green and Inquiry—Common Themes About Predictive Coherence and Adaptation

Holmes (1881, 1917) noted the diverse forms of logic in human reasoning and was rooted in the empirical sciences in the discernment of reason and problem solving. We come with a grab bag of orientations in solving problems that cuts across human experience. Holmes was justly wary of the stranglehold of formal logic or reducing thought to something that it is not.

Holmes was a pragmatist in some abstract sense but never anything particularly philosophic. His sympathies are about carefulness, some allusion to experience and not logic. Holmes was just a smart careful thinker with some grounding in a sense of a community and evolution of social custom, properties and social sanity. He is enough of a formalist, or realist or pragmatist or utilitarian, to fit some in each category in varying degrees. But probably none of these categories really capture him. After all, still today, he remains a mystery and aloof.

Holmes, like Peirce, believed that a judge's judging is rooted in social consequences; meanings are not strictly in the head but tied to context and matters that are embedded in the community. Science, like art, however, as they were for Dewey, is both an end in itself to be enjoyed and tied to our coping and making sense of the world we are embedded in. There is no absolute separation between these disciplines and endeavors in thought. What matters to pragmatists and to Holmes is in line with breaking down dualisms.

Holmes like other pragmatists eschewed traditional philosophical bifurcations. Frozen thoughts are embedded practices well worked out in a community, as Peirce understood and Holmes noted. Dewey made it clear that his corpus was embedded in showing just how porous and malleable the traditional distinctions were. He leaned always towards clarity and science; how much science he knew is not clear to me. Like Chauncey Wright, his inclination was toward biology, some sense of the natural world. Wright was a fan of Darwin and much less so of Herbert Spencer who became a lead voice of social Darwinism. A biological perspective does not lead to a narrow notion such as social Darwinism (Power and Schulkin 2009).

Unlike Peirce, Holmes did not write about logic or the logic of law. Peirce laid out the themes of logic. For both, inferences are embedded in experience; that is why pragmatists like Peirce and Holmes emphasize experience. Holmes sat in on Peirce's discussion on inquiry and discovery and the role of experience in the Metaphysical Club (Wells-Hantzis 1988).

Peirce was and still is the great expositor of human reasoning, human inference, and the logic of discovery or what he called "abduction" or "retroduction" or simply "hypothesis" (1878a, b, 1898/1992). These terms probably bothered Holmes, who was much more austere and economical in his prose and terms. But then Holmes was not the Leibnitz of the US as one eminent scholar has depicted Peirce (Fisch 1986). Peirce was an inventor of symbolic language, contributor to statistical inferences, surveyor of all things scientific and technical and deeply historical in the history of philosophy and science. But Peirce was no poet. Holmes was. Both, though, were pragmatists. And Holmes's outlook and his orientation to foraging for coherence and investigation was "closer to Peirce than to James" (Fisch 1942; Kellogg 2007). For both men, the collective community is what matters, the collective proceeds of the community of investigators.

Induction, deduction, and abduction were three features of reason and judgment that Peirce emphasized. The first two were by the end of the twentieth century traditional, the third not. Abduction was tied to the genesis of ideas, hypothesis formation; all three figure in judgment in any field in judges or the sciences. They are endogenous in the organization of thought and decision making. These are not abstractions but features of thought from which we slide, one to another, in the way we understand the world. These are not strained except when we make such features apparent in consciously solving problems (Fig. 3.1).

Generality and hypothesis are intimate partners in understanding; what Peirce called abduction, the genesis of an idea, and what Holmes understood as generality (Wells-Hantzis 1988). Something like a community of inquirers and something like the community of law-abiding citizens (Rosenblatt 1975) is idealized for both Peirce and Holmes.

Certainly, this is true of Peirce, perhaps less true of Holmes. Peirce understood explicitly the role of prior cognitive expectations, not Cartesian for which Peirce argued against (1868, 1878a, b), and therefore not certain, simple, introspective and indubitable.

Inference and Hypothesis

Deductive or Analytic Synthetic

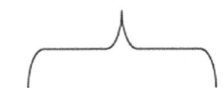

Induction   Hypothesis/Abduction in context & ecology

Fig. 3.1 Flowchart

Moreover, the settled beliefs of the community of inquirers for Peirce rests in the continual interplay of innovation, interest and meaning; meaning is social (see also Wittgenstein 1953). Holmes, of course, is arguing for external measures; unlike their old friend James, who celebrates the psychological. The emphasis is on fixing reference on external objects. Holmes had lower expectations; but then he was perhaps less of an empirical inquirer than Peirce. Both were too dismissive of James and James's brilliance in capturing human experience. They actually wavered on this. Both appreciated James, but both thought he was soft as a thinker. All three were part of the Metaphysical Club in the 1870s, however, and there was a lot of cross-influence. Holmes's views are to be sure facilitated by these interactions (Wiener 1949; Menand 2001). Peirce in 1906 would comment that the young inquirers would at times meet in "my study, sometimes in that of William James." Later, he says, "Mr Justice Holmes however will not I believe take it ill that we are proud to remember his membership" (Peirce, Collected Works 1958; Haack 2011).

As Dewey noted, people "do not [think] with premises. They begin with some complicated and confused case" (1931, p. 134). Action embodied in experience is a signature of this form of pragmatism. Where Holmes meets other pragmatists is on the importance and use of prediction and the logic of statistical reasoning (Hacking 1965).

Holmes mostly talks about law as "recipe making," emphasizing the practical side. Here he is closer to James (much probably to Holmes's chagrin). He is searching for something that works and is predictive. The lawyer is angling to discern what a judge might do. That is what Holmes keyed in on. Lawyers want to win, scientists want to discover, bankers want to invest, prey want to survive and predict the probability of predator occurrence.

Holmes believed in judicial restraint, or at least in restraining his own view to the settled view of the larger community, the commons of law and sentiment. Here he is close to Peirce in glorifying the larger community. Like Dewey and Peirce, he followed science in appreciating "improved machinery and good conduct."

Making sense of law in the context of the lawyer and the judge and what matters is what the judge may do, and is likely to do (Lawyers Law Holmes). Of course, Holmes understood his work as "an experiment as all life is" (see Dewey 1928). Law is continuous and embedded in the larger culture of which it is an important part. Holmes is in the culture of science, prediction and coherence. Law is not ethics and not science, but reflects both. Law is a reflection of our community, the oxygen of our rational behavior, or not.

Law is about continuity and legitimation and accountability. James, in letters to Holmes, urged him to form a group of thinkers to discuss ideas. Holmes was intellectually curious: *what is the nature of law?* It was a question that permeated his pen at the outset of his activities. As Peirce asked: *what is the nature of science?* And James, *what is the essence of psychology?* Holmes would do the same with law.

Theories of the scientific method were running across the face of Europe (e.g., Whewell, Helmholtz) and the empirical/rationalistic traditions were finding more and more moments of contact. Peirce would understand this more than any other in this group, or any group during the nineteenth century, certainly on these shores.

Holmes thought Peirce was pompous, egotistical and oversold in terms of talent; the first two are true, the third is not. Holmes did not like Peirce's cosmic view, his religious sensibility, and thought Peirce overrated.

Peirce was not overrated for the vast majority of thinkers. Peirce, a statistician, set up the first laboratory in psychophysics in America at Johns Hopkins University, before he was kicked out for non-Quaker like behavior. He understood the technical sciences, contributed to symbolic logic and the logic of algebra. He was off the charts in capability and fulfillment in so many fields, including philosophy. But he was self-destructive, conceited, singular and unable to be in any community; much as he glorified the community of inquirers, he was an *n* of 1 (Brent 1993).

Holmes was irritated by Peirce's "pontifical self," but he did respect him, although, as he continues in a letter to Morris Cohen, "he does not move me deeply" (see also Diggins 1994).

Peirce heralded in principle a community of inquirers; settled beliefs in which test, experiment, and rationality would prevail. But Peirce was a Trinitarian; a believer and rather traditional here. He was no Unitarian as Holmes nominally was. Peirce's view of evolution and unity was based on his sense of evolutionary love and the community. This was not the thinking of Holmes.

Holmes, who thought less than Peirce, thought no more of James. Like Peirce, Holmes underestimated his brilliance, capability and profundity with regard to the human condition. The Metaphysical Club (1870–1872) would provide a context to materialize ideas, in science, about the human condition, and in law (Fisch 1986; Menand 2001). James may have been softer than these two, but he was profound in capturing the human condition, better than the two proclaimed giants in this Metaphysical Club.

In the Metaphysical Club, four out of the seven in the group were Harvard trained lawyers. The lawyers were Nicholas St. John Green, Joseph B. Warner, John Fiske, and Holmes, the other three members were Chauncey Wright, Charles Peirce, and William James. One in particular was very influential on Holmes: Chauncey Wright. But so was Green.

Nicholas St. John Green died relatively young in 1876, in his 40s. Like the rest in this Club, his roots were in New England. Green published just a handful of texts. Everyone in the group, though, commented that he understood the modern sense of prediction of events as critical in understanding human cognitive capability and human understanding.

Interestingly, Green (1870) published a piece in the *American Law Review*, where Holmes was the editor, about causation: "Proximate and Remote Cause." This is an important paper highlighting the confusion about causation and its complexity in the human lexicon of understanding. He criticizes Aristotle and the scholastics for their rationalistic conception of causation, a frozen notion. He tied causation to practical activity, what could be understood. Indeed, Peirce himself gave Green credit for like-minded thinking with regard to what Peirce called pragmatism. Peirce, on Green, said that he was "the grandfather of pragmatism" (see Frank 1950; Horwitz 1992).

Both Green and Holmes would reject part of the approach of Langdell (Grey 1989; Winter 2001), and what became in part the study of the logic of case law as an end itself (Holmes 1880). Rejecting formalism is something akin to a kindred spirit of them with W. V. Quine or Wilfrid Sellars. They reject a formalism in a Kantian separation of the analytic and synthetic in empirical assessment (Smith 1970; White 1947, 2000; Brandom 2010).

Green, a major influence on Holmes, demythologizedthe term causation. In his influential article, he states that proximate and remote causes can co-occur at the same time. The separation of causation that may not be causal at all was another issue he highlighted. What Green highlighted was the fact that an event has many senses of causation, and there is a very porous relationship between proximate and remote causes.

Green's (1870) appeal is to experience. He believed that no absolute separation of such causes can be found in the flux of experience and in its application and that it is best not to mythologize and render them abstract and misleading. Unlike Peirce, the mathematician in the group, Green looked on abstractions with suspicion (Wiener 1948; Frank 1930). Green was at the deathbed of his friend, Chauncey Wright, only to die a short term later. Both men never had the public face of many of the others in the famed group of thinkers.

Wright's sense of biology, the appreciation and non-ideological side of embracing new ideas in science, was vibrant and important for Holmes, but his link to Emerson also remained perhaps constant (Levinson 2000; Mendenhall 2015a, b; see also Chapter 6). Green, like Holmes, was intellectually in the tradition of Bacon and Mill, of lawyer scholars (Wiener 1948). Green was closer to Holmes than he was to his classmate Langdell. An interesting fact to note is that Wright is to the philosophy of science what Green, in perhaps more modest terms, is to the philosophy of law (Frank 1930, 1950). Both were major influences on Holmes and all shared the same breakout of science and the continuity with law and the larger sense of human experience in epistemic adventures.

All acknowledge Green as the progenitor, or at least participating in the origins of what became pragmatism (e.g., James, Peirce). All make an appeal to experience, in a large sense of that term. But the appeal about causation is rooted in human experience; something found in James, Holmes and Peirce. Green is knotted to Mill, utilitarianism, social welfare, and to methodological positivism (Kellogg 2007).

Causation in the law is a concept from the Greeks and their law (Frank 1932). It means responsibility, accountability, or guilt. For the law, liability or indemnity are major considerations. For Holmes and many other thinkers, there were no oracles in judicial law making and thinking. They are replaced by hard fought ideas, some of which are better than others. In law, legal reasoning is liberated from anything Euclidian in metaphor (Frank 1950). What is elevated is rhetoric and empiricism, the empiricism that is identified with pragmatism, and the continuity of science and the law. Rationality is not by reason or argument alone, but the hard work of empiricism, the testing of ideas, and their evolution in practice.

Holmes had little scientific training, which he admitted, but he accepted the uncertainty of the human condition and of reasoning. In these senses he was, as Jerome Frank writes, "a completely adult jurist" in 1930. I am not sure what it means to be "completely adult" or if that is achievable, accept for noting the sense of uncertainty that pervades as Dewey noted the knowing process (see also Posner 2008). Holmes was an independent mind, someone who would reason, or try to.

Knowledge spreads across a community of investigators or inquirers. Settled opinion is like common law in that it reflects the testing of time and adjudication or the sense of prediction or "bettabilitarianism" (betting on likely predicted outcomes). The allure of this rhetoric is ripe with imaginative metaphors (see *Schenck v. United States* 1919), but the orientation is towards the future, outcomes that are statistical.

The sage of the New York Yankees, Yogi Berra, would suggest to us: "Prediction is extremely difficult, especially about the future." Diverse pragmatists emphasize the sense of experience, causation as part of experience, and a feel for likely outcomes. Predictive capabilities are tied to our cognitive competence and tied to our biology. What Holmes held fast to was a sense of the probability of a universe. In a moment reminiscent of Peirce and his invention of terms, Holmes would write Frederick Pollock:

Chauncey Wright, a nearly forgotten philosopher of real merit, taught me when I was young ... that we don't know whether anything is necessary or not. So I describe myself as a bettabilitarian. I believe that we can bet on the behavior of the universe in its contact with us. We bet we know that it will be. (Holmes-Pollock letter August 30, 1929; see also Fisch 1988)

Of course, that is fine, but was he a "bettabilitarian" in ways that determined his adjudication of events?

Holmes is the opposite of his old friend James, the one who would write about what makes a life meaningful. Despite Holmes's rather arrogant protestations against James, similar to Peirce's, they both lacked perhaps James's basic human response. Both were arrogant, not an uncommon feature of the talented mind, but a seductive misguided sensibility.

James could write about the mind of the saint or sage with insight. But Holmes understood the travails of war and comradeship; after all, he, and not James, put his life on the line. There is nothing innocent in Holmes; there is some left in James. Holmes accepted the naturalistic evolutionary perspective that runs through much of pragmatism (Wiener 1949; Hauser 2003).

Holmes was unlike his colleague J. C. Gray of Harvard and a fellow member of the Metaphysical Club in generating a global view of the human condition. Although, Holmes alluded to this in a letter to Laski (June 19, 1919), in a "bet" on the West over the East (Duggan 2007).

Holmes has a realist view, an economic view (Posner 2007) of competition and transaction. This is lamentable to his critics who expect more and are disappointed. Of course, economics is not considered the best of all sciences to hang so much on (Leiter 2000). But economics and statistical inference are important in undending the human condition But Holmes would probably say rightly that economics is one tool amongst others. And, in fact, that is right and a rather pervasive position.

Tracking Consequences, Utilitarian Sensibilities, and Staying Anchored to Objects and Social Considerations: A Recurrent Emphasis for Holmes and Other Pragmatists

Holmes, as I have said, was drawn to utilitarian sensibilities (Kellogg 2007). Utilitarianism grew out of an emphasis on the foundational component of the senses in knowing and motivation; the allure of the pleasant and the avoidance of the less pleasant. Two endeavors were at stake in grounding everything in the senses: knowledge and human motivation. As Epicurus would assert, "we must use our sensations as the foundation of all our investigations" (1964, p. 8). Holmes was a "preference utilitarian" (Grey 1989, 1992).

Jeremy Bentham (1782/1970) based his view of utilitarianism on a calculus of pleasures and pangs; on a political theory based on sensations and satisfaction; and on a general principle with some sense of moral sensibility, of the greatest goods. These were not narrow. There were pleasures such as "skill," "power," "benevolence," "expectations;" and the pains were construed broadly too: "regret," "awkwardness," etc. These were not narrow gluttonous pleasures or pains but broad-based human capabilities of meaning and contact.

And we are not very good at the calculus side of predicting what we might want and get pleasure from (Kahneman et al. 1982; Kahneman 2011). Moreover, the calculating side does not capture what we should want. Here Dewey and the road not taken by Holmes is instructive. Dewey (1925/1989) was always in subtle and less subtle ways indicating what we should want, care about, and strive for—namely, the aims of life in trying to adapt to circumstance and context.

Still, it was an advance to suggest. It was just now in some social political context to imagine and then test what produced the greatest good for the greatest number, as John Stuart Mill would expand on Bentham's position, to create a just society of trade, commerce, and minimal interference by government. But this is limited if not set in a context of what we should care about (Hart 1958, 1961; Postrema 2011).

Morality, though as Dewey in *Human Nature and Conduct*, noted, while much appreciative of the insights of Bentham and Mill would argue calculations of a large fit of possible pleasure is no substitute of what counts in life; joint cooperative endeavors. As we now know, our calculating pleasure is not very good anyway. Indeed, we are not very good at it (Kahneman 2011), and diverse regions of the brain tied to pleasure and pain are also tied to a larger social milieu of making sense of the world we inhabit; the brain in suitable contexts is prepared to compute uncertainty relationships into meaningful action (Glimcher 2003; Clark 2017).

Empiricist based approaches would appeal to Holmes. They ground ethical consideration in a consideration of the natural inclinations of human beings, and more broadly construed and based on a calculus of sorts that was expansive of the social and historical context. It was empirical, not mere thought. The calculus was concrete and based on human wants and desires, and it used a set of moral sentiments that seemed roughly agreed upon by the learned secular class.

Indeed, utilitarianism linked the consideration of the human condition to the management of the social milieu and the polity (Sidgwick 1902/1988). Still, it was narrow in the eyes of moral theorists and too narrow about human experience.

Pleasures and pains sprawl across the human conditions of social contact in the hands of Bentham (1782/1970) in particular. The end result is "utilitarian doctrine is that happiness is desirable" (Bentham 1782/1970, p. 307). There is a disinterested part to step back in the consideration of what is virtuous and what is not, what is worthy and what is not. But contact matters; it is human contact in judgment that puts us in touch with phenomenon; essential for law and adjudication.

Much research in neuroscience is about human contact (e.g., Greene 2014): maintaining in development, fostering, sustaining, expanding it, and deepening it. We survived because we could make contact, create bounds of meaning and utility. Utility is one side of human meaning. Because something has utility does not undermine the value of an individual or event. The mistaken view of appreciating something because it is so pure as to be outside the scope of possibility, utility, and meaning is the radical separation that pragmatists like Dewey eschew. Instrumentalism is not incompatible with human worth. The barriers are permeable.

Keeping track of human worth is something within the codification of the law, and the sense of fairness and the moral compass of the dignity of human or the larger animal sense of what we inhabit, depend upon, and need. Our social bonds with each other, as Holmes saw it, are instrumental, but instrumentalism was seen as degrading true value: value on one side, fact or instrumental use on the other. As Holmes suggested in 1914 (*International Harvester Co. v. Kentucky*) "value is the effect in exchange of the relative social desire for compared objects."

Amidst the great praise of Holmes in many quarters, he was and is often criticized for his moral timidity. First was a separation of morality from the law and understood in part of the abuses, misleading, and unintended consequences that can occur. Holmes, like Nietzsche, worried about moralism and human bestiality. Indeed, morals runs through the law as Holmes understood from the body politic in which we live. But if the body politic runs astray, law's role is muted. This has worried some of Holmes's critics (Alschuler 2000; Kelley 1993; Gordon 1982).

Holmes was not bold in these contexts; but he did not envision the law in the context of being bold. It should be careful, clear, and compatible with science. He understood law in limited but obviously important terms, in terms of our cultural evolution. Knowledge was understood in what was settled upon in the community of inquirers. The good was settled in the community of participants. But what about those who are not participating? His responses to these circumstances were soft or mute. He was no firebrand before the war, nor after, just a rational agent and steadfast in making clear his own independent decisions. But the good was collective in the community (Holmes and Frankfurter Correspondence 1912–1933), like science in the community for Peirce.

Holmes would suggest to look to the common person, or criminal, in discerning expectations of adjudication and of future behavior. There was the consideration of something common, an allusion perhaps to a common law or a common faith.

Law is understood in the larger context of the public. The evolution of thought entailed the rigor of codification and empiricism. Yet much human contact, discovery, and understanding is broader than that.

Reasoning and Context: Continuous Monitoring, Coping with Uncertainty

The utter continuity of reasoning within constraints and having on occasion to go beyond settled precedent, moreover the abstract notions of reason and justification, is inherent and can be transparent, or not. But the hodge-podge character of the law as a metaphor and Toulmin's (1950, 1958, 1972) "intellectual ecology" are suitable playmates.

Like biology, law exists in an evolving context. And like biology we may hit on some ideas. But there is continuity, regularity, and predilection built on a capability for problem solving. In fact, regularity of expectations rests upon a continuum; laws are inherent in our expectations of events. There are many meanings of law-like relationships. One law-like relationship is obviously within the profession of law whether in the cultural sense of law-related events or the biological. Both reflect the sense of regularity in a universe of expectations, of coherence in a changing landscape, and of stability and longer-term viability as we make our way through life.

The many meanings of the law are something akin to the many meanings of science, clunks of meaningful material and activities. Looking at the many meanings of the law is an important window of human social binding. One asks, what features of the binding relationships are predictive events?

Laws provide a platform for predicted regularity, a sense of solidity. We come prepared to discern patterns, solidity providing a cephalic system anchored to features of the worlds in which we are in habitation and adapting to—in a word, living. What evolved is the time to reflect; to reflect on what matters. And what matters is what Holmes and others understood as the deepening of human experience. This is a concern for both judges and the rest of us who reflect on human existence and the exigencies of human insecurity in search of stability; it is a sentiment and proposition tied to John Dewey.

Existential angst is another feature of our existence, that wish to placate where we can. We do it in many forms of social structure of which law is one. In an evolving sense of the law—in which foundations are less frozen and where there's a sense of change—propositions and a commitment to viability, fairness, and social action are leading sensibilities for the deepening of human experience. Within each is investigation or inquiry. A locus of interests and diverse topics, and drudgery predominate. There is nothing romantic about laboratory work for most of us: there are labor and disappointments; many assays don't work; many cases are fruitless.

Cognition runs through action, as we sample and cope with our surroundings (Dewey 1896; Engel et al. 2015). Both imagined action and real action can elicit the same neural activities in the brain. I watch you do the action and I don't move, nevertheless my patterns of neural activity will reflect yours. The patterns are not identical, but they are close enough to give us meaning. Imagining action is as important in learning about the world as doing. Imagining is nontrivial in getting a foothold in the life of others, in considering a course of action.

While opportunity is one thing, success is another. Dewey (1925) embraced just about everybody with suggesting possibility, Holmes much less. But both embraced biological and cultural continuity and both understood good judgment as part of matters of great worth, the deepening of human experience. Both men understood "experiment" at the heart of human experience, including for Holmes a living and evolving constitution by the cumulative cultural evolution of us as a nation or species. Holmes looked to the larger community of investigators for insights that might impact the court, the lawmakers and hope for what Peirce called the "community of inquirers" in addition to what Peirce called "fallibilism" in our judgments subject to self-correction.

Holmes (1881, 1917) linked legal reasoning to a context of predictive capability. This core feature of biological design in the evolution of the brain and in learning is about just associating events in time as Hume and others had thought, but also about linking events as predictive. This reaches into the very primitive aspects of cephalic capability, namely predictive events.

After all science (like art) reaches into everything, something both Holmes and Dewey understood. Statistical inference is essential in science, in reasoning, in surviving. The pragmatist just tied to ongoing action (Engel et al. 2015). The parlor or thought discussion without a context of adaptive capabilities can wrangle about association of ideas as Hume did or rationalist clarity as Leibnitz or Descartes did. But the stuff of adaptation that is about prediction of events and the machinery which makes it possible was something Holmes very appreciated.

Leibnitz's sense of science and the law is tied to demonstration and logic. The inventor of calculus was in search of the harmony of reason and jurisprudence (see *Elements of the Law*). Leibnitz understood reason in law was within the necessary sciences; "numerical relationships" and numbers were the key to science. Necessity reigned, not probability. The latter would wait a few centuries to became of the fabric of understood and practiced science.

Indeed, numbers or more generally diverse forms of mathematics underlie human reason, human cultural evolution, of which and what Holmes understood as the "quantitative" is inherent. But Holmes, likes his mentor in the sciences, Chauncey Wright (1877), is tied to the biological and behavioral sciences, of coping and foraging to make sense, of adapting and understanding. For Holmes, there was little edifice to stand on, except hard work and good fortune.

The function of this decision is tied to environments that one is trying to survive, avoid events, approach events, etc. Our notion is tied to brute experience, the sorts of experiences that William James would insightfully adumbrate in *The Principles of Psychology*, one of which is our sense of causation. There is something fundamental in our experience in linking causation to prediction, a precondition in animal sensibility, what Santayana called "animal faith" (Santayana 1967).

A faith that Holmes understood was built on experience, or the direct causation of adaptation and prediction. The issue now was in linking everyday experience and developmental foundations to a larger culture of experiment and self-correction, hypotheses testing, and the competition of ideas. Prediction is the other side of causation. We come prepared to learn causal events and to group events into categories of coherence. Endless opportunistic bootstrapping underlies adaption to events, linking events together, testing, and waiting and determining if events are linked (for instance, of food resources and gastrointestinal distress, the co-occurrence of predators and prey, etc.).

Determining causal relations is not an extra property of our brains but built right into the diverse forms of learning. Causation is both an abstract and incredibly concrete idea. Abstraction is in the hands of the rationalist and the counter intuitive concepts that we live with (e.g., action at a distance). Pragmatists understood that linking the abstract capability of linking diverse events into a predictive event is tied to our understanding of causation. We come prepared to track events, to monitor important and causally efficacious events. We just often have limited theoretical understanding of events.

The twentieth century philosopher Nelson Goodman (1955), in a book about projectable predicates, gave a nice everyday linguistic account of how some arbitrary matter could be couched in common discourse about prediction. It takes a few further steps to link it to causation. Causation is a big term, and there are many disputes about the many meanings of this term. What counts is some notion of adaptation in some form of coherence and capability. That was good enough for Holmes and that was good enough for diverse pragmatists, and it is good enough for most scientists.

Is it totally satisfying? No. But then few things are in epistemic ventures. As Dewey noted, "the union of the hazardous and the stable, of the incomplete and the recurrent, is the condition of all experienced satisfaction as truly as of our predicaments and problems" (1925, p. 54).

Dewey (1925) understood that something like an original condition for our species, straddled between the precarious and the relatively stable, is "the natural and original bias of man is all toward the objective." It is a fairly narrow universe of survival and foraging for some coherence.

The core value whether biological or cultural, narrow possibilities or broad, was an evolving body in an evolving universe with relative stability or viability. As Holmes (1913) put it "we need to think things instead of words" (*Law and the Court*). We need to be anchored to objects with references that fix us to an external world. Holmes's intellectual ear is towards a naturalistic, scientific and realist sense of the human condition.

One core value is that of education impacting changing behavior. Neuroscience provides a rich context for understanding the machinery inside of the head. That is not a replacement obviously for what goes on outside, in the pedagogic context. Progressives like Dewey made a lifetime passion of facilitating educational contexts and then linking those contexts to democratic sensibilities such as an enhanced capability, relatively free of fear and tyranny by laws that decrease human bestial behaviors and enhance human participation in the democratic process.

Neuroscience Considerations: One interesting fact about neural design is the fact that active inference is a feature of cephalic capabilities (Friston 2010; Sterling 2004; Clark 2017). Pragmatists emphasized an active brain, not a passive one. Indeed, the experience is search, appetitive search or foraging for epistemic satisfaction. It is the active of a mind adapting to nuance and niche that we need to take into account when considering an understanding of human agency and accountability in then context of neuroscience and the law. Agency needs to be demythologized. Indeed, it is a very imperfect concept for which we may have a predilection to both attribute it and over attribute it (cf. Altran 2002; Alces 2018). Holmesian sensibility embraces the concept of agency for what it is worth; a critical term in understanding ourselves and our actions, but not without ambiguities, and certainly pockets of epistemic moments where it makes less or little sense to use the term in the law or elsewhere.

Dewey (1925) emphasized appetative and consummatory experiences. Indeed, these events pervade the neural orchestration and neural design (Swanson 2000a, b). Moreover, Holmes like Green and certainly Peirce emphasized statistical inference, an essential element in reason, tied into hypothesis formation. Holmes in the *Path of the Law* defined law as "systematized prediction". Active sampling is a fundamental part of our evolutionary neural design. It figures in most of what we do. Holmes and other early pragmatist like Green were prescient for linking it to reasoning within the law, like we do in science. And indeed, we know understanding the foraging epistemic assessment as utilizing statistical inference (Schulkin 2015).

This view of the brain, of how we reason, sets the stage for understanding other features about us, our sense of agency, of adaptation to diverse forms of social and ecological diversity. The law is critically tied

to a notion of agency (Holmes 1881), and our integration of neuroscience into the law draws out these issues in adjudication and their difficulty (Alces 2018). This problem-solving capability is not just cortical,
it cuts across both neorcortical and diverse regions of subcortical and
motor systems (Schulkin 2004, 2015). Indeed, the neural systems keep
tag on what is familiar and what is not. We clothed expectations within
habits of action that are warranted, adaptive; the habits orchestrated by
neural systems minimize disturbance. This is a core view of the pragmatist view of inquiry. It underlies both our understanding of neural
systems and their evolution.

Conclusion

Predictive capabilities are tied to continuous coherence, or good enough
coherence to capture what matters. Holmes certainly believed that law
is what happens in the courts. The power source is in the larger body
politic, which is always vulnerable to bestial explosion of the irrational.

But settled thought is still within the realm of further inquiry in
which predictive coherence and debated principles predominate in
securing some modicum of rationality and securing forms of peace,
the opportunity to secure wants and needs, and the possibility of participation by individual ingenuity and talent. Holmes's reasoning is
embedded, like Peirce's, in the statistical realm. And what underlies is a
sense of collective experience, and settled opinion through hard fought
inquiry: an inquiry of experiment and experience embedded in cultural
evolution.

Holmes, despite his disputations, valued the search for propositions about the human condition. The search for wisdom underlies
the very idea of the law, of what is fair, judicious; but the law is also
about "taming chance" (Hacking 1990), a phrase from Peirce. Chance
could be understood; chance was a property of the world. Our world is
a world in which probabilities and statistical inference are meaningful
and predictive. And that made all the difference in ushering in modern
inquiry. Statistical records became part of the larger framework of event
prediction.

John Maynard Keynes (1921) would assert that statistical probability is a very primitive feature of our epistemic orientations. It figures in more ancient adaptations such as animal minds rich in tracking events or of determining likelihoods that manifest in surviving. Holmes, the naturalist, would appreciate this feature of the animal mind. Certainly, his mentor in biological sciences, Chauncey Wright, did so. Predictive capability and coherence and background knowledge mattered, for which probability could be measured (e.g., Bayes' theorem). This no doubt excited Holmes as much as it did Peirce. There is one difference between the two: Peirce really understood and contributed to statistics, and added a calculus of probability and a calculus of logical inference.

Indeed, Peirce had much more depth here than Holmes. Holmes was to statistical understanding what James was to experimental studies; they were abstract for them, and not up close and personal. Peirce's work in experimental psychophysics, at the first experimental psychophysics laboratory at Johns Hopkins, his work on statistics of the earth for the geodesic office of the US government (his longest serving occupation after he left Johns Hopkins in part for an affair and in part for his entrenched stubborn behaviors). Unlike Holmes, Peirce did everything not to fit in and be agreeable. Their dispositions and capabilities made them both outstanding individuals, but one was headed for social and public success, the other not. One landed on the Supreme Court and the other penniless and addicted to morphine at periods in his life.

Foundations are malleable for pragmatists; logic and consistency are only part of thinking in the law. Law is set in an historical context; there is no pure logic that reveals the nature of law, there is only logic or logics (as Dewey would put it) in thinking about the law, in thinking in science, etc. The issue about experience, in the famous quote of Holmes, is down playing the excessive sense of logic, mere analytical thinking in Kant's term.

Law is tied to life, continuous with our biological constitution and our evolving sense of the world and our knowledge. Frozen logic is not the metaphor, but instead expanding experience and regulation, maybe minimal regulation to maximize human expression and human choice.

But the pragmatist and Holmes are anchored to objects beyond internal rules. One version of pragmatism, the pragmatism of Peirce and Dewey, is anchored to objects, the internal and external barriers are dissolved in the context of inquiry, probing and testing. What takes precedence is statistical prediction, and conceptual and habitual coherence in the organization of inquiry and action. Brains are operating in a context, many contexts, and are anchored to environments. Our tools extend our brains to enhance our capability in environments. The continuous thread between what is so-called internal and what is external is fluid and permeable, something quite noted for many pragmatists with diverse inclinations.

Mind/brain considerations include the environments of adaption (Clark 2013, 2017). We are embedded in environments trying to make coherence from within and from a view from outside; both are metaphorical (Lakoff and Johnson 1999). Realism and antirealism are dissolved in adaptation, coherence, and purpose. Objects are not reified nor denied; what Peirce called "secondness" or resistance is a constant reminder of what is real or not to varying degrees.

There is a real sense that these intellectual battles are not about external and internal debates or realism and antirealism, but about adaption, significance, and human well-being. Holmes was steadfast in linking his philosophical predilection to external anchors or objects, and to cognitive/neural predictive capabilities.

References

Alces, P. A. (2018). *The Moral Conflict of Law and Neuroscience.* Chicago: University of Chicago Press.

Alschuler, A. W. (2000). *Law Without Values: The Life, Work and Legacy of Justice Holmes.* Chicago: University of Chicago Press.

Atran, S. (2002). *In Gods We Trust: The Evolutionary Landscape of Religion.* Oxford: Oxford University Press.

Bentham, J. (1782/1970). *An Introduction to the Principles of Morals and Legislation: Of Laws.* London: University of London.

Bernstein, R. J. (2010). *The Pragmatic Turn.* Cambridge: Polity Press.

Brandom, R. B. (2010). *Reason in Philosophy*. Cambridge: Cambridge University Press.

Brent, J. (1993). *Charles Sanders Peirce*. Bloomington: Indiana University Press.

Clark, A. (2013). Whatever Next? Predictive Brains, Situated Agents and the Future of Cognitive Science. *Behavioral and Brain Sciences, 36*(3), 181–204.

Clark, A. (2017). *Surfing Uncertainty*. Oxford: Oxford University Press.

Dewey, J. (1896). The Reflex Arc Concept in Psychology. *Psychological Review, 3*, 357–370.

Dewey, J. (1925/1989). *Experience and Nature*. LaSalle, IL: Open Court Press.

Dewey, J. (1928). Justice Holmes and the Liberal Mind. *New Republic, 53*, 210–. In J. A. Boydston (Ed.), *The Later Works of John Dewey, Volume 3, 1927–1938* (pp. 177–183). Carbondale: Southern Illinois University Press.

Dewey, J. (1931). *Philosophy and Civilization*. New York: Minton, Balach.

Dicey, A. V. (1882). Review of Holmes's *The Common Law*. *The Spectator, 3*, 745–747.

Diggins, J. P. (1994). *The Promise of Pragmatism*. Chicago: University of Chicago Press.

Duggan, M. F. (2007). The Municipal Ideal and the Unknown End: A Resolution of Oliver Wendell Holmes. *North Dakota Law Review, 83*, 463–544.

Engel, A. K., Friston, K. J., & Kragic, D. (2015). *The Pragmatic Turn: Toward Action-Oriented Views in Cognitive Science*. Cambridge: MIT Press.

Epicurus. (1964). *Letters, Doctrines and Vatican Sayings*. Indianapolis: Bobbs-Merrill.

Fisch, M. H. (1942/1986). Justice Holmes, the Prediction Theory of Law and Pragmatism. In M. H. Fisch (Ed.) (1986), *Peirce, Semiotic and Pragmatism*. Bloomington: Indiana University Press.

Frank, J. N. (1930). *Law and the Modern Mind*. New York: Brentano Publishers.

Frank, J. N. (1932). Mr. Justice Holmes and Non-Euclidean Legal Thinking. *Cornell Law Review, 17*, 568–588.

Frank, J. N. (1950). Modern and Ancient Legal Pragmatism. *Yale Law Review, 25*, 207–255.

Friston, K. (2010). The Free Energy Principle: A Unified Brain Theory? *Nature Reviews, 11*, 127–136.

Gigerenzer, G. (2000). *Adaptive Thinking, Rationality in the Real World*. New York: Oxford University Press.

Glimcher, P. W. (2003). *Decision, Uncertainty and the Brain: The Science of Neuroeconomics*. Cambridge: MIT Press.

Goodman, N. (1955/1973). *Fact, Fiction and Forecast*. New York: Bobbs-Merrill.

Gordon, R. W. (1982). Holmes's Common Law as Legal and Social Science. *Hofstra Law Review, 10*, 719–740.

Green, St. J. (1870). Proximate and Remote Cause. *American Law Review, 4*, 201.

Greene, J. D. (2014). Beyond Point and Shoot Morality: Why Cognitive Neuroscience Matters for Ethics. *Ethics, 124*, 695–726.

Grey, T. C. (1989). Holmes and Legal Pragmatism. *Stanford Law Review, 41*, 787–870.

Grey, T. C. (1992). Holmes, Pragmatism and Democracy. *Oregon Law, 71*, 521–542.

Haack, S. (2011). Pragmatism, Law, and Morality: The Lessons of Buck vs. Bell. *European Journal of Pragmatism and American Philosophy, 2*, 67–87.

Hacking, I. (1965/1979). *Logic of Statistical Inference*. Cambridge: Cambridge University Press.

Hacking, I. (1990). *The Taming of Chance*. Cambridge: Cambridge University Press.

Hart, H. L. A. (1958). Positivism and the Separation of Law and Morals. *Harvard Law Review, 71*, 593–629.

Hart, H. L. A. (1961). *The Concept of Law*. Oxford: Clarendon.

Hauser, N. (2003). Pragmatism and the Loss of Innocence. *Cognitio, 4*, 197–210.

Holmes, O. W., Jr. (1880). Review of CC Langwell: A Selection of Cases of Contracts with a Summary of the Topics Covered by the Cases.

Holmes, O. W., Jr. (1881/1952). *The Common Law*. New York: Dover.

Holmes, O. W., Jr. (1913). *Law and the Court*. Speech at a dinner at the Harvard Law School.

Holmes, O. W., Jr. (1917). *Southern Pacific Co. v. Jensen*.

Holmes, O. W., Jr. (1919). *Schenck v. United States*.

Horwitz, M. J. (1992). The Place of Holmes in American Legal Thought. In R. Gordon (Ed.), *Legacy of Oliver Wendell Holmes Jr*. Palo Alto: Stanford University Press.

Howe, M. D. (1957/1963). *Justice Oliver Wendell Holmes, Volumes 1 and Volumes 2: The Proving Years*. Cambridge: Harvard University Press.

James, W. (1890/1952). *The Principles of Psychology*. New York: Dover Press.

James, W. (1907/1955). *Pragmatism*. New York: Meridian.

Janack, M. (2012). *What We Mean by Experience*. Palo Alto: Stanford University Press.

Kahneman, D. (2011). *Thinking Fast and Slow*. New York: Farrar, Straux, and Giroux.

Kahneman, D., Slovic, P., & Tversky, A. (1982). *Judgment Under Uncertainty: Heuristics and Biases*. Cambridge, UK: Cambridge University Press.

Kelley, P. J. (1993). Holmes's Early Constitutional Law Theory and Its Applications in Taking Cases on the Massachusetts Supreme Court. *Southern Illinois University Law Review, 18*, 357–414.

Kellogg, F. R. (2007). *Oliver Wendell Holmes: The Legal Theory as Judicial Restraint*. Cambridge: Cambridge University Press.

Keynes, J. M. (1921/1957). *A Treatise on Probability*. New York: Harper & Row.

Lakoff, G., & Johnson, M. (1999). *Philosophy in the Flesh*. New York: Basic Books.

Laski, H. J. (1925). *Socialism and Freedom*. London: Fabian Society.

Leiter, B. (2000). Holmes, Economics and Classical Realism. In S. J. Burton (Ed.), *The Path of the Law and Its Influence: The Legacy of Oliver Wendell Holmes Jr*. Cambridge: Cambridge University Press.

Levinson, S. (2000). Emerson and Holmes: Serene Skeptics. In S. J. Burton (Ed.), *The Path of the Law and Its Influence: The Legacy of Oliver Wendell Holmes Jr*. Cambridge: Cambridge University Press.

McDermott, J. J. (2007). *The Drama of Possibility: Experience as Philosophy of Culture*. Bronx: Fordham University Press.

Menand, L. (2001). *The Metaphysical Club*. New York: Farrar, Straus, and Giroux.

Mendenhall, A. P. (2015a). Pragmatism on the Shoulders of Emerson: Oliver Wendell Holmes Jr.'s Jurisprudence as a Synthesis of Emerson, James and Dewey. *The South Carolina Review, 48*, 93–109.

Mendenhall, A. P. (2015b). *Oliver Wendell Holmes Jr., Pragmatism and the Jurisprudence of Agon*. Lewisburg: Bucknell University Press.

Moreno, J. D. (1995). *Deciding Together*. Oxford: Oxford University Press.

Neville, R. C. (1974). *The Cosmology of Freedom*. New Haven: Yale University Press.

Pappas, G. P. (2008). *John Dewey's Ethics: Democracy as Experience*. Bloomington: Indiana University Press.

Peirce, C. S. (1868). Questions Concerning Certain Faculties Claimed for Man. *Journal of Speculative Philosophy, 2*, 103–114.

Peirce, C. S. (1878a). Deduction, Induction and Hypothesis. *Popular Science Monthly, 13*, 470–482.

Peirce, C. S. (1878b). Doctrine of Chances. *Popular Scientific Monthly, 12*, 604–615.

Peirce, C. S. (1898/1992). *Reasoning and the Logic of Things: The Cambridge Conferences Lectures of 1898 (Harvard Historical Studies)* (K. L. Ketner & H. Putnam, Eds.). Cambridge, MA: Harvard University Press.

Peirce, C. S. (1958). *Collected Papers*. Cambridge: Harvard University Press.

Posner, R. A. (1992). *The Essential Holmes*. Chicago: University of Chicago Press.

Posner, R. A. (2007). *Economic Analysis of the Law*. Alphen aan den Rijn: Wolters Kluwer.

Posner, R. A. (2008). *How Judges Think*. Cambridge: Harvard University Press.

Postrema, G. J. (2011). Justice Holmes: A New Path for American Jurisprudence. In G. J. Postrema & E. Pattaro (Eds.), *Treatise on Legal Philosophy and General Jurisprudence, Volume 11: Legal Philosophy in the 20th Century—The Common Law World*. Dordrecht: Springer.

Power, M. L., & Schulkin, J. (2009). *The Evolution of Obesity*. Baltimore: Johns Hopkins University Press.

Rosenblatt, R. (1975). Holmes, Peirce and Legal Pragmatism. *Yale Law Review, 84*, 1123–1140.

Santayana, J. (1967). *Animal Faith and Spiritual Life*. New York: Appleton-Century-Crofts.

Schulkin, J. (1991). Science and Human Rights. *The Journal of General Evolution, 32*, 243–253.

Schulkin, J. (2004). *Bodily Sensibility: Intelligent Action*. Oxford: Oxford University Press.

Schulkin, J. (2015). *Pragmatism and the Search for Coherence in Neuroscience*. London: Palgrave Macmillan.

Seigfried, C. H. (1996). *Pragmatism and Feminism*. Chicago: University of Chicago Press.

Sidgwick, H. (1902/1988). *Outlines of the History of Ethics*. Indianapolis: Hackett.

Smith, J. E. (1970). *Themes in American Philosophy: Purpose, Experience and Community*. New York: Harper & Row.

Sterling, P. (2004). Principles of Allostasis: Optimal Design, Predictive Regulation, Psychopathology and Rational Therapeutics. In J. Schulkin (Ed.), *Allostasis, Homeostasis and the Costs of Physiological Adaptation.* Cambridge: Cambridge University Press.

Sterling, P., & Laughlin, S. (2015). *Principles of Neural Design.* Cambridge: MIT Press.

Swanson, L. W. (2000a). What Is the Brain? *Trends Neuroscience, 23,* 519–527.

Swanson, L. W. (2000b). Cerebral Hemisphere Regulation of Motivated Behavior. *Brain Research, 886,* 113–164.

Toulmin, S. (1950). *Reason in Ethics.* Cambridge: Cambridge University Press.

Toulmin, S. (1958). *The Uses of Argument.* Cambridge: Cambridge University Press.

Toulmin, S. (1972). *Human Understanding: The Collective Use and Evolution of Concepts.* Princeton: Princeton University Press.

Wells, C. P. (2000). Holmes and William James. In S. J. Burton (Ed.), *The Path of the Law and Its Influence: The Legacy of Oliver Wendell Holmes Jr.* Cambridge: Cambridge University Press.

Wells-Hantzis, C. (1988). Legal Innovation Within the Wider Intellectual Tradition: The Pragmatism of Oliver Wendell Holmes Jr. *Northwestern Law Review, 82,* 541–595.

White, G. E. (2000). *Oliver Wendell Holmes Sage of the Supreme Court.* Oxford: Oxford University Press.

White, M. (1947). *Social Thought in America: The Revolt Against Formalism.* Boston: Beacon Press.

Wiener, P. P. (1948). The Pragmatic Legal Philosophy of N. St. John Green. *Journal of the History of Ideas, 9,* 70–92.

Wiener, P. P. (1949). *Evolution and the Foundations of Pragmatism.* Cambridge: Harvard University Press.

Winter, S. L. (2001). *A Clearing in the Forrest: Law, Life and Mind.* Chicago: University of Chicago Press.

Wittgenstein, L. (1953/1958). *Philosophical Investigations.* New York: Macmillan.

Wright, C. (1877/1971). *Philosophical Discussions.* New York: Burt Franklin.

4

Holmes, Pragmatism and Nature

Introduction

Holmes waffled with some of his philosophical allegiances but was clearly most closely tied to the pragmatists (Wells-Hantzis 1988; Grey 1989) and the larger intellectual tradition of grounded understanding in a naturalist anthropological view about possibilities of nature. Holmes valued utilitarianism and positivism (cf. Pohlman 1984; Kellogg 1984), but neither as ends themselves. Holmes was inconsistent in his views, but expectations and the outcomes from guiding antecedent practices are dominant recurring themes in his consideration of the law.

Prediction and coherence of consequences is the lifeblood of adaptation; in law, in science, in everyday life, in our evolutionary history. The pragmatists were keen on making this paramount. Clarity and exploration, testing and sampling and trying for making things transparent, this is the stuff that makes up inquiry. Holmes, like James, was experimentalist in theory, although neither man was experimentalist in practice. They were expositors. In contrast, an experimentalist is often dirty and murky and the life of the mind is not often clear in the realm of the

© The Author(s) 2019
J. Schulkin, *Oliver Wendell Holmes Jr., Pragmatism and Neuroscience*,
https://doi.org/10.1007/978-3-030-23100-2_4

experimentalist. Holmes, like James, appreciated this concept but did not practice it.

Holmes was eager for law to be associated with science, with values on one side and facts on the other. At least that is one reading of Holmes, and one way to explain his behavior is to suggest that his separation between values and facts harmed his thinking. The separation between values and facts can be too radical.

Holmes understood that there is no one definition of the law. He sought to seek clarity to give perspective and nuance in statutes, and to be predictive. Now it is not obvious that Holmes knew much about the new sciences of statistical inferences as a tool to guide inquiry, to draw meaningful relationships that in principle are predictive. Holmes liked the idea of prediction; but statistical inquiry was as foreign to him as it was to James. Or, more fairly, Holmes did not directly participate in the act of scientific inquiry. Yet he was surrounded by an intellectual air where prediction was a formidable and desired outcome of inquiry.

In writing to Sir Frederick Pollock (fellow of Trinity College, Cambridge University, who himself had written a number of important treatises, amongst other things on *The Genius of the Common Law* as a series of lectures at Columbia University in 1912), Holmes talks about despair coupled with fatalism, or as he put it, and the sense of the "inevitable" (Holmes-Pollock Letters).

In this chapter, I further set Holmes in the context of pragmatism and its origins, and the link to the law, to the self-corrective within the culture of science that pervades many of the pragmatists' positions, along with a prevailing of naturalism but not natural law. This will figure in understanding in seeing the law and inquiry as continuous and the larger body public in which science takes place.

Variants of Pragmatism

There are many variants of pragmatism (Grey 1989; Haack 1998; Kellogg 1992). Holmes might be best called a moderate pragmatist. He had a strong urge to use the tool of utilitarian considerations, as well as a sense of biology and coping with law as one route in the regulation of retribution and minimizing violence. Holmes notes, "law is not science, but essentially empirical." This is a key feature in his philosophy of law and has implications for how we take neuroscience into the context of adjudication.

But rationally is not mythologized and frail. The view is historical, our cultural expressions. But history is contextual and tied to "fashion." As Holmes rather oddly put it in his article on law in science and science in law, "the law of fashion is the law of life". And indeed, fashion, demythologized from the narrow notion of fragrance and appearance, dominates science.

One view of pragmatism is that the philosophy eschews theory (e.g., Atiyah 1983; West 1989). However, that is certainly not the pragmatism that I understood. Instead, grappling with theory in action was the running theme in providing coherence and at times predictive capability within variants of pragmatism. For pragmatists, theory is essential. The question is: how much? And that is contextual (Haack 1998; Grey 1991, 2000; Farber 1988).

The larger point about pragmatism is how intertwined theory was with practice. Theory guides, but it is refined and reevaluated because we want to be clear about what works. And classical pragmatism wants to acknowledge what ends are pursued in the context of what means. Thus, pragmatism is far from anti-theoretical. Theory and practice are not separate as an orientation as ideas (Dewey 1925/1989).

Holmes (1881) understood something about the bounds of reason. He understood the sense of proportionality. He certainly understood the body politic and the importance of debates and integration of issues, such as brain death being a feature of decision making in health care. The continuity of law, science, inquiry and human participation are the ideals that Holmes held up and that Dewey tried to make clear. Holmes, reflecting on the law, wrote that "an ideal system of law should draw its postulates and its legislative justification from science" (Learning and Science 1895; see also Reece 1998).

Like other pragmatists such as James or Dewey, Holmes placed an emphasis on the diverse forms of human experience—what is constant and what is variable, how much plasticity and under what conditions—but he also did this to make consideration of the law less about vengeance and more about what works to reduce harm in the future.

Individuals do not exist in isolation. Our survival is rooted in our connections with others as well as in our ability, through pedagogy, to conform and expand. We are the pedagogical species capable of learning throughout and to the end of our long lives. Holmes himself is a symbol of lifelong learning, but he understood that judges are not professors (1899). Still, he represented the naturalization rather than the myth-making feature of the law in the context of the human drama. It is Holmes's naturalism and stoicism, as well as his realism, that grounds him in his relationships with others (1881). Constraints are there to curtain over human expression and abuse. As Holmes knew firsthand, we are potentially a half-neuron expression away, at any particular moment in time or history, from unleashing our nasty and brutish sides.

Holmes was part of the American intellectual ambiance that embraced evolutionary thinking, diverse as it was, and he understood law in the context of our evolution. In *The Common Law* (1880), he says that "the law embodies the story of a nation's development" (Kuklick 2000). Holmes himself continued to evolve within a context of adjudication and in terms of what constitutes good legal reasoning. He did not veer from his dislike for Kantianism and formalism or from couching a legal context in terms of inherent tendencies in human composition (Vetter 1984). Holmes viewed considerations of growth and development as background metaphors for the law and in understanding human development. He believed that nothing is frozen or fixed; the universe is open ended.

Towards Pragmatism

Dewey embraced the diverse and changing ways in which we can be in the world and make our world (see Grey 1992; Pearcey 2001). Holmes understood what Dewey (1929/1960) called "The Quest for Certainty" as something to be dustbinned to history, asserting that "certainty generally is illusion." What dominates the work of pragmatists like Dewey and Holmes are the themes of coping with precarious events and searching for coherence; nature is alive and evolving. These are dominant metaphors in Holmes's and Dewey's works. In his speech on December 3, 1902 to the Middlesex Bar Association, Holmes asserted that he "tried to see the law as an organic whole." And it is this side of the organic and evolving that puts us in a position to see what a law means in the consideration of a neuroscience, both are culturally tied together. Epistemic quests fueling adjudication and judgment.

There is a side of Dewey, particularly in his *A Common Faith*, that digs deep into the human condition and embraces the common bonds found in social solidarity. Holmes admired the book, *Experience and Nature*, and the naturalistic sensibility that pervades the text. But Dewey could be obscure, which was unnerving to Holmes's delicate pen and tendency toward detached pontification. Where Dewey was in the trenches and in close touch with the plight of persons, Holmes's view tended to be more detached and intellectual. He certainly wasn't consistent on free speech or rights (Rogat 1964; Rogat and O'Fallon 1984). Holmes was also contextual, nuanced, unsettled (Grey 1992) as well as seductive and provocative. But a sense of common bounds, except in war, was not something he would later reveal.

The point for pragmatists like Holmes and Dewey is that culture is not on one side and nature on the other. What made Holmes a pragmatist, in part, were his external standards for anchoring adjudication as well as his sense of science and the law as continuous. There is also the emphasis he placed on the precarious and the stable and, finally, the idea that ends and means are continuous.

Oliver Wendell Holmes Jr. and Pragmatism

Holmes was ambivalent about his relationship to pragmatism (e.g., Fisch 1942; Grey 1992; Kellogg 2004). Although he has been forever linked to the pragmatists by proximity and interaction, he could also have been linked to utilitarians like John Stuart Mill and his father, James Mill. Toward the end of his life, Holmes singled out John Dewey and the book *Experience and Nature* as a work to which he felt particularly close. To me, this puts him in line with pragmatist thinking.

Pragmatism has several key features. One is the active sense of experience that pervades inquiry. Another is discerning consequences. One without the other lessens the impact and importance of pragmatism, so Holmes held on to both. A third feature is the endlessly and open-ended continuity of nature and culture including experimental inquiry.

Holmes could be quite brutal in his correspondence with others (Posner 1990). He called James's pragmatism "an amusing humbug" (Letter to Einstein, June 17, 1908; see also his letters to Pollock 1942, 2015). He would say of Dewey's book *Human Nature and Conduct* that "I seem to know the fundamentals before" (June 14 Letter to Harold Laski).

As it did for other pragmatists, the topics of nature and science figured importantly in Holmes's thoughts. Just as prominent was the subject of science and the law. Like most thinkers, Holmes straddled the various topics with inconsistency, moving between positivism and pragmatism, individual concerns and the social good and intended and unintended consequences of decision making and social policy. Not surprisingly, he also bounced between bad faith and the facing of consequences and also between aspects of the human recognition of diverse forms of fatalism and human fortitude and hope. Hope glimmers within survival, a survival that takes its toll, a wear and tear on the human body. Holmes struggled from a detached perch in the consideration of the common law. This reflected, in some historical contingent context, not only the cultural evolution and continuity of science, but also the law as a competition of ideas. Holmes asserts that "the common law is not some brooding omnipresence in the sky" (1917).

Importantly, Holmes often focused on the sense of the community, and the larger body politic (1880, 1917). All thought is embedded in a social context. For Holmes, that context was the larger community that one fixes on when trying to make sense of what is right and what isn't. He clearly didn't believe in the reduction of the individual to the community but accepted the fact that an individual is embedded in one and having been thrown into a world that challenges their capabilities as an individual, still strives to survive. Holmes's wartime experiences were a palpable reference throughout his life whereas neuroscience is part of the larger epistemic and translational community.

Holmes was ambivalent about pragmatism, in part, because he thought James was soft and Peirce conceited. It was Dewey, whom he knew less well that he would identify with in the end, particularly the Dewey of *Experience and Nature* (see *Collected Works* in Posner, p. 71). And the one philosopher that he liked the most from his early days in The Metaphysical Club was Chauncey Wright (1877, 1878; Menand 2001).

Wright never identified himself as a pragmatist. He died before Peirce and then James, and—much later—Dewey made the diverse forms of pragmatism more of a household philosophical term. Wright appreciated the biological sciences, evolutionary thought and our capabilities when it comes to survival, adaptation, and consciousness. He was an outspoken evolutionist and corresponded with Darwin, whom he looked to and admired as someone who spoke the language of the new sciences in biology.

Good Enough Reasoning, Practice and the Law: Legal Pragmatism

The pragmatism dear to Dewey understood participatory democracy and the evolution of our cultural enterprises as knotted to science. Science does not lead the way; human meaning, experience, community and purpose do (Smith 1970). Importantly, the knowing process is quite active, not passive, in the assessment of sensations or rationalistic flair of great clarity. There is just a sense of empiricism where you have to battle for your ideas and engage in activity. An active mind is the utter sense of pragmatism and not mere logical inference, although inferences are absolutely essential.

Posner (2003) nicely depicts a long thread of pragmatists and the law, a thread that may run from John Marshall through Joseph Story through Holmes. It is a view with reasonableness in adjudication and not just consequences. Pragmatism incorporates worthy ends to be fought over within a marketplace of ideas. As Posner put it, "legal pragmatism is forward looking" and empirical (2003, p. 60). It is a variety of problem solving with purpose that contributes to a conception of law and our cultural evolution. But Posner is narrow, and dismissive and gives no sense of the greater meaning of the human condition. Pragmatism that Homes found in the end in Dewey; has meaning at the heart, standards and participation in a community.

Deliberative democracy goes hand in hand with pragmatism (Dewey 1931; see Sullivan and Solove 2003). The evolution of the intelligence reflects "distributive intelligence." Here a sense of realism is tied to consequences (as in contract law). Here Holmes is less sanguine about our prospects; support for democracy is one thing, optimism about the human condition quite another. Holmes is not Dewey, the optimist. Holmes was a skeptic about reason and our human prospects. We are bound by our limitations, of being swept away by the body politic. The war experience was always close.

Law is tied to custom, context, and adjudication and the practical means for cooperative social transaction. Holmes, as Posner (2003) notes, is "the most influential expositor of legal pragmatism." But legal pragmatism means more than one thing (Barzu 2018). What dominated the discourse of judicial rationality and realism is being systematic, reasonable, anticipatory, ecological, and deeply embedded in argument that includes the art of rhetoric (Posner 2003). Legal reasoning is part of practical reasoning. Posner (*Harvard Magazine* interview 2015) thinks the adversarial emphasis is compromised, because it is too much about pushing the client's case and not enough about truth seeking. Of course, truth is faint and difficult a motive and an outcome to discern.

Of course, legal pragmatism is in part caught in debate within the philosophy of law: more theory or in praise of theory (e.g., Dworkin 1987), too little or more dismissive of theory (Posner 1990), or at least "thicker" than proceduralism (Posner 2008) or just theory enough when it counts (Grey 1989). We certainly want to retain our idea of getting things right (e.g., women's right, rights of minorities, expanding participation, etc.). The question for a pragmatist like Holmes is having enough theory, theory that matters, that captures the phenomenon with an attention to consequences. And here we look to Dewey; and Dewey (1916) tied democratic sensibility to habits of acculturation and education. Legal pragmatism of this sort is through enhancing participation in our culture. And one part is making decisions about the sort of neuroscience or other forms of science that takes place, gets supported, cultivated and integrated into the larger body politic, including law.

Holmes is indeed part of legal pragmatism, but, as I have indicated, there are many meanings of pragmatism (Rosenfeld 1996). With regard to the law, many leading interpreters of Holmes have tied pragmatism to minimal theory (Grey 1989; Kellogg 1984, 2007, 2018) and to an economic view of competition and free exchange (Posner 2003, 2007). Free exchange and competition is a common currency. But, in pragmatism, all things must be within the bounds of reason. For example, hate speech, though allowed, can also have limitations in contexts that might do harm. Pragmatism is nuanced; practical implications are imperative.

A major interpreter of Holmes is Appellate Judge Richard Posner of Chicago. Posner is an avowed legal pragmatist, who provides an economic perspective on the law. He is an editor of Holmes, a biographer of Benjamin Cardozo, and a prolific author. In a book on judges, he starts with the obvious: "judges are not moral or intellectual giants … nor calculating machines" (2008, p. 7). They are mostly pragmatic, or they arrive at a solution over time.

Legal pragmatism, Judge Posner (2008) asserts, is quite separate from pragmatism proper, and law puts it in the backdrop of decision making for the judge. Posner echoes Holmes in positing that law is, in part, what judges do when acting like judges. Strict legalists look behind them, but legal pragmatists, like other pragmatists, also look forward. They expand the horizon (Posner 2008; Domnarski 2016).

Holmes and other legal pragmatists tend towards strict textualism. The Constitution for them was an historical document, not a given revelation of a plan for the universe. In this, they resemble the historical-critical approach to the Bible, not coincidentally developed over the spread of Holmes's career. Realism, or rather "critical realism," is a core feature of a variants of pragmatism (Schulkin 2012). The realism is earned by the struggle earned by competition, or just being lucky to hit on the right idea—what Peirce called abduction depicted in Chapter 2.

Still the legitimacy is fought over, even though the generating idea might be quite spontaneous. Pragmatists tend to be weighted to experienced events, as opposed to only method and logic. That is all the difference that really matters. Anchored to events is the calling card of a critical realist such as Holmes.

Legal pragmatism, with a minimally expected theory within the law, aims towards some sense of what Peirce called "critical common sensism" (see also Reid 1764). It is anchored to some settled goals in principal of value and substance (Sullivan 2007). And it is tied to our social milieu, the oxygen required to participate and the means to participate in the wider polity.

Pragmatism as Peirce understood it is tied to critical common sense aims—not naïve realism but critical realism (Schneider 1946/1963). But all things, including perhaps common sense, critical though it may be, can be limited (Blumenthal 2016). After all the "common man" is a Holmesian phrase, and what to expect may be limited to a set of individuals and not others. The pool of others is larger than the pot that Holmes was thinking about. And what is common can be a shield to keep others from participating.

The degradation of pragmatism is the reduction to mere method, which for the most part it was not. Nor was it the anti-intellectual rejection of theory with which it was associated in places practiced or understood this way. For Peirce (1877, 1898/1992), who labored forever to keep theory and inquiry essential for pragmatism, it was not, and for many others it was not.

At its worst, in the hands of Bertrand Russell or T. S. Eliot, "the great weakness of pragmatism is that it ends up being no use to anybody." But pragmatism has always meant more than one thing. Certainly, over the last 50 years, the link to nature has been diluted and the sense of conversation in communities has been highlighted (Rorty 1979). This occurred within the philosophy of law and practice (e.g., Smith 1990).

Pragmatists, like Holmes do emphasize experience (McDermott 2007; Smith 1978; Neville 1974). A running theme in pragmatism is an active sense of experience ripe with orientation and, in some contexts, self-correction and continual assessment.

What many commentators on pragmatism and science misconstrue is the thin rush for seeing pragmatism as power politics and a vulnerability to variants of scientism and as only methodological (e.g., Hollinger 1990). Of course, this common view is understandable and there are many variants. There is definitely a worry about science and instrumentalism and the loss of the individual in the prediction of events.

Of course, that was not the view of any of the classical pragmatists. Indeed, being instrumental is not antithetical to the loss of the plight of the individuals, their concerns, pain and ill fortune. Indeed, the normative goal of the most recently heralded pragmatist, Richard Rorty (1990), is to endlessly be mindful of not causing harm or discomfort, a very lofty ideal and worthy normative goal.

Holmes entered a modern world, a pragmatic one in which valuation is embedded in adjudication (Dewey 1939). Although not always consistent, Holmes rejected the modern separation of facts from values, of nature from culture, which misses that they are more porous, permeable and continuous with one another. Here he is consistent, namely a broad view about what is essential to the natural state of humans is eschewed. The many variants of natural rights as part of natural law (Finnis 2002) is replaced with a sense of nature and of our bodily politic and the utter continuity of one with the other.

The many meanings of pragmatism suggest the dissolution of philosophical disputes (e.g., realism vs. anti-realism), rendering the issue a practical matter and a matter embedded in experience. Not the so-called "pure experience" that James occasionally alluded to, but the sense of "casual efficacy" (Whitehead 1925) that underlies a sense of transactions with anticipation of future outcomes knotted to the sense of transitions. Indeed, James would say "life is in the transitions" (Koopman 2009), what James thought of "the basic stuff of experience" and the "continuity of experience" (James 1890) that mark experience in our lives. This is what some pragmatists emphasize and link to Emerson and genealogical conception. Pragmatists emphasize the transitions in life (Levin 1999; Koopman 2009, 2011; Albrecht 2012; Stuhr 1997).

The Emersonian emphasis is of "vitality"—the zest for life in the battle for survival (see Chapter 5). Change is endemic to our experiences.

We experience as James masterfully understood; coping with it, enhancing it, directing it, accepting, modifying, etc. Transitioning is another way to refer to the breakup of habits, and expectations are again elicited and change is occurring. Learning is one feature in the context. The pragmatist view of next expectations is a result of new habit or inference, or behavior being warranted. The prediction and the external anchors for Holmes's sense of adjudication and the law fit within this perspective.

Two themes stand out. First, the emphasis on experience and the utter continuity of nature and culture which are rooted in transactions tied to transitions in meaningful events. Second, interpretative sensibility pervades the whole sense of experiences with a normative goal towards self-corrective inquiry. Holmes is part of this.

Holmes, like Peirce or Dewey, literary and sophisticated, was looking for anchors in a world of adjudication, in a world of objects. He ran too far perhaps based on an outdated notion of the subjective and the objective. But an anchor to real entities was his normative goal. It just was the acknowledged porous relationships between seeing and doing, acting and observing, means and ends, theory and practice, fact and value.

Preserving our notion of experience is keeping track of people's wants, desires, goals, pain, etc. Moreover, in theory, formation we are guided by the disappointing fact of being wrong; something called "secondness" in the world of hypothesis formation. Coupling this sense of theory formation and rejection is what James and Whitehead referred to is the experience of causation "causal efficacy." Perhaps pragmatism frees one from the desperate need to have a deep theory in the law (see Grey 1989; Rorty 1990). But we want theory where we can get it; we need tools, perspective, and worthy ends as a necessity for everyday life, in law and elsewhere. What is theory mean here in part; what are the goals we are striving, what are the means at hand; drilled down more local; how do we get greater participation in evaluating evidence, how do we get oriented towards an experimental sensibility, how do we reduce human suffering and enhance human well being? Neuroscience matters here.

Neopragmatism

Rorty (1982) moved his variant of pragmatism away from the naturalism that was part of Holmes or certainly Dewey. Rorty's (1979) battle was against a form of epistemic representationalism, but that is not an issue for most pragmatists today. Today, pragmatism is tied to reliable coherence, prediction, sense and inference, the stuff that permeates Holmes's intellectual sensibility. Holmes was endlessly fallible, inconsistent and "all too human," a phrase of Nietzsche.

Rorty (1982) talks about language and not experience. After all, he coined the phrase "the linguistic turn." And indeed, talking about language is easier and more informative perhaps than talking about experience. Rorty, like the historian Hollinger (1977; see also Jay 2005), pokes fun at how little came from the pragmatist talk of experience, which was sometimes even confusing and misleading and definitely not informative. Moreover, leaving aside a naturalistic frame of mind is part of the linguistic feature. We are at the table of language; it is natural language, the language of everyday discourse that roots out our vulnerability to philosophical perplexities (Wittgenstein 1953).

But Rorty, in what became known as the "linguistic turn," leaves behind something important: experience. This was intentional. Language is something one can be anchored to. Experience can be too ambiguous. Rorty (1979, 1982), a neopragmatist, makes this clear. The pragmatist perspective of all things relative to background perspective, anticipation, and embedded and understood in the larger body of epistemic concerns or inquiry is maintained. Just the emphasis on experience is diluted.

But Rorty knows this and feels adding talk about experience is confusing and not helpful. Rorty (1990) does argue for one fundamental fact: alleviating others' discomfort is a primary normative goal of human ethical behavior and the celebration of democracy. But Holmes (1918) was very ambivalent about the prospects for the human condition, skeptical about "majority vote," and, of course, acknowledging that "majority vote can lick all others."

Variants of legal pragmatism, from Holmes to Benjamin Cardozo, permeated the philosophical air. Legal pragmatism is construed in narrow terms: a little theory, a lot of technique (Farber 1988; Warner 2010). Holmes was a master of technique and frankly the tool side of pragmatism is a real property of anyone's pragmatism. The degree of theory varies. In Peirce, it is quite pronounced and profound. In most others, it is much less so in philosophy.

Law by definition can be narrow or not, technique-dominated and craftsman-oriented. But pragmatism is an organic philosophy, where testing ideas predominates. Holmes resonates with the organic sensibility of the law and life. After all, the lifeblood of inquiry is the generation of ideas and the testing of ideas, and then there is the beauty of ideas and their integration in the context of what it leads to. Such separation of ideas from technique is narrow thinking indeed. After all, the ideal of the law is something about fairness and justice, but in this case issues of realism and antirealism as well. But pragmatists also tend to dissolve this distinction. There is continuity between nature and culture, not hard foundations about realism. This is something that resonates with Holmes's sensibility. Of course, he never wrote a systematic treatise, so we do not really know.

Realism pervades theory as we bang into what Peirce called "secondness" and the rejection of our ideas by brute facts of the matter (Haack 1998). Holmes, like Peirce, is not looking to what is often referred to as the subjective stance (Horwitz 1992). Holmes is anti-subjectivist. But subjectivism is an ambitious term. The contrast should rather be put into what is potentially part of the culture of self-corrective inquiry, subject to testing, hypothesis formation, etc. Realism, legal or otherwise (Holmes, Pound, Klewellyn), is tied to action, events, communities and perhaps what Cardozo (1921) called "the method of sociology." Holmes was grounded; but the tension between the terms that are inward and those outward, subjective vs objective, predates and outlived him.

Whether it is the early Holmes tied to uncovering custom and inherently common-sense rationalism, or Holmes, the more practical adjudicator and more knowledgeable talking about prediction, completion of ideas and the body politic, his issues lie with tying to external events and objectivism (Horwitz 1992). Legal formalism is replaced with law being tied to basic problem solving within the larger culture. It becomes a range of tools used in legal reasoning, as in other forms of problem solving and reasoning.

Indeed, many of us never wrote about pragmatism in the narrow way of anti-theory or anti-intellectual or antirealism, but of course this came to dominate many people's understanding of pragmatism. No doubt this was part of Peirce's loathing rejection of James and his "what works" identification about pragmatism, and then turning around and renaming what Peirce understood pragmatism to be about.

This discussion in the law about theory vs no theory or less theory is a reflection of this tension. The same surrounds science, including neuroscience. Pragmatism is associated with technique; and yes that is a true reading, but pragmatism is broader than that. The many variants of pragmatism is one thing and is a fact; the practice of inquiry is quite another. The practice of inquiry is rich in theory whether explicit and detailed or not. Theory permeates in practice, in lives lived, in practicing the law or science. That was certainly important for Dewey. Such separation between theory and technique is to be eschewed.

Keeping law pure (Langdell 1880) is keeping technique elegant and theory minimal (Pound 1908, 1921a, b), which is a worthy but misguided goal. The model of science for which "Langdell's orthodoxy"— named after the Dean of the Harvard Law School in the late nineteenth century (Grey 1983; Wells-Hantzis 1988)—and the law was modelled is misconceived, for science itself is not a set of theorems. Of course, this is nineteenth century fantasy and some twentieth century fantasy, but a recurrent fantasy. Formalism is the end-all with a set of rules. Rules are internal to legal judgment. This is part of legal reasoning but misconstrues the law and its link to the wider culture in which law is housed. Holmes understood the seduction of formalism and also the very useful and essential features of rules and formalism and inference from within.

Holmes, in elegant and slightly devious prose, would ridicule Langdell, calling him the "greatest living theologian"—quite a slight— and chiding the vulnerable emptiness of his "logical integrity" in reviewing one of his books (Holmes 1880). He was not always fair about Langdell. Experience also mattered to Dean Langdell as it did for Holmes (Grey 1983; LaPiana 1994) It was a book about case law amongst others. But the emphasis was on logical inference and rigorous methods. Holmes like many others after him within pragmatism would revolt against the absolute or mere logical clarity and the foundational footing this would hold.

Indeed, the orientation is nested beliefs, or what Dewey would call "warranted beliefs." These are ideas that have competed and have come to matter amid a backdrop of a democratic sensibility of self-corrective inquiry, an experimental spirit. Law without foundations is not without warrant, coherence and inherently valuable. Law provides a fundamental bedrock for continuity and clarity. Logic matters, always, but experience is not something other or outside logic. And argument varies depending upon context, circumstance and capability.

And so, Rorty's (1979) version of pragmatism took hold in the later part of the twentieth century for many people. His version of pragmatism is about conversation, rhetoric and community akin to the later investigation. The emphasis as it became in the philosophy of law is on practice and interpretation (Patterson 1990) and artisanship with elegance. Of course, elegance in law is tied to theory, to orientation. We want technique that matters, that makes a difference to the lives of individuals, to maintain peace, and to be functionally responsible and accountable for behaviors, under the law for all. But the metaphor of the "world well lost" (Rorty 1979), the argument of detached representations is an anthema to the pragmatist. Detached representations is one thing, but losing a sense of realism is quite another. Holmes is neither but never spells anything out.

Of course, he is not John Dewey: programmatic and hopeful about the human condition. Dewey, in contrast to Holmes, rejects fatalism and puts his faith in collective action, the critical intelligent action, and self-corrective inquiry. Dewey's emphasis is in capturing control where one can, and forging coherence and worthy ends for a collective sensibility of worth. Democracy is an achieved worth as the polity is opened to others. Holmes is more pessimistic than Dewey with regard to human possibilities and expression. Perhaps that war made all the difference.

But Rorty's blend of Dewey, Heidegger and Wittgenstein, intellectual acumen and knowledge of the philosophical tradition helped him expand beyond a narrow view of philosophy, and indeed to give up on a lot of it. He often searches in literature for the profundity required to take account of our existence and human understanding. Dewey was one of his intellectual heroes as were Wittgenstein and Heidegger. Neither the experimentalist investigation of Peirce and Dewey nor the Jamesian depiction of human experience is bracketed in language; at one level it is all language.

Unlike Rorty, and more like Holmes and Dewey, I hold on to the older language of experience (Schulkin 1982; see also Smith 1970; Neville 1974). But not the mythologized language of experience; sometimes there is just not a lot to say. Rorty comes out on the right side of social hope, a phrase linked to Dewey, meaning that there is one important normative goal, to work to avoid human discomfort and pain.

Dewey's road in contrast to Holmes is tied to social hope. Social hope entails something about what we owe each other. Social hope is at the heart of the progressive sensibility. It pervades what Dewey (1934, p. 1070) suggested in *A Common Faith*: a natural/atheist, religious sensibility of others, a natural piety (Shook 2010).

As Dewey (1934/1970) noted, "there is but one sure road to access the truth – the road of patient, cooperative inquiry operating by means of observation, experiment, record and controlled reflection," which is something Holmes understood. Holmes was a rational agent, for whom these features are part of being a rational agent. This is quite different from being a rationalist.

Both Holmes and Dewey, I think, would concur that progress here is the opportunity for shared decision making, or the right to decline to participate; the range of options, including not to participate, is part of the progress. The other part of the body politic in progress is the evolution of knowledge, shared knowledge.

For Holmes, nothing was off limits in a conversation about knowing, reading, or writing. In a world of letters and practice, his conversations were within a context of the culture of mind. Rarified as it sounds, this was Holmes's early ambiance at breakfast, lunch, and dinner; he sat at a table where knowledge acquisition was valued.

When Holmes depicts the inconsistent features of the law, when he tirades against logic, he perhaps overstates his case. Logic is a tool, like its sister field, mathematics. Logic is not an end and not a reduction to its form to have meaning in science as the positivist. When as a tool, logic's use is clear or becomes clear. What Holmes was probably attacking was the view of thought as only about logic. The many meanings of logic are perhaps what Wittgenstein called participation in diverse language games. Wittgenstein rendered thought external and social. For Wittgenstein, there was no "language of thought" for which the brain was a common element. There was no consideration of the precondition to render action understood in the context of cognitive systems.

What perhaps Wittgenstein did not consider enough, in part, was the contribution of what we impose on the world. The common sense that Peirce wrote about and associated with Thomas Reid was familiar and the sense of science associated with Chauncey Wright and Darwin were close to Holmes. These were pockets of progress to what Holmes recognized. An important component of this sense of science and experience is rooted in keeping track of objects, or external markers of events. With these markers, there is a viable sense of realism to anchor oneself (Daniels 1974/1989).

Neuroscientific Considerations: Pragmatist orientation, as I have indicated, is naturalistic but not reductionistic (Dewey 1925; Kitcher 2012), it is integrative in which diverse forms of social sensibility are woven into the fabric of understanding (Schulkin 2000). Understanding human agency and human accountability, a cardinal feature of the law is understanding the active sense of foraging for some coherence that underlies human adaptation, and epistemic urges quelled by satisfaction (Hohwy 2013).

The neural systems are designed were selected to enhance this process and to promote good enough problem solutions (Gigerenzer 2000). Regions of the brain that capture the gist of events, for example the amygdala, and basal ganglia underlie diverse problem solving are the staple of neural activity, underlying exploration and habit formation (Schulkin 2015).

The active sense of experience, that pragmatist emphasized is tied to tracking causal events (James 1890/1952) that enhance neural fluidity and coherent action. It is this tracking of causal events, and tying it to some aspect of experience that reflect something about the coherence on active neural systems. It also reflects the development of the brain, the budding of agency overtime, and the building of responsible agents, or not (Moreno 2003; Racine 2010).

All of this is to be understood in the culture we live; a culture of neuroscience. And the embrace of neural science is within a consideration of our evolutionary past.

Conclusion: Pragmatism and Naturalism, *Not* Natural Law

While Holmes (1918) was a naturalist, he was not a natural law theorist. In fact, Holmes was thought to be immoral in his rejection of natural law—something that is still regarded as a powerful and mainstream approach. Of course, natural law can mean more than one thing but a moral order of one sort or another is a large part of the concept and Holmes has often been attacked by his detractors for rejecting natural law and the sense of moral order (e.g., Finnis 2002; George 2004). A number of Jesuits (e.g., Gregg 1942–1943; Alschuler 2000) took aim at him for his vulnerability to human barbarity, lack of moral foundation, and his separation of the law from morality. These critics believed Holmes's jurisprudence would dissolve into moral ineptitude and a lack of moral bearing, with no inherent meaning in the universe, and no natural law to follow, to be part of, no city of God, no moral transcripts to eternally follow, and, worse, no foundational eternal verities.

During his correspondence with Sir Frederick Pollock, a major interpreter of the law and common law (see Duxbury 2004) as well as a close colleague, Holmes was reminded by Pollock "that in the middle ages natural law was regarded as the senior branch of the divine law and therefore had to be treated as infallible" (December 20, 1918; Pollock-Holmes Correspondence 1942, 2015). But with regard to divine law, infallibility is not a factor in the knowing process and both Peirce and Holmes would assert that infallibility is the very thing one gives up in a culture of inquiry. In devaluing the natural human dignity that goes with the natural law (Lucey 1941), Holmes became vulnerable. Natural dignity was something he understood from the feeling of solidarity he experienced during the war as well as from the battle of ideas and the social ties of human meaning—but not from natural law.

Holmes's version of pragmatism applied to a world without extra-human meaning. Counter to what some suggested, a Nazi state does not result from Holmes's orientation (Gregg 1942–1943) as the Nazis drew from all quarters (Sherratt 2013). Most upsetting to Holmes's critics was that in referencing a world in which we are insignificant in the cosmos, he resorted to comments about power (cf. Posner 1992; Luban 1992; White 1993; Alschuler 2000) and of being swept up in the moment—a will to power within an historic moment (see also Nietzsche 1886; Vanatta and Mendenhall 2016).

Pragmatism was (and is) regarded as anti-intellectual in many quarters. Holmes was an atheist but not anti-religious; by contrast, Peirce remained a Trinitarian and argued for the existence of God. Though not anti-religious, William James was anti-foundationalist and Dewey claimed having a "Common Faith" as noted (see Smith 1970, 1992; White 2015). These men traced their sense of wonder at the world back to the experimentalism of Jonathan Edwards, the sense of nature and the growing body of science. It was a social philosophy that eschewed "social absolutes", demythologized human and judicial decision-making and understood humanity's imperfect core *with its features that pervade our legal decision making and system.* Our legal system is historical, not eternal, not frozen (Laski 1931; Frankfurter 1923, 1939).

When Holmes sees inconsistencies between logic and experience, Dewey asserts "the undoubted facts which Justice Holmes has in mind do not concern logic but rather tendencies of the human creatures who use logic which a sound logic will guard against" (1931, p. 131).

Common law, like common faith, the former for Holmes and the latter for Dewey, is grounded in historical common contact. There is no separation of the natural from the cultural—only endless permeable relationships. Like Dewey, Holmes placed an emphasis on "growth" or "organic development" (White 1971). In much of his correspondence, he talked about the growth and organic features of the law (e.g., correspondence with Frederick Pollock 1874–1932).

The law is designed to enhance human safety and entrepreneurial activity. In Holmes's mind, law was a reflection of the larger social milieu and the body politic, both of which were expanding and embattled. He believed that the state of nature was no clue to the state of law except in the desire to avoid harm and be free of fear. On this point, his thinking did not differ from that of many others—most notably Hobbes and Spinoza. To Holmes, there was no higher power, only the powers of adaptation and desire. Nor were there saviors in Holmes's mind. He had a good sense of the anthropology of human existence, which pervades the pages of his major treatise on the law. Though not scholarly or deep, the treatise is attentive to the human condition, and pertinent to its understanding over which Holmes would preside as a judge and have to make decisions, including about the consideration and understanding of brain function.

The naturalization of the human condition was as familiar and as common a topic as the evolution of common law. The two became entwined with each other as the search for clarity in the law that wouldn't sacrifice our human sense of wonder continued. Holmes's sense of wonder may have been depleted by war and then later, by the overwhelming number of cases he had to judge, but he nevertheless managed to retain it. How else could he discover a kindred spirit in Dewey's *Experience and Nature*?

With no bifurcation of nature and culture, we get some sense of Holmes the pragmatist. Comfortable with regard to the naturalization of human knowing and surviving, unlike Dewey, Holmes did not emphasize human social contact—an evolving capability tied to evolution and the development of our social health and well-being. His personal correspondence reveals how truly important contact, both intellectual and catholic in scope, was to him. While Holmes had little direct contact with Dewey, he elevated Dewey to represent his (own) views on the metaphysics of nature and of human struggle; he just left out the ends and the means. "I have tried to see the law as an organic whole" said Holmes (1902), adding, "I also have tried to see it as a reaction between tradition on one side and the changing desires and needs of a community on the other." Both Holmes and Dewey were conscious of the endless continuity of culture and nature. They knew that we are a species that is social by orientation, by survival, by cephalic enlargements and by capability.

Holmes's perspective on naturalism was rooted in Emerson but also in Charles Darwin and Chauncey Wright. Naturalism pervades the sense of experience in the tradition of which Holmes is a member (Richardson 2007, 2014).

Nature and law are continuous and not inherently opposed. Laws restrain behaviors that are harmful. Laws, which promote, for instance, the right to know about chemical toxins, facilitate two fundamental ideas for Holmes: making information available (in this case, about toxins and companies that produce them), and integrating that knowledge with human action (Sarokin and Schulkin 2016).

A keen sense for nature and survival is one thing. For natural law, quite another. Holmes was a lifelong opponent of natural law. He said of it, "it seems to me this demand is at the bottom of the philosopher's effort to prove that truth is absolute and of jurist search for criteria of universal validity which he collects under the heading of natural law" (Holmes 1918).

For Holmes, nature had no higher purpose than the brutal whiff of survival. Darwin provided a language to describe this purpose. Herbert Spencer was in the intellectual milieu. Perhaps the poetic side dissipates with the deprivation of the war that he participated in and survived.

Holmes, the naturalist, labored mostly in the abstract while observing the constant influx of science and the law. For him, law was informed by the sciences and the behavioral sciences in particular. He would have appreciated the behavioral outcomes of our vulnerability to drawing inferences that are not always the best explanation but, in fact, reflect a vulnerability to making mistakes. Nor would he have been surprised that this vulnerability to drawing inferences is hard to correct (see Kahneman 2011)

References

Albrecht, J. M. (2012). *Reconstructing Individualism*. New York: Fordham University Press.
Alschuler, A. W. (2000). *Law Without Values: The Life, Work and Legacy of Justice Holmes*. Chicago: University of Chicago Press.
Atiyah, P. S. (1983). The Legacy of Holmes Through English Eyes. *Boston University Law Review, 63,* 341–380.

Barzun, C. L. (2018). Three Forms of Legal Pragmatism. *Washington University Law Review, 95,* 1003–1034.

Blumenthal, S. L. (2016). *Law and the Modern Mind: Consciousness and Responsibility in American Legal Cutlure.* Cambridge: Harvard University Press.

Carodozo, B. N. (1921). *The Nature of the Judicial Process.* New Haven: Yale University Press.

Daniels, N. (1974/1989). *Thomas Reid's Inquiry.* Palo Alto: Stanford University Press.

Dewey, J. (1916). *Democracy and Education: An Introduction of the Philosophy of Education.* New York: Macmillan.

Dewey, J. (1925/1989). *Experience and Nature.* LaSalle, IL: Open Court Press.

Dewey, J. (1929/1960). *The Quest for Certainty.* New York: Capricorn Books.

Dewey, J. (1931). *Philosophy and Civilization.* New York: Minton, Balach.

Dewey, J. (1934/1970). *A Common Faith.* New Haven: Yale University Press.

Dewey, J. (1939). *A Theory of Valuation.* Chicago: University of Chicago Press Press.

Domnarski, W. (2016). *Richard Posner.* Oxford: Oxford University Press.

Duxbury, N. (2004). *Frederick Pollock and the English Juristic Tradition.* Oxford: Oxford University Press.

Dworkin, R. (1987). *A Matter of Principle.* Cambridge: Harvard University Press.

Farber, D. A. (1987–1988). Legal Pragmatism and the Constitution. *Minnesota Law Review, 72,* 1331–1377.

Finnis, J. (2002). Natural Law: The Classical Tradition. In J. Coleman & S. Shapiro (Eds.), *Jurisprudence and the Philosophy of Law.* Oxford: Oxford University Press.

Fisch, M. H. (1942/1986). Justice Holmes, the Prediction Theory of Law and Pragmatism. In M. H. Fisch (Ed.) (1986), *Peirce, Semiotic and Pragmatism.* Bloomington: Indiana University Press.

Frankfurter, F. C. (1923). Twenty Years of Mr Justice Holmes's Constitutional Opinions. *Harvard Law Review, 36,* 909–939.

Frankfurter, F. C. (1939). *Mr Justice Holmes.* Cambridge: Harvard University Press.

George, R. P. (2004). Holmes on Natural Law. In J. De Groot (Ed.), *Nature in American Philosophy.* Washington, DC: Catholic University of America Press.

Gigerenzer, G. (2000). *Adaptive Thinking, Rationality in the Real World.* New York: Oxford University Press.

Gregg, P. L. (1942–1943). The Pragmatism of Mr Justice Holmes. *Georgetown Law Review, 252,* 263–295.

Grey, T. C. (1983). Langdell's Orthodoxy. *University of Pittsburgh Law Review, 45,* 1–53.

Grey, T. C. (1989). Holmes and Legal Pragmatism. *Stanford Law Review, 41,* 787–870.

Grey, T. C. (1991). What Good Is Legal Pragmatism. In M. Brint & W. Weaver (Eds.), *Pragmatism in Law and Society.* Boulder, CO: Westview Press.

Grey, T. C. (1992). Holmes, Pragmatism and Democracy. *Oregon Law, 71,* 521–542.

Grey, T. C. (2000). Holmes on the Logic of the Law. In S. J. Burton (Ed.), *The Path of the Law and Its Influence: The Legacy of Oliver Wendell Holmes Jr.* Cambridge: Cambridge University Press.

Haack, S. (1998). *Manifesto of a Passionate Moderate.* Chicago: University of Chicago Press.

Hohwy, J. (2013). *The Predictive Mind.* Oxford: Oxford University Press.

Hollinger, D. A. (1977). Review: The Culture of Experience: Philosophical Essays in the American Grain. *Transactions of the C. S. Peirce Society, 13,* 312–315.

Hollinger, D. A. (1990). Free Enterprise and Free Inquiry: The Emergence If Laissez-Faire Communitarianism in the Ideology of Science in the United States. *New Literary History, 21,* 897–919.

Holmes, O. W., Jr. (1880). Trespass and Negligence. *American Law Review, 14.*

Holmes, O. W., Jr. (1881/1952). *The Common Law.* New York: Dover.

Holmes, O. W., Jr. (1895). Scientific Proof and Relations of Law and Medicine: Learning and Science. *Boston University Law Review, 26.*

Holmes, O. W., Jr. (1899). The Theory of Legal Interpretation. *Harvard Law Review, 12,* 417–420.

Holmes, O. W., Jr. (1902). *Twenty Years in Retrospect.* Speech at a Banauget of the Middlesex Bar Association.

Holmes, O. W., Jr. (1917). *Southern Pacific Co. v. Jensen.*

Holmes, O. W., Jr. (1918). Natural Law. *Harvard Law Review, 32,* 40–44.

Horwitz, M. J. (1992). The Place of Holmes in American Legal Thought. In R. Gordon (Ed.), *Legacy of Oliver Wendell Holmes Jr.* Palo Alto: Stanford University Press.

James, W. (1890/1952). *The Principles of Psychology.* New York: Dover Press.

Jay, M. (2005). *Songs of Experience.* Berkeley: University of California Press.

Kahneman, D. (2011). *Thinking Fast and Slow.* New York: Farrar, Straux, and Giroux.

Kellogg, F. R. (1984). *Formative Essays of Oliver Wendell Holmes Jr.* London: Greenwood Press.

Kellogg, F. R. (1992). Who Owns Pragmatism? *Journal of Speculative Philosophy, 6,* 67–80.

Kellogg, F. R. (2004). Holistic Pragmatism and Law: Morton White on Justice Oliver Wendell Holmes. *Transactions of the Charles S. Peirce Society, 40,* 559–567.

Kellogg, F. R. (2007). *Oliver Wendell Holmes: The Legal Theory as Judicial Restraint.* Cambridge: Cambridge University Press.

Kellogg, F. R. (2018). *Oliver Wendell Holmes Jr. and Legal Logic.* Chicago: University of Chicago Press.

Kitcher, P. (2012). *Preludes to Pragmatism.* Oxford: Oxford University Press.

Koopman, C. (2009). *Pragmatism as Transition.* New York: Columbia University Press.

Koopman, C. (2011). Genealogical Pragmatism. *Journal of the Philosophy of History, 5,* 533–561.

Kuklick, B. (2000). *A History of Philosophy in America.* Oxford: Oxford University Press.

Langdell, C. C. (1880). *Summary of the Law of Contracts* (2nd ed.). Boston: Little, Brown.

LaPiana, W. P. (1994). *Logic and Experience.* Oxford: Oxford University Press.

Laski, H. J. (1931). The Political Philosophy of Mr. Justice Holmes. *Yale Law Review, 40,* 683–695.

Levin, J. (1999). *The Poetics of Transition: Emerson, Pragmatism and American Literary Modernism.* Durham: Duke University Press.

Luban, D. (1992). Justice Holmes and Judicial Virtue. *Nomos, 34,* 235–264.

Lucey, F. E. (1941). Natural Law and American Legal Realism. *Georgetown Law Review, 30,* 493–533.

McDermott, J. J. (2007). *The Drama of Possibility: Experience as Philosophy of Culture.* Bronx: Fordham University Press.

Menand, L. (2001). *The Metaphysical Club.* New York: Farrar, Straus, and Giroux.

Moreno, J. D. (2003). Neuroethics: An Agenda for Neuroscience and Society. *Nature Reviews, 4,* 149–153.

Neville, R. C. (1974). *The Cosmology of Freedom.* New Haven: Yale University Press.

Nietzsche, F. (1886/1972). *Beyond Good and Evil.* Mineola: Dover.

Patterson, D. M. (1990). Law's Pragmatism: Law as Practice and Narrative. *Virginia Law Review, 76,* 937–991.

Pearcey, N. R. (2001). Darwin's New Bulldogs: Scopes and American Legal Philosophy. *UL Review, 13,* 483–514.

Peirce, C. S. (1877/1992). The Fixation of Belief. In C. Kloesel & N. Houser (Eds.), *The Essential Peirce: Selected Philosophical Writings Volume 1* (p. 115). Bloomington: Indiana University Press.

Peirce, C. S. (1898/1992). *Reasoning and the Logic of Things: The Cambridge Conferences Lectures of 1898 (Harvard Historical Studies)* (K. L. Ketner & H. Putnam, Eds.). Cambridge, MA: Harvard University Press.

Pohlman, H. L. (1984). *Justice Oliver Wendell Holmes Jr.* Cambridge: Harvard University Press.

Posner, R. A. (1990). What Pragmatism Has to Offer the Law. *California Law Review, 63,* 1653–1670.

Posner, R. A. (1992). *The Essential Holmes.* Chicago: University of Chicago Press.

Posner, R. A. (2003). *Law, Pragmatism and Democracy.* Cambridge: Harvard University Press.

Posner, R. A. (2007). *Economic Analysis of the Law.* Alphen aan den Rijn: Wolters Kluwer.

Posner, R. A. (2008). *How Judges Think.* Cambridge: Harvard University Press.

Pound, R. (1908). Mechanical Jurisprudence. *Columbia Law Review, 8,* 609–610.

Pound, R. (1921a). A Theory of Social Interests. *American Sociological Society, 15,* 16–45.

Pound, R. (1921b). Judge Holmes's Contributions to the Science of the Law. *Harvard Law Review, 34,* 449–453.

Racine, E. (2010). *Pragmatic Neuroethics: Improving Treatment and Understanding of the Mind/Brain.* Cambridge: MIT Press.

Reece, H. (1998). *Law and Science.* Oxford: Oxford University Press.

Reid, T. (1764/1997). *An Inquiry Into the Human Mind.* Edinburgh: Edinburgh University Press.

Richardson, J. (2007). *A Natural History of Pragmatism.* Cambridge: Cambridge University Press.

Richardson, J. (2014). *Pragmatism and American Experience.* Cambridge: Cambridge University Press.

Rogat, Y. (1964). The Judge as Spectator. *University of Chicago Law Review, 31,* 213–256.

Rogat, Y., & O'Fallon, J. M. (1984). Mr Justice Holmes: A Dissenting Opinion. The Free Speech Cases. *Stanford Law Review, 36,* 1349–1406.

Rorty, R. (1979). *Philosophy and the Mirror of Nature*. Princeton: Princeton University Press.

Rorty, R. (1982). *Consequences of Pragmatism*. Minneapolis: University of Minnesota Press.

Rorty, R. (1990). *Philosophy and Social Hope*. New York: Penguin.

Rosenfeld, M. (1996). Pragmatism, Pluralism and Legal Interpretation: Posner's and Rorty's Justice Without Metaphysics Meets Hate Speech. *Cardozo Law Review, 18,* 97–152.

Sarokin, D. J., & Schulkin, J. (2016). *Missed Information*. Cambridge: MIT Press.

Schneider, H. W. (1946/1963). *A History of American Philosophy*. New York: Columbia University Press.

Schulkin, J. (1982). *The Pursuit of Inquiry*. New York: SUNY Press.

Schulkin, J. (2000). *Roots of Social Sensibility*. Cambridge: MIT Press.

Schulkin, J. (2012). *Naturalism and Pragmatism*. London: Palgrave Macmillan.

Schulkin, J. (2015). *Pragmatism and the Search for Coherence in Neuroscience*. London: Palgrave Macmillan.

Sherratt, Y. (2013). *Hitler's Philosophers*. New Haven: Yale University Press.

Shook, J. R. (2010). Dewey's Naturalized Philosophy of Spirit and Religion. In *John's Dewey's Philosophy of Spirit. With the 1897 Lecture on Hegel*. New York: Fordham University Press.

Smith, J. E. (1970). *Themes in American Philosophy: Purpose, Experience and Community*. New York: Harper & Row.

Smith, J. E. (1978). *Purpose and Thought*. New Haven: Yale University Press.

Smith, J. E. (1992). *America's Philosophical Vision*. Chicago: University of Chicago Press.

Smith, S. D. (1990). The Pursuit of Pragmatism. *Yale Law Review, 100,* 409–449.

Stuhr, J. (1997). *Geneological Pragmatism*. Albany, NY: SUNY Press.

Sullivan, M. (2007). *Legal Pragmatism*. Bloomington: Indiana University Press.

Sullivan, M., & Solove, D. J. (2003). Can Pragmatism Be Radical? Richard Posner and Legal Pragmatism. *Yale Law Journal, 113,* 687–739.

Vannatta, S., & Mendenhall, A. (2016). The American Nietzsche? Fate and Power in the Pragmatism of Justice Holmes. *UKMC Law Review, 85,* 187–205.

Vetter, J. (1984). The Evolution of Holmes, Holmes and Evolution. *California Law Review, 72,* 343–368.

Warner, R. (2010). *Legal Pragmatism: A Companion to Philosophy of Law and Legal Theory*. New York: Wiley.

Wells-Hantzis, C. (1988). Legal Innovation Within the Wider Intellectual Tradition: The Pragmatism of Oliver Wendell Holmes Jr. *Northwestern Law Review, 82,* 541–595.

West, C. (1989). *American Evasion of Philosophy.* Madison: University of Wisconsin Press.

White, G. E. (1971). The Rise and Fall of Justice Holmes. *University of Chicago Law Review, 39,* 51–77.

White, G. E. (1993). *Justice Oliver Wendell Holmes: Law and the Inner Self.* Oxford: Oxford University Press.

White, J. W. (2015). *Lincoln on Law, Leadership and Life.* Naperville, IL: Source Books.

Whitehead, A. N. (1925/1997). *Science and the Modern World.* New York: Free Press.

Wittgenstein, L. (1953/1958). *Philosophical Investigations.* New York: Macmillan.

Wright, C. (1877/1971). *Philosophical Discussions.* New York: Burt Franklym.

Wright, C. (1878). *Letters of Chauncey Wright with Some Account of His Life* (J. Thayer, Ed.). Cambridge: Press of John Wilson and Son.

5

Duty, Surviving, Social Contact

Introduction

In 1864, Holmes was fatigued. His letters to his parents reveal a person literally at the edge from the war experience that is ravaging his brain and body. He became a survivor; survivors have at least one mode, namely social contact and maintenance of human dignity. He did this in part by reaching out to others. He was to be rejuvenated within a short period of time and would return to an intellectual compass and a sense of Emersonian self-reliance. Holmes understood the meaning of friendship and cultivated a diverse and wide-ranging coterie. Emersonian discipline and a keen sense of nature and history were his backbone. Holmes understood something about what Emerson would suggest, namely "self-trust is the essence of heroism" (1855, p. 226).

Holmes was surrounded by a rich breadth of access. He was talented and fortunate (talented to have skill and fortunate to be motivated). Holmes was ambitious, consuming volumes of legal and more

© The Author(s) 2019
J. Schulkin, *Oliver Wendell Holmes Jr., Pragmatism and Neuroscience*,
https://doi.org/10.1007/978-3-030-23100-2_5

broadly intellectual history; fortunate and talented to became involved in a project that was at the heart of the development of American law. Holmes was recruited for the editing of the 12th edition of Kent's *Commentaries on American Law*. Chancellor and then Professor Kent of Columbia College wrote a very learned book on the law which was hugely successful and extremely influential. Holmes made his legal name working on this book and then became editor of the *American Law Review*. Holmes was well placed to be at the heart of American law. He was also fortunate to be able to travel after his studies and became friends with his legal colleagues in the United Kingdom. Holmes did not generously give credit with regard to how he came to his views (White 1993, 2015).

In this chapter, we begin with a discussion of duty, a core feature for Holmes, and social contact, a primary adaptive feature and a life-long feature of Holmes. And he is an important factor in the cultural evolution of law in the US. But it is Dewey who links pragmatism, the law and the promise of social evolution to a meaningful philosophy of law with ideals worthy of pursuit; human well being.

Duty to the Republic: Marshall, Jefferson, Lincoln, Holmes

Holmes was a federalist. He fought to preserve the union. The union was as sacrosanct as anything could be for him. Duty, stoicism, rationality, and consistency were all tied to his sense of the republic. John Marshall, one of Holmes's (1901) heroes, was as he put it in a "fortunate circumstance," in the right place at the right time to make a difference. Neither Holmes nor Marshall were radicals or reactionaries on race; they were thoroughgoing moderates on most things.

Holmes was a great fan of Marshall as he was of Montesquieu. Of Marshal, he would say in a speech honoring him that "the theory for which Hamilton and he decided and Webster spoke and Grant fought and Lincoln died is now our cornerstone" (1901, p. 270, *Collected Legal Papers*), and he suggested that Marshall "stands for a new body of jurisprudence"—a union of statutes across states.

Battling for Federalism Within the Bounds of Reason: Social Viability

Marshall, who, like Holmes, fought and understood battle, was a staunch defender of the nation, a federalist. Marshall was a country lawyer and a greater admirer of Washington, with a temperament to tie together the differences and squabbles of his peers (Adams, Jefferson Hamilton, etc.).

Marshall emerges from the politics of Virginia as a dedicated states-men and the attacks from his fellow countrymen, Madison or Jefferson, were mainly about government: how much or how little, which have helped define who we are. These were worthy fights about nationhood and identity; about safety, and about law (Newmyer 2001). John Adams nominated Marshall to the Supreme Court. He was admired by the fire-brand from Massachusetts.

Marshall was the consummate lawyer, a modern man looking for a cultural evolution on matters that counted (Ellis 2007). Holmes was envious a bit for Marshall's being present at a pivotal moment in the formation of the United States; but Holmes had his own historical moments. In *Maybery vs Madison* (1803), Marshall was the defender of the court, Jefferson the cultural dissenter; Marshall won. Jackson, also an ardent nationalist, asserted "Marshall has made his decision now he has to enforce it" (McDowell 2010). Jackson also hated Marshall.

Marshall was understood perhaps by Holmes as a pragmatist of some sort like himself, but then so was Lincoln (Posner 2003), contextualized by a great moment in history: the forming of a nation. But what then of pragmatism, it means everything so as to mean too little? Both Marshall and Holmes were formidable. They were ardent nationalists, fighters in wars to persevere. Marshall, like other founders following Locke, held the pillars of independence to be three independent branches of govern-ment. An ardent nationalist, Marshall, like Holmes or Lincoln, believed in preserving the union; strengthening it was a primary motivation (Ellis 2007). But what emerged from the second revolution of the civil war was a second sense of rights (McPherson 1991), expressed in the 13th, 14th, and 15th amendments.

Holmes would comment on the fact that "we live by symbols." Marshall and Lincoln were symbols of the coherence of a union of differences into a more perfect (if endlessly imperfect) union. But so is Holmes, symbolic as a major judicial figure (Ferguson 1988).

Holmes understood the two senses of freedom that Isaiah Berlin (1969) would write about. What Holmes and others called "negative freedom," freedom to be left alone, what Berlin called "unobstructed freedom." Maximal freedom with few governmental or law-like enforcement mechanisms. The second sense of liberty according to Berlin is to be one's own "master." What we later enforced legally, freeing African Americans in principle by amendments to the constitution.

Lincoln evolved from a repulsion of slavery to an active voice to abolish and to begin a process of expanding rights (McPherson 2008; Foner 2011). Lincoln went beyond where he was, and he came to represent what was possible—an evolving historically dependent cultural evolution in our sense of rights. Not by natural rights built on necessity nor a historically linear expression of cultural evolution on some plan, but rather hard-fought moments under adversity.

While Holmes at whatever level of enthusiasm was always anti-slavery, Peirce's family was pro-southern (Brent 1993)—probably why Peirce, a New Englander, stayed out of the war entirely. Thoughts about slavery reflected time and context. Holmes is no different; he is entirely a man of his time. While he did oppose slavery, on the Supreme Court, however, he was no particular friend of African Americans and their plight in the South (Baker 1991; but see Moore vs Dempsey, 1923). Here local custom prevailed, even when local custom was unjust and unfair, as he would acknowledge and argue for a new trial. Both Holmes and Peirce were independent thinkers. Neither showed great empathy for those not born to privilege and circumstance as they were. Both were self-enveloped. It was their other colleague that emphasized that trait central to our social evolution: thinking about others, what we might owe them.

Duty and Rules

Rules, regulations, and customs are substantiated in the fabrics of communities. They are the oxygen of everyday consumption and transaction with one another. Getting a foothold in the world is learning the rules and regulations, and following them or not.

While there is no inherent rule in the trajectory of the universe, as Holmes certainly and consistently thought, a set of common law principles and a propensity for problem solving is essential to our survival in nature and continuous expression in our cultural evolution—an evolution that Holmes recognized he was part of and, indeed, fought for.

Few of us can say that we fought for something as he did; the war to hold a union together and to eliminate slavery had its price. He would write home about the lice, the wounds, the agony. He would write home about "the determination of the south" (*Touched with Fire*, p. 80). Holmes was fighting less like Captain Shaw of the 20th and more through happenstance and duty. Duty eventually to his fellow soldiers, of both sides of the conflict he would admire, and of which the battle of ideas would materialize in a bloodbath with lots of incompetence and wanton slaughter.

In writing his father, Holmes reminded him of the reality of this war "…if it is true that we represent civilization… [it] is in its nature, as well as slavery, diffusive and aggressive …" As if he is doubting the validity of the slaughter still, he says, "I am to be sure heartedly tried and half worn out body and mind by this life, but I believe I am as ready as ever to do my duty" (*Touched with Fire*, p. 80).

For Holmes, duty, following the law, and reason pervade his orientation to the world in which he lived. Duty is the common vernacular for Holmes; but so is intellect, independence, and competition. As he put in a letter to his parents on May 30, 1864, the "wear and tear strain" has meant that "I am not the same man (may not have the same ideas)." Holmes's beliefs were pitted against a world of nature as he put

it "indifferent to us." What was not indifferent was human social contact in duty. Duty was elevated into a kind of mythical connection;
an "overrsoul" along the lines of Emerson. Holmes straddles between
Emersonian mysticism, variants of Nietzsche and perturbations of the
will (see Luban 1992; Posner 1990).

Law and duty were bright spots. After all, the law represented the
possibility of social cohesiveness. Holmes bleeds mystical with references to "vital forces" to features of life in which "life is a roar of bargain and battle, but in the very heart of it there rises a mystic spiritual
tone that gives meaning to the whole. It transmutes the dull details into
romance."

Vitalism in duty runs through Holmes. The battle wounds, the battle roar, and the sense of life in battle all ruminate through his prose
of self-disclosure grounded in mystical metaphors of common battle
and the larger public; the language of perseverance, necessity and duty
(Luban 1992). But a majority rules on Holmes and the sense of what is
right runs thin in a language of duty and competition and force. Like
Nietzsche (1878, 1886), there is a will to power beyond morality that
underlies the brute competition, duty, that can look very unenlight-
ened. Of course, Holmes, like Nietzsche, was not about sugar coating
statements.

There was also little abstract "sentiment of rationality" in Holmes
as his friend James would write about, or it was hard fought realization about human finitude, his own. James, who dedicated his book
on pragmatism towards the close of the century (1897) to Peirce, who
neither fought in that war nor whose family thought it worth dissolving slavery (see Brent 1993), would write that "moral skepticism can
no more be refuted by logic than intellectual skepticism" ("The Will to
Believe," p. 22). Willing to believe, or simply persevering despite the
entropy that prevails, therein lies a sense of duty persevering when there
is less of one to be found. He thought "duties precede rights" (*Codes and
the Arrangements of the Laws*). But, like all things, it has limits. Duty is
tied to perseverance.

Social Contact, Holmes and Friends: Keeping Contact Through Correspondence Over a Lifetime

Holmes, except for what seems like absolute solidarity with his fellow soldiers, kept contact through correspondence and friendship. He has broad tastes. But such contact was a big part of his life as an adaptation to the trauma of being left for dead in a field of battle, scars across a brain, punctated by human contact. Human contact fundamental in ordinary life and particularly in combating scars of trauma.

Frederick Pollock, Harold Laski, Morris Cohen, and Patrick Augustine Sheehan are only a few names in an impressive corpus (Menand 2001). Holmes's relationship with Lady Pollock, Lady Clare Castletown and Nina Gray spanned decades. With Lady Clare Castletown, he had the occasional tone of flirtation and intellectual romance. Holmes was known as a big flirt with women (White 1990; Monagan 1988). Holmes's marriage to Fanny Dixwell, the granddaughter of Justice Jackson, was a long one. But Holmes appeared to find some outlets of intimacy within an intellectual acumen with Lady Castletown (White 1990).

Holmes was acutely aware of the irrationality running deep in the expression and imagination in all of us, an instinctual sense of survival at all costs and the utter expression of the will to power, and its impact on the law and the larger body politic (Dailey 1998; Posner 1992, 2007). In a letter to James in 1896, he wrote of being able to still "sympathize deeply" with a romantic spirt of what is ideally possible for human expression.

Of course, there is the romanticism of Emerson, Keats, or Wordsworth that is less about power and the unconscious in contrast to the so called rational and the conscious. Of course, that sort of separation is the sort of thing that pragmatists such as Dewey would never make. Holmes is not Dewey; and he certainly let that be known.

Holmes, on the one hand, wanted to separate the practice of law from the consequences that may occur; but, on the other hand, he understood the larger social body politic that underlies law. Holmes is moved in life by intellectual friendships, and his relationship with Harold J. Laski is an important one. They had little in common, in background, in belief, in politics.

Holmes could hold others close intellectually and could write with passion to them about ideas, about being in the world, about frailties. His letters and his responses to Laski are one amongst others. And one core issue is freedom of speech in a democratic world, in a local environment fearful of socialism (like the United States). Laski was a socialist, a political theorist, an academic. What makes evident Holmes's big spirit is his embrace of someone like Laski; even more so as Laski is savagely attacked for his socialism (*Harvard Magazine* article), a view that Holmes did not share and was angered by (Healy 2013).

It was the same with Louis Brandeis. These were forms of intellectual friendships, colleagues of intellect and affection. The mandate of free speech was one the issues that continued a long-standing relationship between Laski and Holmes, especially when the issue is about ideology, and socialist Red scares in the US (*Schenck vs United States*, 1919). The rhetoric of "clear and present danger" set off a vital ensemble of reflection, of thought. It still does, ripe with elements of ambiguity (Stone 2004), freedom of speech in an evolving democracy. When there are no absolutes, as there are none for Holmes, context and nuance matters. Herein lies vestiges of rationality that are nurturing, in prose and profundity.

While he couched legal understanding in terms of experience, Holmes also looked for the biggest integrative bang for abstractions, while acknowledging the underdetermination of applications of general laws. As we would put it, general principles do not prove the concrete (Menand 2001, 2002). Holmes is many things: a thinker in transition and influenced by positivism, formalism, and, of course, pragmatism, and yet not strictly of any one of these "-isms."

Holmes's rejection of mere logic or method and of foundations predates the separation of the dilution of the analytic and synthetic distinction that would later dominate some of the philosophical disputations in philosophy. The boundaries are porous and evolving in analytic and synthetic propositions or sentences; that is one of pure logic and ones of synthetic, empirical, or creative implications (e.g., Kant). Emphasizing is not degrading logic, method, or clarity (White 1947).

Holmes eschewed abstractions, though he himself was quite abstract as a person in some ways. But cases are situated and not axiomatically decided by deduction. There is an ambiguity factor in Holmes' thinking (White 2002; Grey 1989) that made him vulnerable to attack or misunderstanding (Hart 1958). For instance, the predictive value of events in judicial reasoning is just one factor. The issue of pragmatism is the porous boundaries and at times less clarity than an analyst would want. Even Holmes often ridiculed pragmatists.

A Promised Land: Evolving Law and What Holmes Symbolized for Some

Morris Cohen, a City College student off to Harvard at the turn of the twentieth century to study philosophy in the golden era (Hollinger 1977; Kuklick 2000) with James and Royce, was a dominant figure when he returned to City College to teach for the duration of his career till the early 1940s. He dominated generations of students at City College. Importantly, Cohen edited the first Peirce volumes which appeared in 1917. His student, Paul Weiss, who would also go on to Harvard, would, along with Charles Hartshorne, go on to produce the first multivolume collected works of Peirce, a major achievement in the early 1930s when they were both graduate students.

Cohen would become immersed in the Cambridge group and carry that with him to City College for others to breathe, like Paul Weiss, whom I knew and studied with one summer in 1975 and got to know during the late 1980s till his death at 101. Cohen's interest was in everything. A rationalist Aristotelian, who was encyclopedic, his range covered just about everything from science to law to an understanding of his new country where he arrived at the end of the nineteenth century from Russia. Holmes represented that new country par excellence, the intellectual Brahmin class of Cambridge Massachusetts. Yankee blood brewed in this blood and breath.

Holmes had many relationships with Jewish intellectuals, of whom Cohen was one (Hollinger 1975, 1996), and was quite accepting of the large cohort of left leaning Jewish intellectuals, unlike some of his contemporaries (e.g., Henry Adams, Henry James).

Cohen, like Holmes, prided himself in the imagery of logic. Cohen would write a number of books on logic and indeed started out as a professor of mathematics, statistics and logic like Peirce. Predictive capability was at the heart of reasoning for all three. But Cohen was much more of a traditionalist than Holmes. Cohen was an essentialist with regard to knowing and proportions. Holmes was more rooted in growth and the expansion of knowledge and less about the Platonic essences, if ever at all, and the struggle for knowledge.

Holmes understood mathematics as a tool and expressed that to Cohen. As I indicated earlier, Holmes alludes to his father as the one brought up in science. Holmes Sr., a physician, was known for first discussing the concept of "crib death" in newborns (what we now call SIDS or sudden infant death syndrome).

Cohen and Holmes really had a meeting of the minds when it came to nurturing emerging scientific kinds of evidence that aid the law and the larger aims of justice. Cohen wanted Holmes to be a progressive, like him. Perhaps this wanting to see Holmes clouded his view of who Holmes was—someone who had much less hope in the human condition and one for whom human irrationality was a powerful feature of human expression (Grey 1992).

Holmes (1920) does concede that "science and philosophy are necessities in life" and not just pragmatic necessities; what Peirce worried about was when people over-identified pragmatism with what James characterized simply as what works. Cohen lines up with Peirce's version of meaning and pragmatism. But he again reiterates in a letter to Cohen in 1923 (September 14th) and acknowledges Peirce's originality but not "his self-satisfaction… and when it comes to religion it reflects what we want to believe and not his devotion to logic."

Holmes's letters to Cohen often end with a sense of stoicism and grappling with life, morality being one way in which to grapple with life. Holmes (1920) would suggest that "morality is simply another way of living." Morality, for Holmes, is contextual, built on circumstance and chance, but also history.

Liberalism and Pragmatism

Morris Cohen, the philosopher and friend of Holmes, wrote a book on "The Faith of a Liberal." His friend, Felix Frankfurter, would become more conservative as he sat on the bench (Feldman 2010). But many of Holmes's close colleagues were liberals. Holmes though was not.

In the letters of Cohen and Holmes, you can see that Holmes appreciated philosophical interactions. Frankfurter, a graduate of City College of New York, like Morris Cohen, was also his close friend and his roommate at Harvard. Both men would become intellectual interlocutors with Holmes. Frankfurter would be instrumental in introducing Cohen to Holmes (see Cohen-Holmes Correspondence). Cohen, an avowed liberal, would try to link Holmes to liberal sensibility. But it was a stretch. Holmes was an instrumentalist or pragmatist to Cohen's rationalism. In a letter to Cohen (September 10, 1918), Holmes would write that "mathematics is a tool with which to work on given premises. The premises are a matter of insight." Holmes valued tools, and mathematics was just one of them.

Both Holmes and Frankfurter would understand the constitution as an evolving document and a fundamental instrument of government and order. Frankfurter would begin as an advocate, a defender of labor, a foe of injustice (see Sacco and Vanzetti trial and execution) and ended up arguing close to Holmes for judicial restraint on the bench. Perhaps the removal from the arm pit of humanity in a brazen and bustling city to the rarified air of the Supreme Court played some role in this change (see Holmes and Frankfurter Correspondence 1912–1934).

Holmes admired Frankfurter perhaps the way he admired Abbott during the war. Both men were courageous in battle. One survived, the other did not (see Chapter 1). In contrast, Holmes did his duty, both in service and in law.

Holmes expresses closeness in his letters. He writes to Frankfurter in January 1918, "except in the news of trouble in your family your letter gives me great pleasure" (January 5, 1918). Frankfurter, who was later tasked to write a biography of Holmes, wrote in a letter, "it is a wonderful thing for an old fellow to find that he is not lonely when pretty much all of his contemporaries and early friends are gone" (Holmes and Frankfurter Correspondence 1920).

Frankfurter facilitated Holmes's relationship with Cohen. Holmes's clarity, not necessarily his consistency, and his way of dissolving philosophical disputes and misleading dichotomies comes out in his correspondence with Cohen. Cohen would eventually dedicate a major work on what he took to be the scientific method, "To Mr. Justice Oliver Wendell Holmes: The courageous thinker and loyal friend."

Cohen was a modern rationalist about science. That is, he believed in foundational systems of rational thought, based on a large sense of science. But Cohen had an idealized version of the scientific process of data (Hollinger 1975).

Despite these differences, both Cohen and Holmes had a long and detailed correspondence and spent time together along with their respective spouses. Judging from this correspondence, they seem close (see Cohen-Holmes Correspondence). With his student, Ernest Nagel, Cohen would write an important book on logic and would speculate about logic and the law. Cohen (1931) was looking for grounding and like many others found it in the structure of reasoning (Hollinger 1975). Cohen thought there was a logic that underlies human reasoning across topics including the law, and there is. It is, just as Dewey said, not one formal frozen piece of logic.

Both Holmes and Dewey believed in the methodological reflexive use of rules. Holmes sees judicial decision making as the application of rules, statutes, and law. The basis of making laws has been the experience of individuals and the competition of ideas. This is what Holmes called "the felt necessities of the times, the prevalent moral and political theories, intuitions of our public policy, avowed or unconscious, even the prejudices which judges share with their fellow men, have had a good deal more than the syllogism in determining the rules by which men should be governed" (*Common Law*, 1881).

The issue about liberalism and pragmatism would be central to Dewey. Liberals of the twentieth century were progressives. Liberals like Dewey had a wide view of human experience and a pervasive sense of experiment and correction. The liberal view under Dewey is the release and development of positive human potential. Perhaps there is a mythology associated or naiveté associated with the expansion of human potential and the deepening of human depth.

Meliorism and Human Progress: Dewey Not Holmes

Dewey (1928), when writing about Justice Holmes and the liberal mind, is just fanciful. Morris Cohen, defender of Holmes, is also a little fanciful, as is Harold Laski. Dewey is trying to argue against those who say Holmes "has no social philosophy." Dewey defines modern liberalism as indicative of the orientation towards "experiment, the method of intelligence and problem-solving."

Dewey quotes Holmes in "Justice and Holmes and the Liberal Mind" (1928), as saying "Your business as thinkers [addressing judges] is to make plainer the way from something to the whole of things; to show the rational connection between your fact and the frame of the universe."

The larger sense of experimentalism is foremost in the American sensibility, e.g., Franklin and those who helped found the American Philosophical Society, a society that represents "useful knowledge, and where the philosophical is practical in implications." Both men are tool oriented. Indeed, Dewey's philosophy has often been identified with instrumentalism and with technology. But our capabilities are tied to creative use of tools to enlarge our adaptive capabilities and to expand our horizons (Hickman 1992). We are just not quite there with many of our tools. But the convergence of tools and theory is participatory.

Dewey (1925, 1948) is associated with a philosophy of technology and instrumentalism. Dewey was clear about human progress and the role of science. Often cited by his faith in science, even Dewey was more nuanced than his critics allowed (Niebuhr 1932/1960). He knew our bestial capabilities. Our human condition is fraught with abuse, both Dewey and Holmes engage these palpable facts. Certainly, Holmes understood that; but his outlook was blurred from a war that "changed everything" for him.

Holmes said "the constitution is an experiment," and the experiment is tied to experiences in transitions from one event to another. It is hard to satisfy everyone, but the normative goal of a participatory democracy is greater participation of an informed citizenry (Sarokin and Schulkin 2016; Butler 2017).

Dewey, of course, believed in human progress much more than Holmes. Meliorism (Koopman 2009) is faint in Holmes. Certainly not after a war that left him wounded, his neural circuits reverberating with what we do to each other.

Pragmatism is put in this larger context of deriving all possible consequences in determining an idea: experimentalism. It is a method related to the ideas (pragmatism) essential to human meaning and human understanding. Within the law, it is essential for evolving epistemic orientations (Ansell 2015).

Holmes travels between positivism, legalism, and realism but tends towards modest theory in pragmatism. In correspondence, Holmes thrives amidst the discussion of ideas. He also understood the impact of force and the reality of human bestiality that is ever-present in the human condition. While the evolution in culture may reflect the allure as Whitehead put it from force to persuasion, the reality of "just good enough" decision making must take account of human propensity for violence. Holmes understood the hodge-podge of the law and human decision making.

Holmes saw no necessary steps in progress; but diverse forms of progress are plain. The idea of progress is an important element. There are no cosmic nor necessary steps, but a cultural evolution is a major theme of pragmatism as it has always been a major theme in the history of diverse strains of pragmatism (Schneider 1946/1963).

For Holmes, life was the embodiment of struggle; and it is. As a result, stoical responses are necessary to persevere. But you can be tough minded and have normative goals. Law, in part, embodies the normative goals of our culture (Diggins 1994). Dewey was too optimistic for Holmes some of the time; but he still chose Dewey for representing the larger sense of being in the world as found in Dewey's *Experience and Nature* (1925). Commenting on Dewey, Holmes would assert that he has … "to have more of our cosmos in his head than any philosopher I ever read" (Holmes correspondence with John CS H Wu. see Fisch 1942). And what Dewey emphasized there was "the precarious and the stable" in a context of worthy ideals where ends and means are permeable in the everyday matters of life and transactions with one another. Law blends into the fabric of social life as do bits of science. Neuroscience becomes part of this fabric.

Transition underlies heightened experiences. It is transition that underlies memory and adaptation between events, since experience is a key factor in learning and inquiry. Struggle lay at the heart of human existence. The world is at once the precarious amidst bouts of stability (Dewey 1925/1989).

Holmes belittled Dewey, and called his writing "unreadable." On Dewey's constant reminder of the downtrodden masses, Holmes said, "He talks of the exploitation of many by man – which rather gets my hair up" (Letter to Laski, see also Diggins 1994). Holmes was interested in the law but not the intentions of the individual committing a crime but about the probability of the crime occurring again. The meanings of things, what we should pursue, belongs to the body politic, the legislative branch and not the judicial branch. Here Holmes is consistent and some might consider him even to be conservative (Mendenhall 2013).

Neuroscientific Considerations: Human social contact is at the heart of human viability, something Dewey (1925) emphasized. We are a social species; our evolution is tied to cooperative endeavors. Indeed, cortical function is tightly linked to social contact (Dunbar 1996, 2016). The greater the social contact, the greater cortical function. Further, a good deal of the brain is neocortical in our species; about a third of the brain (Swanson 2000). The tagging of social contact, the friendships (and those not), the transactions of the everyday that mark our experiences are couched within cortical regions, both old and new cortex.

Discussions of neuroscience and the law are set in part by small groups at institutions evaluating whether an experiment can go forward;

institutional review boards (IRM committees) are a very first step in the larger legitimation process of what we decide together, towards eventual use of evidence that might be used in a courtroom from discoveries about the brain (Moreno 1995). This is a key thing; we are in this together, directly and indirectly. And one of the issues is about trauma and the brain, fear and the brain, uncomfortable experiments; loss, depression, frustration etc. And of course, exposing trauma, exploring it; war time trauma that impact the brain.

I began the book, with capturing Holmes Jr. experience of the war, his view about it, and the impact of that war might have had on his brain. We know more about the impact of war on the brain, and the longer-term considerations. These epistemic considerations of neural systems are gradually being integrated in how we reason, with regard to human agency, reasoning and accountability and the law (Churchland 2011; Kolber 2014), in everyday life, for understanding human development and human devolution, both in the courts and not (Hirstein et al. 2018).

The war experience for Holmes' was also couched in the context of duty. But his longer-term viability perhaps is in the diverse social forms of contact that he facilitated through his diverse friendships over his long life (Cheney and Seyfarth 2007). And friendships are key to survival, or at least a sense of well-being. They can decrease the information molecules like cortisol and CRH noted in the first chapter that are also tied to energy consumption (Schulkin 2011, 2017). These basic forms of social contact placate agitated systems, sustain longer term viability from over use and degeneration (Sapolsky 1992; McEwen et al. 2014; Schulkin 2017).

Dewey, Holmes's Better Half or "The Road Not Taken" by Holmes: Our Social Nature

Dewey himself did a dissertation at Johns Hopkins University mostly having to do with Hegel. But Dewey shed the Hegelian suasion of his dissertation. The sense of human unity in *Experience and Nature* written some 50 years later depicts not disunion so much as continuity and overlap. But Dewey shed the Hegelan part and concentrated on nature and culture as something we adapt and create and where there is much continuity. Dewey also emphasized our social progress, as we are a social species. Holmes rescinded from this.

Holmes believed in science and revered the law (1899) and the competition of ideas, and felt inevitability of outcome. He did, like Dewey, tie social context with empiricism was and is a trend in the right direction. That is what mattered is the direction of the engagement, and not just the competition. Ends mattered as they merged into means or instruments of engagement. The fluidity was a feature of the human experience, and now could be linked to law, to what mattered to enhance deliberative participation in the social milieu, small and large.

Holmes was evolving as he grappled with law and knowledge as he understood it being that of science. His view of science was narrow. His view of human suffering was up close from his experience in the war and his view of the mysticism in nature was vast as it was for his revered senior statesman of the Cambridge community—Emerson. Emerson as we will see in the next chapter, provided a sense of place, an appreciation of nature, and the poetic sensibility of being awed by nature, and being part of it. No separation of us and a budding development of a pragmatic orientation of tools and use, ends and means and dilution of dividing lines in epistemic orientation between action and thought.

Holmes never developed a full-blown sense of positivism or utilitarianism. Of course, he was influenced by this and was inconsistent. The desire for objective measures as a worthy pursuit is independent of both positivism and utilitarianism. Separating facts from values, morals from rules, that was a common malady and not consistent with the pragmatism that Dewey would later develop. A pragmatist orientation is one which value underlies action and action is embedded in thought.

Dewey was wide eyed and looked naïve to some (e.g., Bertand Russell). Holmes never did. Holmes was a progressive in passing, namely never really one or simply passing was the sense in which science can become dominant and where the law is buttressed by the discoveries in science. Dewey was a progressive in substance; the pragmatism rooted in law was not simply one of instrumental conveni-ence and prediction but towards larger rewards of human uplifting.

When it came to cases involving labor laws, for instance, limited hours that can be worked or worker protections (*Lochner vs NY* 1905), Holmes sided in favor of contract and freedom of capability rather than worker protection, nor did he evoke any form of natural rights, or any particular economic theory being tied to the constitution. Holmes was careful about over reading too much into the constitution (George 2000; Baker 1991; Wellington 1990; White 2015). As Holmes would assert in *Lochner vs New York*, namely that rights lay in the "majority to embody their decisions in law."

Holmes was inconsistent, but then he is contextual (Grey 1989; Posner 1992). The many variants of utilitarianism and the link to hedonic evaluation has taken many forms. In more modern forms, it is tied to the larger calculus of pleasure. But pleasure is not a calculus—certainly not exact—but then what is? Certainly not that; we are just left to consider the facts about us. Our reasoning under diverse con-ditions with regard to calculating is mired in flaws (Kahneman et al. 1982). But certainly, the calculating device is good enough under con-ditions of survival, viability, longer-term survival.

Holmes embraced a version of utilitarianism, though not stated as such and housed in a context of an ethic of capitalism. Holmes was no socialist. For him, there was no loss of an individual in the greater com-munity. Here he was quite different from Dewey, whose progressiveness was in light of a community. Holmes deferred to the community and the politics of the community. This was independent of his position on the Supreme Court. This was a property of him, a kind of aloof-ness from others (Rogat 1964). Others were more abstract, or, at least, it seemed that way. He has been characterized as a positivist, as a pejo-rative term in the sense of the loss of a moral fabric of what matters. An anchor in metaphysics and theology was just what positivism was trying to eradicate.

Holmes was a wavering positivist, but he was heavily influenced by the tools of social or legal positivism. Holmes was anchored to many kinds of explanations. The body politic was palpable in all things including the law.

Now the issue about subjectivity vs objectivity is an issue that perhaps has outlived its usefulness. There are degrees of objectivity depending upon the phenomenon being studied, is what many of us would say now. Degrees of inquiry and objectivity pervade most of what we do when we investigate; the demarcation of the objective from the subjective is much more fluid, permeable and continuous. Again, pragmatism, unlike naïve positivism, acknowledges and has, since Peirce, undercut such distinctions between observation and theory, invention and discovery—or, as Dewey would have said, with regard to fact and value.

References

Ansell, C. (2015). *Pragmatic Democracy: Evolutionary Learning as Public Policy*. Oxford: Oxford University Press.

Baker, L. (1991). *The Justice from Beacon Hill: The Life and Times of Oliver Wendell Holmes Jr*. New York: Harper & Row.

Berlin, I. (1969). *Four Essays on Liberty*. Oxford: Oxford University Press.

Brent, J. (1993). *Charles Sanders Peirce*. Bloomington: Indiana University Press.

Butler, B. E. (2017). *The Democratic Constitution*. Chicago: University of Chicago Press.

Cheney, D., & Seyfarth, R. M. (2007). *Baboon Metaphysics*. Chicago: University of Chicago Press.

Churchland, P. S. (2011). *Brain Trusts*. Princeton: Princeton University Press.

Cohen, M. R. (1931/1959). *Reason and Nature: An Essay on the Meaning of the Scientific Method*. New York: Dover Press.

Dailey, A. C. (1998). Holmes and the Romantic Mind. *Duke University School of Law, 48*, 429–510.

Dewey, J. (1920/1948). *Reconstruction in Philosophy*. Boston: Beacon Press.

Dewey, J. (1925/1989). *Experience and Nature*. LaSale, IL: Open Court Press.

Dewey, J. (1928). Justice Holmes and the Liberal Mind. *New Republic, 53*, 210–212. In J. A. Boydston (Ed.), *The Later Works of John Dewey, Volume 3, 1927–1938* (pp. 177–183). Carbondale: Southern Illinois University Press.

Diggins, J. P. (1994). *The Promise of Pragmatism*. Chicago: University of Chicago Press.

Dunbar, R. I. M. (1996). *Grooming, Gossip and the Evolution of Language*. Cambridge, MA: Harvard University Press.

Dunbar, R. I. M. (2016). *Human Evolution*. Oxford: Oxford University Press.

Ellis, R. E. (2007). *Aggressive Nationalism*. Oxford: Oxford University Press.

Emerson, R. W. (1855/1876). *Nature, Addresses and Lectures*. Cambridge: The Riverside Press.

Feldman, N. (2010). *Scorpions*. New York: Hachette Books.

Ferguson, R. A. (1988). Holmes and the Judicial Figure. *University of Chicago Law Review, 55*, 506–548.

Fisch, M. H. (1942/1986). Justice Holmes, the Prediction Theory of Law and Pragmatism. In M. H. Fisch (Ed.), *Peirce, Semiotic and Pragmatism*. Bloomington: Indiana University Press.

Foner, E. (2011). *The Fiery Trial: Abraham Lincoln and American Slavery*. New York: Norton.

George, R. P. (2000). *Great Cases in Constitutional Law*. Princeton: Princeton University Press.

Grey, T. C. (1989). Holmes and Legal Pragmatism. *Stanford Law Review, 41*, 787–870.

Grey, T. C. (1992). Holmes, Pragmatism and Democracy. *Oregon Law, 71*, 521–542.

Hart, H. L. A. (1958). Positivism and the Separation of Law and Morals. *Harvard Law Review, 71*, 593–629.

Healy, T. (2013). *The Great Dissent: How Oliver Wendell Holmes Jr. Changes His Mind and Changes History of Free Speech in America*. New York: Henry Holt.

Hickman, L. A. (1992). *John Dewey's Pragmatic Technology*. Bloomington: Indiana University Press.

Hirstein, W., Stifford, K. L., & Fagan, T. K. (2018). *Responsible Brains: Neuroscience, Law and Human Culpability*. Cambridge: MIT Press.

Hollinger, D. A. (1975). *Morris R. Cohen and the Scientific Ideal*. Cambridge: MIT Press.

Hollinger, D. A. (1977). Review: The Culture of Experience: Philosophical Essays in the American Grain. *Transactions of the C. S. Peirce Society, 13*, 312–315.

Hollinger, D. A. (1996). *Science, Jews and Secular Culture*. Princeton: Princeton University Press.

Holmes, O. W., Jr. (1881/1952). *The Common Law*. New York: Dover Publications.

Holmes, O. W., Jr. (1899). The Theory of Legal Interpretation. *Harvard Law Review, 12*, 417–420.

Holmes, O. W., Jr. (1901). One Hundredth Anniversary of the Day on Which Marshall Took His Seat on the Bench.

Holmes, O. W., Jr. (1905). *Lochner vs New York*.

Holmes, O. W., Jr. (1918). *Natural Law*. Cambridge: Harvard Law Review 32.

Holmes, O. W., Jr. (1920). *Collected Legal Papers*. London: Constable and Co.

Holmes, O. W., Jr., & Frankfurter, F. (1996). *Correspondence 1912–1934* (R. M. Mennell & C. L. Composton, Eds.). Lebanon: University Press of New England.

James, W. (1897/1927). *The Will to Believe and Other Essays in Popular Philosophy*. New York: Longmans, Green and Co.

Kahneman, D., Slovic, P., & Tversky, A. (1982). *Judgment Under Uncertainty: Heuristics and Biases*. Cambridge, UK: Cambridge University Press.

Kolber, A. J. (2014). Will There Be a Neurolaw Revolution. *Indiana law Journal, 89*, 808–845.

Koopman, C. (2009). *Pragmatism as Transition*. New York: Columbia University Press.

Kuklick, B. (2000). *A History of Philosophy in America*. Oxford: Oxford University Press.

Luban, D. (1992). Justice Holmes and Judicial Virtue. *Nomos, 34*, 235–264.

McDowell, G. L. (2010). *The Language of Law and the Foundations of American Constitutionalism*. Cambridge: Cambridge University Press.

McEwen, B. S., Gray, J. D., & Nasca, C. (2014). Recognizing Resilience: Learning from the Effects of Stress on the Brain. *Neurobiology of Stress, 1*, 1–11.

McPherson, J. M. (1991). *Abraham Lincoln and the Second American Revolution*. Oxford: Oxford University Press.

McPherson, J. M. (2008). *Tried by War: Abraham Lincoln as Commander in Chief*. Princeton: Princeton University Press.

Menand, L. (2001). *The Metaphysical Club*. New York: Farrar, Straus and Giroux.

Menand, L. (2002). *American Studies*. New York: Farrar, Straus and Giroux.

Mendenhall, A. P. (2013). Justice Holmes and Conservatism. *Texas Review of Law and Politics, 2*, 305–314.

Monagan, J. S. (1988). *The Grand Panjandrum*. Lanham: University Press of America.

Moreno, J. D. (1995). *Deciding Together*. Oxford: Oxford University Press.

Newmyer, R. K. (2001). *John Marshall and the Heroic Age of the Supreme Court*. Baton Rouge: Louisiana State University Press.

Niebuhr, R. (1932/1960). *Moral Man and Immoral Society*. New York: Charles Scribner.

Nietzsche, F. (1878/1996). *Human All Too Human*. Cambridge: Cambridge University Press.

Nietzsche, F. (1886/1972). *Beyond Good and Evil.* Mineola: Dover Publications.

Posner, R. A. (1990). *Cardozo: A Study in Reputation.* Chicago: University of Chicago Press.

Posner, R. A. (1992). *The Essential Holmes.* Chicago: University of Chicago Press.

Posner, R. A. (2003). *Law, Pragmatism and Democracy.* Cambridge: Harvard University Press.

Posner, R. A. (2007). *Economic Analysis of the Law.* Alphen aan den Rijn: Wolters Kluwer.

Rogat, Y. (1964). The Judge as Spectator. *University of Chicago Law Review, 31,* 213–256.

Sapolsky, R. M. (1992). *Stress: The Aging Brain and the Mechanisms of Neuron Death.* Cambridge: MIT Press.

Sarokin, D. J., & Schulkin, J. (2016). *Missed Information.* Cambridge: MIT Press.

Schneider, H. W. (1946/1963). *A History of American Philosophy.* New York: Columbia University Press.

Schulkin, J. (2011). *Adaptation and Well Being.* Cambridge: Cambridge University Press.

Schulkin, J. (2017). *The CRF Signal: Uncovering and Information Molecule.* Oxford: Oxford University Press.

Stone, G. R. (2004). *Perilous Times.* New York: Norton.

Swanson, L. W. (2000). What Is the Brain? *Trends Neuroscience, 23,* 519–527.

Wellington, H. H. (1990). *The Supreme Court and the Process of Adjudication: Interpreting the Constitution.* New Haven: Yale University Press.

White, G. E. (1990). Holmes as Correspondent. *Vanderbilt Law Review, 43,* 1707–1760.

White, G. E. (1993). *Justice Oliver Wendell Holmes: Law and the Inner Self.* Oxford: Oxford University Press.

White, J. W. (2015). *Lincoln on Law, Leadership and Life.* Naperville, IL: Source Books.

White, M. (1947). *Social Thought in America: The Revolt Against Formalism.* Boston: Beacon Press.

White, M. (2002). *A Philosophy of Culture: The Scope of Holistic Pragmatism.* Princeton: Princeton University Press.

6

Emersonian Sensibilities

Holmes in a speech at Harvard Law School (February 15, 1913) commented on human thought and noted that "most men think dramatically, not quantitatively." Holmes was depicted as detached, as a spectator, and perhaps some of that is due to his surviving the horror of war (Rogat 1964; Grey 1992). He was the "Yankee from Olympus" (Bowen 1945), and, of course, an elitist. He has been called a lot worse, as I have indicated. He has been called a lot of laudatory things as well. Our expectations of Holmes exceed the reality. Holmes was a well-qualified, well-educated, battle-tested, ambitious intellectual with a pedigree of experience to match any office.

Like John Marshall, Holmes's legacy is built into the fabric of American jurisprudence and language (e.g., "Clear and Present Danger," Leonard 2006; Budiansky 2019; Blasi 1999). John Marshall was in the right historical place, the founding of the republic. But Holmes was as well, having survived a bloody war of meaning that would resonate in him over his long life. The times were utterly human, interesting, and tied to the intellectual milieu of America. Holmes stayed close to the value of free expression, external and not internal machinations, and a culture of competition. Holmes is anchored to a sense of duties and not rights (Atiyah 1983).

© The Author(s) 2019
J. Schulkin, *Oliver Wendell Holmes Jr., Pragmatism and Neuroscience*,
https://doi.org/10.1007/978-3-030-23100-2_6

As I have indicated at the onset of this book, Holmes is not as consistent and as predictable as his supporters initially wanted and expected. He is duplicitous with regard to ideas and where they might come from. He breathed the same intellectual air as the Metaphysical Club, and sought to further his own development by his association with it, and in his diverse correspondence with so many people and across a wide range of cultures and classes. And that is impressive. What is less impressive is his detachment towards the plight of the everyday person. There is a lack of empathy in his tone towards the plights of others.

Posner (2008), writing about what he calls "sensible pragmatists" versus the "short sighted pragmatist" (p. 239), reminds me of Peirce's distinction between James's pragmatism and his own. One is narrower than the other. And narrowness is one of the critiques of pragmatism in general, including legal pragmatism (Warner 1993; Grey 1989; Butler 2002). Another orientation more Dewey-like, view of legal pragmatism ties it to the resolution of conflict with an eye toward social justice (Butler 2002), and which matters for a philosophy of law that integrates neuroscience, a view shared by this author. One feature of social justice is evolving a culture of concern for others, one of the normative goals of a variant of Deweyian deliberative participatory democracy. And it strikes this author as a very worthy though difficult one. But this is far from Holmes, and even further from Posner. But it is close to Emerson. And Emerson is important for Holmes.

And Holmes is also tied to Emerson, a formidable figure of intellect next to his father. Emersonian themes and a sense of nature and poetry run through his work with a vibrant pulse of nutrients, meaning, and sustenance. Emersonian independence was a vital source of intellectual sustenance for Holmes. Emerson would say in his *Natural History of Intellect*, "my belief in the use of a course on philosophy is that the student should learn to appreciate the miracle of the mind; shall learn its subtle but immense power…" (p. 6). This is something that Holmes embodied over a lifetime.

Holmes however, did not regard Emerson as a "thinker" as he would say in one letter (Letter to Patrick Sheehan, October 27, 1912). Emerson for him was primarily a "poet" who revealed truth. And it is

true that Emerson did not argue in a manner that Holmes perhaps preferred. But Holmes knew that Emerson depicts what is important in
human life and, in a letter to Patrick Sheehan, he acknowledged that.

Emerson was more "starry eyed" than Holmes. Emerson's transcen-
dentalism was tied to idealist concepts of perfectibility. Holmes was
not an idealist. His poetry transformed into practical rules of guidance,
methods of reasoning that ground us. Nature was not lost, but there
was little "oversoul" as Emerson would have it and Holmes would not.
But the sense of tragedy which Emerson also wrote about would be a
stoic feature scrawled and sewn into Holmes's sensibility.

Holmes was no reformer either; nor a walker in the woods or trees to
get his bearing in a Walden pond as Thoreau suggested, at least after the
war. He was thrown into and had the ability to succeed in, a profession
in which he fit very well, and in a context so fertile with intellectual
excitement and emboldened by a war. And yet he would never be reduc-
ible to a single "-ism."

In this chapter, we return again to a pragmatist theme: tagging events
and tracking and predicting events within the larger cultural themes in
which Holmes and indeed many of us experience. The roads not taken
by Holmes are highlighted along with his ties to Emerson.

Senses of Nature

In a universe in which thinking is naturalized and demythologized to
the human condition, the law has played an important role in the cultures we have evolved in. Holmes inherited this and contributed to an
Enlightenment perspective modified by an evolving sense of nature. This
sense of nature is traced across culture to native capabilities (Atran and
Medin 2008). Cognitive capabilities underlie action in pragmatism, an
understanding of the brain (Schulkin 2000, 2012), and certainly the
pragmatism of Peirce, Dewey, and Holmes.

Biological organs like the brain are both determined and labile. There
is a solid form of synaptic structure that is predictable and scientifically formidable, and there is endless epigenetic regulation of change.
This brain is not the machine of the sort Holmes might have imagined.

And we re-envision the meaning of determining also by our inventions; the tools that we use that are extensions of our brains. And here the room is endless, or so it seems.

Nature is no abstraction, and we come prepared to make sense of natural kinds of objects. We seek to determine causal links; there is nothing abstract about this but it is a very basic predilection, along with spatial knowledge, kin relationship, prediction of events, facial expression, to name a few, and, of course, language and the facilitation of social contact.

One feature is the fundamental distinction of objects that are alive, and those that are not. The brain is prepared to detect this. Animate and inanimate distinction is perceived early on in babies. They come prepared to understand this and to see others as motivated, or intentional where they can. The brain may not be fully developed, but this distinction about nature alive or not is basic to us. And it should be.

Of course, concepts like bacteria blur lines of living and non-living and notions of immortal cell lines. But for core dividing features in what we evolved in, categories that set the precondition of understanding this distinction are fundamental along with a number of other cognitive predilections that are embedded in neural function and manifest in behavioral expression.

We quickly (in evolutionary time) turned from a species only concerned with the religious to the philosophical being about wonder and systematic connections and clarity of events. Indeed, there are forms of what might be called "native pragmatism" (Pratt 2002) that have roots in knowing nature (Wilshire 2000), grounded in a form of primal responsiveness inherited from an evolutionary past that situates us as the human primate that survived the other hominoids that existed and preceded us, who all died by some 30,000 years ago.

Sense of Place

Both Emerson and Holmes convey a sense of place. The sense of place builds connections of meaning within the storehouse of knowledge. Wisdom is mostly about knowing what matters; intelligence is the means to achieve ends that are worthy. Wisdom is the flame of possibility amid wonder (Midgley 1989) and the possible and actual realization of what we owe each other—a kernel of possibilities of others co-inhabiting a place together (deciding together, Moreno 1995).

Of course, the moments of insight about place and being in Confucius, Mencius, Lao Tsu, and Wang Ming require that a completion of oneself is tied into others, piety, and regard for nature (Chan, Lao Tsu). Glimmers of possibilities about nature and self-development are so easily distorted into the struggles of everyday human strife and limitations, connecting heaven and earth, the possible with the actual. Enhanced attention flickers amidst noise.

Of course, there is the omnipresent danger of predation, decay, and entropy amidst growth and opportunity. Thoreau understood this and put it in the context of identity and protest; the self in search of the pond of wellbeing—the link of heaven and earth. Emerson, reaching out in verse to a poetry of self-reliance, developed a sense of what we owe each other; our survival as dependent upon one another, but our individual sense of duty forming a whole. The bringing together of collective wills of choice is something akin to Rousseau.

Nature and poetry are a way of life, not separate; there can be a continuity of nature and culture with no bifurcation. This would resonate with Holmes Jr. as his father, the polymath physician, was drawing close to the inspiration of Emerson (Menand 2001; Mendenhall 2015a, b). The heavens, or the ideal pursuit of what is worthy, mingling endlessly with the toils of the soil and a world soul expanding, to jolt them into some balance or harmony. Emersonian sensibility pervades a life-long literary theme for Holmes.

The important theme is the continuity of the arts with the sciences; the poetry of the verses of Emerson with the sense of duty and self-reliance. This was an American version of stoicism sprinkled by the Enlightenment and the sense of science, investigation, and invention. But Emerson and a real strand of later pragmatists (Dewey 1925; Neville 1974) would emphasize the aesthetic dimension of thought and of experience, something Holmes understood.

Emerson would emphasize "first experience," and aesthetics would sculpture thought alive, something close to Holmes. From Emerson through Dewey, the emphasis is on nature, but culture is not something foreign. We just keep evolving, while there are constraints of natural objects. Emerson, like Thoreau, captures our rootedness in nature. Holmes inherited this.

The softer ministerial side of Emerson, the teacher of human embrace, is not lost to Holmes, but it is not expressed rampantly except through his attachments to comrades in arms and his intellectual friendships (e.g., Laski, Frankfurter). Holmes was silent with regard to Emerson's ability to reach for Buddha bliss. There was not much bliss to be found after a war that left its scars on him. Emerson was disgusted by slavery; as a literary critic and a sage of the human condition he represents the "better angels" of the US. Holmes knew that. But he was no Emerson.

Emerson the poet was rooted in a sense of nature, of history, and a spiritual embrace of the unknown with an evolving theology. This theology was rooted in some form of unity while also tied to what has been called "Scottish common-sense realism" (Reid 1764; Hamilton 1853; Emerson 1858; Richardson 1995). This sense of realism emphasizes that we are anchored to objects, in nature and our cultural evolution, and historical sensibilities are continuous with nature. The universe is a rich one, endlessly evolving and ripe with human freedom and expression. Holmes's fatalism, though, tends to emphasize what he calls the "can't help" (see his essay on Natural Law). We can not help but responding in such and such a way, or making such a decision.

Holmes was the poet of his Harvard class and remained close in some sense to what he learned from one essential mentor, namely Emerson. Emerson's sense of language, nuance, and depth pervades the pulse of Holmes (Mendenhall 2015a, b). Emerson, a romantic pragmatist (Goodman 1990), cultivated a worthy scholarly sense of developed human experience. Similarly, aesthetic is how many law scholars have described the pen and prose of Holmes (e.g., Posner 1992; Budiansky 2019; Grey 1989; Frankfurter 1939). Judge Posner (1992) would describe Holmes's in one of his dissenting opinions (*Lochner vs NY*) as a "rhetorical masterpiece."

Emersonian sensibility, rich in Holmes, is expressed in verse. Emerson, the literary critic, expositor and practitioner, expressed the very genius he valued (Bloom 1983, 2002). Holmes, like Emerson, would be influenced by the essays of Montaigne for their sagacity. And, as Emerson would say, "Montaigne is the frankest and most honest of all writers" (Montaigne or the skeptic).

The range of rhetorical devices and the generations of transitions of events with energy is a core feature of pragmatism (Levin 1999). The rhetoric of Emerson and Holmes captures this in Emerson's rhetoric and Holmes's dissent—dissent being a maker of energy, of alternative, of differing consequences (Mendenhall 2015a, b). Holmes has often been characterized as a great dissenter, although perhaps less than when compared to some other Supreme Court jurists. But he raised the level of discourse on the court considerably. His use of metaphors and his literacy are part and parcel of his Emersonian rhetorical energy. The many faces of pragmatism, of which one is found in these literary and rhetorical transitions, are anchored to a First Amendment sacredness in which rhetorical presentations tie probable inferences to other probable inferences (Danisch 2008). This rhetoric comes alive in the making of a case.

Nature is a running theme in human life, our embodiment within it, our cultural evolution of self-discovery, to discover a will to persevere. A sense of place, of being, is what we come to experience in nature: either the good and the bad while demythologizing the sense of the romantic. This is true for me, and I would suggest for Holmes, who is not particularly romantic but who had a respect for diverse forms of human original conditions.

Insight by Any Means Necessary: Metaphors and Wallace Stevens

Pragmatists, like Holmes and Dewey, see reason and problem solving as tool oriented; mathematical inference is a tool embedded in action, in foraging for coherence. So are metaphors an aid to thought, and can either aid or hinder investigation and discovery. We use metaphor in science, in engineering, and in law. Metaphor figures in making a case, in facilitating understanding. Try to hold a conversation for five minutes without using a metaphor; you will find it to be an almost impossible endeavor. The literary Holmes was also a pragmatist; tools that aid reason, argument and understanding are used. Metaphors can or not be aids to thought, just like other tools that aid problem solving. Pragmatist legal reasoning, or otherwise as we will see in neuroscience, does not denigrate the use of metaphor in human information processing.

Metaphor can also be misleading, and dangerous; "metaphorical framing is not a game" (Johnson 2007). And yes, they can be fictional and misleading and nonsensical (Cohen 1935). But they need not be, and never as Holmes would attest dangerous enough to ban it, or to ban poetry, which relies heavily on metaphor. Whatever aids understanding is to be used to aid epistemic endeavors (Boyd 1993; Lakoff and Johnson 1999). Cognitive systems run through metaphor as they do with other forms of human reasoning and the sense of action is endemic in cognition (Lakoff and Johnson 1999).

Holmes wants to stay anchored, and as Wallace Stevens evoked "to think things not words" (Grey 1989). "Things" are anchors, external events in which cognitive systems that run through action and are tied to transactions with others (Dewey 1925/1989).

Empiricism is hard, yet pragmatism provides enough context, enough theory; poets like Wallace Stevens, a student of William James, give a feel for what it is like to be a pragmatist (Richardson 2007). Other poets lament what is often thought "the great weakness of pragmatism is that it ends by being of no use to anybody." (T. S. Eliot, cited by Posner 2003). But pragmatism, certainly the way Holmes understood it, in so far as he reflected on it, was good for an adaptive form of self-corrective inquiry. Something brains like our own are prepared to do, which is why science is extended problem solving that reflects and extends and expands basic cephalic capabilities that evolution selected.

Human problem solving as Holmes and other pragmatists like Dewey emphasized was problem solving in context, with background capabilities, expectations, all within the larger cultural milieu. What is important is the evolving side of it, but not evolving towards any predetermined end.

Law is about clarity, but literature and its study are endlessly instructive. Where the poem is about nuance and possibility, in Holmes it is the recognition of this in forms of prose in the law. But a pragmatist need not reject "that there are objective criteria of truth" contra to some characterization (Posner 1988/2009); one does not and we do not through objective measures. We just do not exaggerate foundations and note the fluidity of the epistemic endeavors. And that is just it, endeavor is plural. One of the points of pragmatism and the larger body politic and the continuity of science and other endeavors like the law is the groundedness that one senses. We are linked to objects in inquiry: we are linked to shared experiences.

Whatever helps is in the tool box for reasoning. So, if poetry is helpful then use it. Reasoning by metaphor is still reasoning. The issue is the anchor to what is relevant, what must be decided, what opens up a horizon of understanding and judiciousness. Poetry is about the imaginative, and almost all thinking requires that. The issue is staying grounded in what counts, in what is relevant. Wallace Stevens was a lawyer, by trade, a naturalist by predilection and expression perhaps a pragmatist in orientation (Grey 1990, 1991). He was someone where experience may provide a sense of that with the acknowledgement of the important role of metaphor and then the loss of metaphor to a piece of understanding laid bare by our labor—a labor tied to consequences and meaning (see also Levin 1999).

Returning to the allure of the poet, or poetry, when Grey (1990, 1991) makes a case for the poetry of Wallace Stevens to be useful to legal thinking, in part it is a plea for the diverse forms of reasoning, of which strictly logical form is just one. Grey suggests that Stevens, the pragmatist poet (cf. Winter 2001), breaks up the rigidity or dichotomy between worlds, metaphor and reality being one, and perhaps diluting the binary rigidity that is inherent in legal thought (Posner 1988/2009; Winter 2001). Grey suggests that Stevens, the lawyer, could serve as a "therapist" for ameliorating rigidity.

Poetry, may be bridge the mind to what might be relevant, enticing and certainly beautiful by broadening our horizon. Holmes, endlessly literary, "was an artist of a high order" and his opinions indeed remind one of "the poems of Wallace Stevens" (Novick 1992, p. 1248). Holmes, as Posner says (1988/2009), was the most well-read of those that served on the Supreme Court, and sought to dilute rigidity and understand that fewer words can be enlightening and that an argument of persuasion uses diverse forms of expression. The point here is that metaphors can be helpful or not; and they pervade thought whether poetic or not.

Holmes, in his presentations to audiences, could go poetic (see *Law and the Court*) and instill faith and fears of human possibilities. Wallace, a situated pragmatist (Levin 1999) of what is above us or beneath our feet, dilutes the separation of the theoretical and the practical, the imaginative and the real, the valuation and the cognitive (Grey 1992; Richardson 2007; Winter 2001). Here he is close to Dewey and most pragmatists of diverse persuasions.

As Stevens suggested, "the poem refreshes life." Poetry embodies "the first idea," "a response to the daily necessity of getting the world right" (Levin 1999). Of course, there are plenty of instances where this is not the case. Perhaps under an Emersonian light, toward self-reflection, self-respect, self-reliance, and respect towards others in a worldly sense, and perhaps something Holmes understood, that poetry is an ingredient in reasoning which can aid thought. There is nothing necessary or mythical. Just looking for what aids and what does not, and being faithful to phenomena in understanding. Maybe Holmes understood this, kept this close to him, his sense of science tempered and softened by poetic sensibility grounded in a naturalized epistemology that understood heuristics, or good enough problem solving.

As Stevens put it, "we seek nothing beyond reality" (Grey 1990, 1991) and Holmes understood this to be the purpose of the law: where are we and what are predictive implications and possibilities.

As Stevens suggested, "In the world of words, the imagination is one of the forces of nature." But then imagination runs through all epistemic endeavors. Holmes understood this, as did William James.

Problem solving runs through pragmatism. As the poet, Stevens understood as a poet with pragmatist sensibilities is just that, an understanding that nature and culture are continuous with one another (Richardson 2007). A running thread within pragmatism is capturing, attending and being respectful to human experience. Indeed, the literary conception of pragmatism for which Holmes might have felt quite an affinity found quite a bit in conveying human experience under the literary guise of a pragmatic turn (Levin 1999; Poirier 1992; Richardson 2007). Wallace Stevens noted, "the soil is man's intelligence."

Aesthetics, Reasoning and Anchors

Alfred North Whitehead in his book *The Adventures of Ideas* would link beauty as one feature in the adventures of ideas. Dewey in *Art as Experience* would link aesthetics to everyday problem solving. Aesthetics and beauty can aid inquiry and investigation. There is no descent downward by beauty or aesthetics inhering in the everyday—the beauty of an argument, an idea, a piece of understanding. And, of course, aesthetics runs through the law (Schlag 2002; Butler 2003), just as aesthetics does in most human endeavors (Shusterman 1992).

There is nothing not real or less real or less cognitive about it; like most tools it can be helpful or not. The key is when is something helpful. Look at the diverse cognitive tools that enhance our capabilities. These are perhaps frozen modules but interactive helpful tools in adaptation, in making sense of our surroundings, in predicting events.

Indeed, the cognitive revolution writ large has impacted our understanding of reasoning in general and reasoning about law related events (Winter 2001; Johnson 2002). The pragmatist recognition, particularly in Dewey, is the dissolution of diverse forms of dualisms, of which cognition and embodied experience is one; they run together as forms of adaptation in attempts to have coherence in action, in understanding, etc. As Dewey put it, experience is fundamentally tied to doing and "knowledge is…involved in the process by which life is sustained" (1920, p. 87).

Mark Johnson, the philosopher, reminds us "Law is a human creation of human minds dwelling in human bodies, in human societies, operating with human cultural practices" (Johnson 2002). Some activities may be way more dignified or important than others; but interpretation is endemic to inquiry and discovery. A world without foundations of mythic proportions is not a world without solidity and footing.

Most forms of human reasoning are tools. The issue is that there is no one panacea for reasoning judiciously, getting things rights, having the motivation or desire, or fortitude to try to get things right. Let alone being capable of it.

A view about coherence and adaptation undercuts debates about internal and external points of view. There is no denying the existence of external world, not in most instances of adjudication as Holmes well knew. Indeed, the acknowledgment of felt effects of what Peirce called secondness or resistance is omnipresent in movement and action of thought. Context and nuance are conditions of decision making. They are inherent in judicial decision making.

Human reasoning is aided by the body, not abated; no foundationalist rationalist perspective, just hard slogging predictive coherence, adaptation to action. Metaphors run through thought and the depiction of action (Johnson 2007; Boyd 1993). Metaphor is not a bad word, a less objective expression than something that is not. It is a tool that aids or does not aid thought and action. And that is all empirical, and in the mix, with permeable foundations, aims, and capability. The law breathes within this milieu. At least, that is as Holmes understood such events.

Holmes understood that settled beliefs dwell in the conception of ideas, the habits, in an evolving community of inquirers as Peirce hypothesized. Ideas for Holmes like Peirce and the later Wittgenstein (1953) are embedded in a social milieu (see also Mead 1934) or shared meaning and transactions (Schultz and Luckman 1973). Holmes understood this palpably from his war experiences.

They function as social instruments or just instruments. It is the ideas that matter and their free expression for Holmes (Danisch 2007). Free speech, no harm, and competition of ideas are the triad of necessity in a democracy of individuals etching out a life of meaning and possible prosperity or not. Dissent matters (Sunstein 2003) as it calls into question a majority, and the link to history and the larger body politic towards a future. Holmes's rhetoric is future-oriented, which is like other pragmatists.

Perhaps fittingly in giving the Oliver Wendell Holmes Jr. lecture at Harvard, Cass Sunstein (2003) makes clear in citing diverse forms of experimental data is our vulnerability to conformity, experiments on obedience to authority (e.g., Milgram 2009) that undermines the independence of thought for which Holmes is clearly an exemplar. Without much rebuke, what Holmes represents is independence of thought. Dissent is but one expression of this against clan conformity so vital in a democracy.

Holmes perhaps overly identified with his dissenting opinions on the court became something that was also followed by judges, or copied by other judges (Posner 2002; Grey 1995; Sunstein 2005). Hearing differences and considering dissent are at the heart of thought in a democracy. In a crowd of political correctness, here Holmes, though imperfectly so and thus without mythology (e.g., *Abrams vs US*, 1919), is a signal against the tyranny of thought, the rape of free expression even for "the expressions we loathe" and the possibilities "fraught with death." Holmes is a symbol and reality of independence of thought; a contrarian with principal. Holmes here rightfully stakes a legitimate claim, but not consistent and without mythology of some form of Olympus vector or thought; just human all too human (Nietzsche 1886).

But Holmes's prose can be poetic, more Jamesian or Emersonian than his fellow justices. His poetic sensibility never faded from the schoolboy influence of his mentor Emerson. Dante was another poetic luminary Holmes greatly admired (Ethel Scott, February 18, 1910).

The range of literary expression across the justice systems around the globe in history is ancient literary art embedded in the larger reflective culture, some of which was written by individuals who themselves were lawyers (e.g., Kafka). And that, of course, takes us through the circumstances of why law evolved with us, offended us, protected us, and continues to be a vital part in what is anchored to our evolving sensibilities.

Rhetoric, not just argument, is part of the expressive and persuasive phrases of law, a law embodied in Holmes's sense of capturing what is valuable in the law. How a word sounds matters. Moreover, the metaphors are power-oriented. A link to Nietzsche has been suggested (Posner 1992), but, unlike Nietzsche, the metaphors are about temperance in the face of excess, constitutional or otherwise. Here Holmes is perhaps close to Emerson in his poetic pragmatism (Goodman 1990; Poirier 1992).

Tragedy and pain pervaded the war experiences. His world is adversarial. Such is the nature of law. Maybe that is why he tended to undervalue the cooperative side of our survival; the socially bonded part of our necessity, outside the necessities in battle.

It is the "free trade of ideas" and the competition to achieve those ends in an evolving culture that values this participatory democracy. Rhetoric matters as much as logic in making a case (Danisch 2007; Fisch 1986). William James, the writer, lamented that Holmes was "cold blooded," egotistical, and conceited. No doubt he was after the war. Cold bloodedness may be somewhat adaptive. Nonetheless, Holmes (1873) would talk at times of an evolved culture as the "spread of an enriched sympathy." The rhetoric is thick; the spread of what we might owe each other is perhaps no more precise.

But Holmes was ambiguous and contradictory when considering the plights of others. He understood what it took to win, and winning an argument could require rhetorical skills of diverse sorts, and a consideration of the larger self or more social other (Mead 1934), in one case, the juror or the judge.

One of the many paradoxes of Holmes was his "blind acceptance"— the stoic sensibility that pervaded his experience of men in battle, or more generally blind commitments along with an evolving sense of constitutional law and the sustained value of free speech (Menand 2001; Healy 2013). But he also understood that, though frail, our cultural evolution is tied to democracy of which law is an integral part (Posner 2003). Dewey well understood this, and cephalic capabilities were continuous with this realization. Our biology is not one-sided and our cultural achievements on the other; the ugly side, human irrationality and wanton destructiveness, is also always omnipresent, too much so.

Evolution and Cognitive Fluidity

From an evolutionary point of view, cognitive flexibility or fluidity underlies our cognitive and technological capabilities (Mithen 1996; Rozin 1998). Our evolution reflects the expression of specific adaptions, one of which in our case is our cognitive fluidity. One feature in the evolution of neural function is greater access to specialized systems and use across broader and broader domains of problem solving.

And superior fluidity can under some conditions enhance greater cognitive/neural capability in marking the transitions in experience and across events. One view about our evolution is the fluidity between diverse forms of capabilities, one of which is emotional fluidity, fast information processing about the states of others. And another perhaps is what some have characterized as "super fluidity" (Poirier 1992; Mendenhall 2015a, b): enhanced capability in drawing connections in experiences, in transition. Indeed, Emerson would note that "nothing is secure in life but transitions, the energizing spirit."

Of course, super fluidity can also and easily be excessive; the super fluidity of the over-attribution of intentionality to objects, to people inappropriately, without justification, or minimal justification, which is misleading (e.g., Atran and Medin 2008; Frith and Wolpert 2003). Emerson is admired for his poetic vision, and in search of what matters to the human condition, after satisfying basic needs of comfort and human attachment—a core feature of pragmatism (Richardson 2007).

Inherent in these literate expressions are appraisals, and giving value to events in what diverse regions of the brain do all the time. They are mostly not reflective, and subcortical and even when more corticalized it is not necessarily transparent. But good judgment is getting things right, and that of course varies with the subject, context, expectations, resources, etc.

Sagacity is the normative goal for Emerson. Human frailty, the endless extreme, and the horrors of war and wanton destruction lie within the living structure of neural tissue. This is the reality of Holmes. And yet the world screams of possibilities, perhaps Emersonian.

Emerson runs deep towards something about moral perfectionism and attunement to human experience; he writes about things that matter: friendship, a sense of nature, self-reliance, freedom, self-expression (Cavell 1990; Goodman 1990). A romantic feature permeates this strain of pragmatism and variants of American thought. It is found easily in Emerson, James, and Dewey (Goodman 1990). Of course, now pragmatism runs the risk of meaning more than what is manageable; perhaps the many strains of American thought are more suitable, of which Holmes ingested and was part of, and are more accurate.

Holmes, though not consistent, is certainly not rooted in variants of romanticism (Berlin 1999). Neither was Kant with his emphasis on human freedom and self-determination, although Rousseau was a major influence on Kant. Holmes was more sanguine or more pessimistic about the human condition than Kant. And in part this has to do with his notion of experience, philosophical education and acumen. Of course, there are many variants of pragmatism and American sensibility. But Holmes is part of it all (Grey 1989).

Holmes ends up later in his life close to Nietzsche, whose rallying cry is beyond morality; morality that binds people, survival and perseverance, rejection of religion, and the expression of aesthetics as primordial. Holmes, the distant jurist, is at times grounded in an epic ecstasy entered event, a civil war that he never forgot, and the evidence of irrationality and sheer power.

The other feature of Holmes is of the rational agent in addition to the survival side and the amoral sense that appears in some of his writing. He is close to some of the ideas in which belonging to the larger public takes precedence.

The blindness and indifference from a common plight aside from free speech and opportunity is what bothers Holmes's many detractors. He betrays an indifference to common frailty, or those not surviving or thriving. The less fortunate amongst others is not close to perhaps a hardened heart to the frail, along with a stoic realization that necessity and competition prevail and vitalism and action are what one leans on for sustenance. War is rampant with fear, and it is harsh and deadly.

Perhaps this is unfair. After all, Holmes was not consistent. But the Dickensian world of harsh plights is not tempered in his prose about others. Perhaps he concealed that part of himself, or he lost that part in an abstract sense of a mystical world of "a world soul" that Emerson would write about, and certainly the strife and will to persevere against the odds.

The thing about Holmes, in opposition to Emerson, is his level of detachment, perhaps a consequence facilitated by a war where wanton destruction was rampant. Holmes in his first piece on the Law titled "Primitive Notions in Modern Law" would appeal to a predilection towards animism (Santayana 1967; Atran 2002).

Holmes, in a letter to fellow justice Felix Frankfurter in 1924, would acknowledge that "his [Santayana's] way of thinking and mine have much in common" (Holmes-Frankfurter Correspondence). What they had in common was a rooted sense of being a part of nature. They were rooted in some kind of animal faith, a sense of objects, of others. For them, the universe has no subjective idealism, which was fashionable in their time and at other times as rapid constructivism, subjective or collective. Holmes's views about us are rooted in biology, but a limited view about biology, and of course the biology of his times. And even then, he did not know much about science.

Holmes reached out to Emerson in a letter where he describes law as a window into our cohesion from such primitive tendencies as retributive justice to codification in common vernacular and understanding (see Novak 2007). Common threads about democracy run through the art and verse of Emerson, Church, Whitman. It is in the context of stoic voyages of self-understanding (Novick 1992).

Holmes would write to Emerson when this paper appeared in Press:

> It seems to me that I have learned after a laborious and somewhat painful period of probation that the law opens a way to philosophy as well as anything else, if pursued far enough, and I hope to prove before I die. Accept this little piece written in that faith, and as a light mark of gratitude and respect I feel for you who more than anyone else started the philosophical ferment in my mind. (see also Baker 1991)

Mostly, people want to survive and to insure their families' survival even under the best of circumstances. But law without value is harsh, and it is not Holmes's approach, at least not entirely (cf. Alschuler 2000; Grey 2014).

The Nazis may be our touchstone for this, but in fact there are many villains, a litany throughout history is too easy to construct, including in our own times. It is far too easy, in any culture, to lean on "common sense" without question. What philosophers sometime refer to as givens are just well worked practices.

Critiques of Holmes often focus on his apparent moral divorcement, or abstraction from the struggles of individuals (Alschuler 2000). Holmes was steeped in the history of law and also had a profound effect on American law (Horwitz 1992). Many scholars have noted that

Holmes liked the law but had no grand illusions about it and traces of nihilism and self-reflective suicidal ideation appeared with his writings. He did know something about blood and guts, pain and despair, abject fear. His war experience taught him that. Nihilism or fatalism permeates his sense of being in the world, within a context of duty.

Like Emerson, Holmes adumbrates some spiritual whole for individuals, again something of a "world soul," but it is elusive. The war clouds all it seems. At least that is one plausible hypothesis. Holmes had more faith in the deepening of human experience, rather than in the foundations of thought or science. Holmes would remember Emerson and would recall "the early firebrand of my youth that burns to me as brightly as ever is Emerson" (see also Levinson 2000). Of course, nature is alive, brimming with important civilizing features for the human experience that underlie a mystic sense of democracy as expressed by Walt Whitman and decency in the poetic voice of Emerson and at the cradle of Holmes.

Something Holmes first experienced is what Aristotle described: "the noblest death is the soldiers' for he meets it in the midst of the greatest and most glorious dangers" (*Ethics*, p. 94). Holmes understood the frail decay of mortal wounds in battle. He watched others die, and he almost died himself. Holmes may be a distant New Englander, but he was an ardent correspondent with what became long friendships.

This reaching out was a fundamental adaptation—reaching out and letting others reach out to him. Emerson would assert "I do not wish to treat friendships daintily, but with the roughest courage. When they are real, they are not glass threads of frostworks but the soldiers' thing we know" ("Friendship," p. 192).

History underlies all human life, local, global and then universal. We are historical, but there is no historical inevitability as there is, say, to eye color; though even eye color is ripe with probability and so are historical events. Necessity is a thing of logic, something Hegel and others misunderstood as they thought about history.

Holmes, no less than Emerson, celebrated history as the backdrop of understanding in just about everything, and certainly in law and the sciences. The interpretative is not just for the humanities or history, it is the bedrock of all epistemic quests. It was once couched as the historical versus scientific, and the "new sciences" as Vico suggested is something different certainly than the range of Enlightenment science for mechanistic explanations.

Holmes makes it clear, in *The Path of the Law*, that "the rational study of law is still to a large extent a study of history." Knowing ourselves is knowing about this history, a history of the human condition. History is grounded, not in the heavens, but on earth. It is the separation between what we imagine is possible and build from our groundedness with nature. This separation would become part of the larger separation of invention from nature as opposed to the utter continuity of the two.

Pragmatism is often criticized as being ahistorical, but there is plenty of historical context in pragmatism (Koopman 2011; Stuhr 1997). Both Emerson and Holmes are historical enough. Of course, so is Peirce and Dewey.

Vico goes on to describe the Greek philosophy of law as depending on the state; what he called "pragmatics i.e. sheer practitioners" (1709, p. 49). This sounds like legal formalism in more modern times. Greek philosophy, certainly with Socrates, Plato, Aristotle, and Seneca, is about knowledge as wisdom. Law aided these endeavors, both divine and secular. Here Vico emphasizes the history of law in cultures: "jurisprudence was transformed into the science of the just in an art of equity" (1709, p. 56).

The movement is historical. What Vico called the "law of nations" became more of a matter of capturing civilization from the past, and then with a more comparative eye to common forms and diverse forms of human expression and the richness of experiences within the larger human family. Understanding the law was putting this form of knowledge in the class of wisdom.

The larger point about the point of Vico is not reducing all knowledge to mathematics or later logic. What Holmes meant by stressing experience and not logic is the larger sense at least of what constitutes the human experience. The only issue is that logic runs through reasoning, it is not outside it in some pure form. Holmes worried about the narrowness of logic, which is why he emphasized experience like Dewey and many others. The same held for Vico in his critique of Descartes.

Thinking Back and Ahead

We are a species that thinks ahead. The consideration of future generations, a virtue of what we owe each other, is expanded, serving as both
a real discernible measurable feature and a metaphor for future binding
and more open options in considering future engagements.

Holmes thought the framers of the constitution lived in a time of
moral courage. Holmes in some ways envied the revolutionary decision
makers because they worked in a time when their choices mattered.
These were radical ideas for their time (Wood 1993). They were based
in part on the Enlightenment perspectives of liberation, greater participation, and possibilities within the polity along with the reality of contrary sentiments and rules against the very ideas they espoused.

Holmes inherited "radical enlightenment" (Israel 2001). While
he himself was not a radical, he was associated with the progressive
end—though his progressivism was tinted by a dim view about the
human condition and its possibilities (not perhaps as dim as Hobbes
or Machiavelli, but certainly no romantic like Rousseau). In particular,
Holmes shared an Enlightenment perspective with regard to science and
non-foundational feature of legal sensibility, that is historically laden
and ultimately tied to the body politic.

Emerson and Thoreau were part of the larger ambiance of intellectual
air that Holmes was surrounded by when he was developing his own
ideas. Holmes incorporated and inherited Emerson's and Thoreau's rev-
erence and respect for nature as a habitat we share with others, a crucial
part of the landed sensibility of the American persona (Novak 2007).

Emerson turned experience, self-reliance and duty into a mystical
fervor; this sentiment certainly influenced Holmes. Holmes was able to
ground this sense of Emersonian thought, and sense of nature, vitality,
and mystic wonder, into a metaphysics of core experiences. Holmes did
not forget Emerson's high ideals for human existence.

While Thoreau himself was a real radical, Holmes was not. Thoreau
displayed utter disgust toward slavery. Thoreau, like Martin Luther King
Jr. over a hundred years later, proclaimed the value of civil disobedience,
although perhaps not to the level of John Brown's insurrection of violence at Harper's Ferry.

There is a strand of pragmatism that finds roots in a religious sensibility towards nature (Pratt 2002; Dewey 1925). But Holmes eschews any such sentimentalism. He was never an abstract rationalist except by disposition and stoic duty.

Holmes understood the law as not isolated or "self-sufficient" (see also Pound 1921), but entirely embedded in the larger culture, the body politic, the social frameworks and epistemic orientations. Roscoe Pound, a Ph.D. in the biological sciences who would go on to be the Dean of the Harvard Law School, would go on to emphasize the sociological factors and sciences so relevant to the law.

Holmes tended to look down on a "qualitative," as opposed to a quantitative way of thinking. *Qualitative* was a pejorative term for him; though, of course, qualitative studies run continuously with quantitative studies. Dewey, quoting Holmes, noted "the growth of education is an increase in the knowledge of measure. It is a substitution of quantitative for qualitative judgment..." (Holmes in Dewey 1928). What Holmes aims for is the "essence of improvement that we should be as accurate as we can" (Holmes in Dewey 1928, p. 181). For Holmes, this meant a dependency on quantitative ways of thinking. As a result, science is a natural tool; science leans on measurement and prediction. The brain comes prepared to do just that (Clark 2013).

It is not clear that Holmes recognized the social ends that mattered, aside from a sense of duty and free competition of ideas. An informed set of qualitative descriptions of jury members we know to be meaningful, to be predictive, could be part of our arsenal of knowledge. In the rush to look scientific, we sometimes talk about the quantities separate from their qualitative components, e.g., fact versus value. But, for pragmatists, and I am one of them, science is the act of valuing diverse forms of data gathering and monitoring, replicating this data, and extending the data further. Empiricism is hard; Holmes was in a chair removed from data gathering, and the quantitative side that he so valued was still an abstraction. The qualitative runs through the quantitative in a permeable livable fashion. These are the sorts of dualisms that pragmatism looked to dilute (Dewey 1920/1948; White 2002).

Neuroscientific Considerations: And the dualism that are diluted matter; cognition and action, perception and cognition, sensation and cognition, mind and nature, nature and culture figure in how one understands the organization of human action (Dewey 1925; Lakoff and Johnson 1999) and certainly in understanding the neural basis of agency, the sense of responsibility (Dewey 1925; Neville 1974; Schulkin 2000). Indeed, the naturalism that runs through pragmatism is a non-reductive understanding of ourselves (e.g., Wahman 2015), within a broader understood connection to neuronal expression and neural connectivity (Schulkin 2015).

Our nature is the building of culture; the sense of experience is continuous. And that is the point about pragmatism, the pragmatism that runs through Holmes or Dewey; themes of inquiry cohere in our epistemic drives at satisfying our inquiry, or not. We are grounded in place and time in nature and the culture one inhabits, but the thread is not bifurcated into epistemic distortion of dualisms of many sorts (Dewey 1920/1948).

And this pervading sense of naturalism is one feature of our cultural continuity with science, an out growth of core epistemic capabilities selected as good enough problem-solving orientations (Gigerenzer 2000). The neural systems are the conduit for the realization of the organization of action and our epistemic drives (Schulkin 2015).

The brain is surely a piece of matter; but the brain is also an idea; what the amygdala is or where it is a reflection of the culture one participates in, how one carves it up; the amygdala and the frontal cortex to name just two regions of the brain is simply not just there (Swanson 2000a, b). The epistemic application is inherently conceptual; concepts run through epistemic orientations and action. They are not divorced; they exist in the objects we build and the objects we build extend us further. As they provide the expansion of our epistemic. Culture and nature are continuous with us as a species.

And so epistemic affairs are grounded, in practice, in satisfaction. Key epistemic orientations are whether something is understood as alive or not (Whitehead 1938; Atran and Medin 2008). The Emersonian experiences of place, or that of Thoreau of the familiar is continuously expanded into the unfamiliar as we migrate to new places

(Rozin 1976, 1998). Nature is never other as we migrate as we both think back and think ahead; the brain is prepared for these migratory flights of movement, as we capture the past in the present in the expanding present so nicely captured by William James in his magnificent treatise Principles of Psychology.

Metaphors run through problem solving legal or otherwise (Lakoff and Johnson 1999) like cerebrospinal fluid bathing the brain; it is all pervasive. The issue is the epistemic urge and its satisfaction. Fluidity and neural accessability is one feature of our neural design (Rozin 1976, 1998). And within that design are the diverse forms of linguistic and non-linguistic expression that facilitate social contact.

Conclusion: A Sense for What Matters

Emersonian sensibilities pervade Holmes and his pragmatism. It is the softer side of Holmes and perhaps the wiser. Poetic glimpses of human compansion and wisdom grounded in place, nature and human solidarity.

Holmes understood that human meaning is human made. Human meaning is knotted to meaningful social contact into what matters to our species and to our future generations. But there may be no universal law or justice, except the frail sensibility of human empathy and understanding. A hard part about the law is capturing what is right or wrong. Surely that it is at the heart of the search for wisdom; what is essential is what Plato proselytized as philosophical wonder. As a result, we do not want to be glib, as Holmes can seem, with regard to considerations of justice and connections to others and historical depth and sensibility.

We know that the brain is constantly adjudicating, generating appraisals of events. The brain's structure requires the diverse activation of appraisal, valuation of events that matter for survival at its basic level, for cultural ascent, and for what matters with regard to what we owe each other and the expanding participation of others in the larger body politic.

Supreme Court judges like Holmes are placed in a lofty position for adjudicating in the highest matters and are often the last recourse. Yet the judges represent views. There are no blank slates of innocence for any of us. The balance of judgment requires knowing what is right within the law, a sense that the law is historical; but something can be historical and not simply be contextual. We hit on things beyond the context, even though without context, there is no possibility of hitting on what might be better for the human condition. Wisdom is valuing this capability and the many variants of human search and quest (Midgley 1989): the meaning found here on earth, what we owe each other, what we can value, and what we will not tolerate and should not tolerate in one another.

What matters is that lives matter. This is a common religion of human bounds, a common ground of human meaning (Jaspers 1913/1997) that is real and frail and sacred. It is a valuation embodied in facts, in judgment, in the larger body politic, and in the great breath of meaning. Perhaps a variant of "holistic pragmatism," something associated with Morton White (2002), would contribute to an understanding of Holmes (Kellogg 2004).

Like other pragmatists, for Holmes the values of things are inherent in the appraisal systems that have evolved in culture and that are embedded in diverse regions of the brain. Pragmatism is rooted in the considerations of the social milieu (Joas 1993). And Holmes's view on nature is as he understood Dewey's treatise *Experience and Nature*, as ill written as he has suggested it is in at least one of his sets of correspondence.

Holmes over his long-life savored experience. The deepening of human knowledge and judgment was the cornerstone of his pragmatism. This deepening required historical sensibility and the logic that runs through practice and problem solving. Holmes is rooted in sentiment in science, though not in practice. Common law is at the heart of our common experience deepened by test and further practice; but the emphasis is on a collective sense, a community sense of what matters (Parker 2003). Like Peirce or Dewey or later Mead, the collective sense of what matters is captured in the body politic, in our laws for better or for worse. Holmes, though unlike Dewey, eschewed the progressive activism of many; perhaps his sense of the individual was not far from that of James. It is just that James was courageous in looking inward. Holmes was not, but perhaps the pain and the scars of war, his temperament, due to the war still lingered within him—a "bleakness" with regard to the human condition and human prospects (Wilson 1994).

The war experience is the critical underlying theme in Holmes's life and his understanding of "the felt necessities of the time." He would elevate "enlightened skepticism" (Path of the Law, 1897) as critical to judicial sensibility and wisdom within a context of a "spacious present" (James 1890) of looking forward and being anticipatory and looking backward and understanding historical context (Dewey 1920/1948) and shared experiences (Tushnet 1977).

Holmes certainly experienced mindfulness, but the mindfulness is elliptical and not always straightforward. Holmes's (1913) prose was always elegant, the class poet emerges here and there: "But then I remembered the faith that I partly have expressed, faith in a universe not measured by our fears, a universe that has thought and more than thought inside of it, and as I gazed after the sunset and above the electric lights there shone the stars" (Law and the Court).

Holmes is trying to keep track of what matters here (Posner 1988/2009). Of course, what matters to Holmes is the sense of duty; but the duty of the judge is good judgment and independence much to the consternation of many. Holmes was on all sides at one time or other during his time on the Supreme Court (Novick 1992; Alschuler 2000). Holmes's philosophical stance is within the broadening of human experience. He understood it that way, despite his flaws, his limitations, and his own inability to look inward.

His philosophical journey is riddled with a deep sense of matter and wonder. Matter in this case is that the world is that of human meaning, nature, and matter. Coupled with that is a poetic sense of wonder, muted by war, stoic sensibility, and judicial and intellectual engagement. Holmes, like Hobbes, though tempered by Emerson, had an abiding pervasive sense of economics.

The endless bombardment of alienation from what matters was a sign of the times. What, Dewey (1925), or Jaspers (1913/1997) for instance, emphasized was human contact, human solidarity, human responsibility, the acknowledgement of human fragility, our need for some form of acceptance, and our goal to stand alone and to make judgments. Judgment is not something that distorts our sense of the object even when the object is delivered in various degrees. Clarity of mind, enhanced focus to what counts, eliminating the unnecessary. Enhanced attention on the detail that matters is the normative goal of the practice and one end along with something called wisdom; it is about knowing what matters. Of course, knowing is everything in judging, in assessing, in understanding. Techniques and orientation that enhance knowing are tied into understanding, and neural function and the larger environment.

Holmes, who one colleague would call "the mystic philosopher," (Farnum 1937) said, "faith in a universe not measured by our fears, a universe that has thought and more than thought inside of it, and as I gazed after the sunset and above the electric lights there shone the stars" (Harvard Law, NY address). Like Kant and others, the sense of nature pervades with wonder and power and awe. But unlike Leibnitz or Kant and their sense of what constitutes human natural inevitability and the role of law, reason and science, for Holmes, the science of the mind is located indeed in what can be known, and what might be known, and what might never be known with no inevitable plan of reason, and the palpable fact of the limitations of human reason (Luban 1992).

Holmes understood the "incommunicable" features of our shared experiences, or our being thrown onto a battlefield of irrationality and brute force. As he expressed for a long duration of a very long life, "life is a roar of bargain and battle," but he goes on to breathe Emersonian with "but in the very heart of it there rises a mystic spiritual tone that gives meaning to the whole" (The class of 61, 1913; see also Luban 1994).

The Holmesian emphasis is on the further fluidity of science in the context of the law (Haider 2006), in ethics, in human decision making. But, like everything, this connection should be demythologized and not oversold. Decision making should be within the bounds of reason. Indeed, the whole tenor of the pragmatists is the dissolution of diverse forms of pernicious dualism where ethics breathes in the law, cognition runs through practice, and values are inherent in facts.

Freedom of expression, variants of fatalism, self-expression, self-restraint, and self-reliance are the moral elements of life that interest him. Holmes was not much for metaphysics, but there is a metaphysics that borders on libertarian sensibilities that underlies his philosophy of judicial restraint. He is more tied to what was considered normative in the nineteenth century than what one might associate with a liberal in the twentieth century (Luban 1992, 1994; Burton 1979).

References

Alschuler, A. W. (2000). *Law Without Values: The Life, Work and Legacy of Justice Holmes*. Chicago: University of Chicago Press.

Atiyah, P. S. (1983). The Legacy of Holmes Through English Eyes. *Boston University Law Review, 63,* 341–380.

Atran, S. (2002). *In Gods We Trust: The Evolutionary Landscape of Religion*. Oxford: Oxford University Press.

Atran, S., & Medin, D. (2008). *The Native Mind and the Cultural Construction of Nature*. Cambridge: MIT Press.

Baker, L. (1991). *The Justice from Beacon Hill: The Life and Times of Oliver Wendell Holmes Jr*. New York: Harper & Row.

Berlin, I. (1999). *The Roots of Romanticism*. Princeton: Princeton University Press.

Blasi, V. (1999). Reading Holmes Through the Lens of Schauer: The Abrams Dissent. *Notre Dame Law Review, 72,* 1343–1360.

Bloom, H. (1983). *Agon*. Oxford: Oxford University Press.

Bloom, H. (2002). *Genius*. New York: Warner Books.

Bowen, C. D. (1945). *Yankee from Olympus*. Boston: Little Brown.

Boyd, R. (1993). Metaphor and Theory Change. In A. Ortony (Ed.), *Metaphor and Thought*. Cambridge: Cambridge University Press.

Budiansky, S. (2019). *Oliver Wendell Holmes: A Life in War, Law, and Ideas*. New York: Norton.

Burton, D. H. (1979). *Oliver Wendell Holmes Jr.: What Manner Liberal*. New York: Krieger.

Butler, B. E. (2002). Legal Pragmatism: Banal or Beneficial as a Jurisprudential Position. *Essays in Philosophy, 3,* Article 14.

Butler, B. E. (2003). Aesthetics and American Law. *Legal Studies Forum, 1,* 203–220.

Cavell, S. (1990). *Conditions Handsome and Unhandsome.* Chicago: University of Chicago Press.

Clark, A. (2013). Whatever Next? Predictive Brains, Situated Agents and the Future of Cognitive Science. *Behavioral and Brain Sciences, 36*(3), 181–204.

Cohen, F. S. (1935). Transcendental Nonsense and the Functional Approach. *Columbia Law Review, 35,* 809–849.

Danisch, R. (2007). *Pragmatism, Democracy and the Necessity of Rhetoric.* Columbia: University of South Carolina Press.

Danisch, R. (2008). Enthymemes, and Oliver Wendell Holmes Jr. and the First Amendment. *Rhetoric Review, 27,* 219–235.

Dewey, J. (1920/1948). *Reconstruction in Philosophy.* Boston: Beacon Press.

Dewey, J. (1925/1989). *Experience and Nature.* LaSalle, IL: Open Court Press.

Dewey, J. (1928). Justice Holmes and the Liberal Mind. *New Republic, 53,* 210. In J. A. Boydston (Ed.), *The Later Works of John Dewey, Volume 3, 1927–1938* (pp. 177–183). Carbondale: Southern Illinois University Press.

Farnum, G. F. (1937, May). Holmes—The Solitary Scholar. *Journal of the Law Society of Mass, 1,* 2–9.

Fisch, M. H. (1942/1986). Justice Holmes, the Prediction Theory of Law and Pragmatism. In M. H. Fisch (Ed.) (1986), *Peirce, Semiotic and Pragmatism.* Bloomington: Indiana University Press.

Frankfurter, F. C. (1939). *Mr Justice Holmes.* Cambridge: Harvard University Press.

Frith, C., & Wolpert, D. (2003). *The Neuroscience of Social Interaction.* Oxford: Oxford University Press.

Gigerenzer, G. (2000). *Adaptive Thinking, Rationality in the Real World.* New York: Oxford University Press.

Goodman, R. B. (1990). *American Philosophy and the Romantic Tradition.* Cambridge: Cambridge University Press.

Grey, T. C. (1989). Holmes and Legal Pragmatism. *Stanford Law Review, 41,* 787–870.

Grey, T. C. (1990). Hear the Other Side: Wallace Stevens and Pragmatists Legal Theory. *Southern California Law Review, 63,* 1569–1595.

Grey, T. C. (1991). *The Wallace Stevens Case: Law and the Practice of Poetry.* Cambridge: Harvard University Press.

Grey, T. C. (1992). Holmes, Pragmatism and Democracy. *Oregon Law, 71,* 521–542.

Grey, T. C. (1995). Molecular Motions: The Holmesian Judge in Theory and Practice. *William and Mary Law Review, 37,* 19–45.

Grey, T. C. (2014). *Formalism and Pragmatism in American Law.* Boston: Brill Press.

Haider, A. (2006). Roper vs Simmons: The Role of the Science Brief. *Ohio State Journal of Criminal Law, 3,* 369–377.

Hamilton, W. (1853). *Discussions in Philosophy and Literature.* Edinburgh: MacLachlan and Stewart.

Healy, T. (2013). *The Great Dissent: How Oliver Wendell Holmes Jr. Changes His Mind and Changes History of Free Speech in America.* New York: Henry Holt.

Holmes, O. W., Jr. (1873). The Gas-Stokers Strike. *American Law Review, 7,* 582–583.

Holmes, O. W., Jr. (1913). The Class of 61. Harvard University.

Holmes, O. W., Jr. (1919). *Abrams vs United States.*

Horwitz, M. J. (1992). The Place of Holmes in American Legal Thought. In R. Gordon (Ed.), *Legacy of Oliver Wendell Holmes Jr.* Palo Alto: Stanford University Press.

Israel, J. I. (2001). *Radical Enlightenment.* Oxford: Oxford University Press.

James, W. (1890/1952). *The Principles of Psychology.* New York: Dover Press.

Jaspers, K. (1913/1997). *General Psychopathology* (Vols. I & II, J. Hoenig & M. W. Hamilton, Trans.) Baltimore: The Johns Hopkins University Press.

Joas, H. (1993). *Pragmatism and Social Theory.* Chicago: University of Chicago Press.

Johnson, M. (2002). Law Incarnate. *Brooklyn Law Review, 67,* 949–962.

Johnson, M. (2007). Mind, Metaphor, Law. *Mercer Law Review, 58,* 845–868.

Kellogg, F. R. (2004). Holistic Pragmatism and Law: Morton White on Justice Oliver Wendell Holmes. *Transactions of the Charles S. Peirce Society, 40,* 559–567.

Koopman, C. (2011). Genealogical Pragmatism. *Journal of the Philosophy of History, 5,* 533–561.

Lakoff, G., & Johnson, M. (1999). *Philosophy in the Flesh.* New York: Basic Books.

Leonard, G. (2006). Holmes and the Lochner Court. *Boston University School of Law.* No. 06–38.

Levin, J. (1999). *The Poetics of Transition: Emerson, Pragmatism and American Literary Modernism.* Durham: Duke University Press.

Levinson, S. (2000). Emerson and Holmes: Serene Skeptics. In S. J. Burton (Ed.), *The Path of the Law and Its Influence: The Legacy of Oliver Wendell Holmes Jr.* Cambridge: Cambridge University Press.

Luban, D. (1992). Justice Holmes and Judicial Virtue. *Nomos, 34,* 235–264.

Luban, D. (1994). Justice Holmes and the Metaphysics of Judicial Restraint. *Duke Law Journal, 44,* 449–523.

Mead, G. H. (1934). *Mind, Self, and Society.* Chicago: University of Chicago Press.

Menand, L. (2001). *The Metaphysical Club.* New York: Farrar, Straus and Giroux.

Mendenhall, A. P. (2015a). Pragmatism on the Shoulders of Emerson: Oliver Wendell Holmes Jr.'s Jurisprudence as a Synthesis of Emerson, James and Dewey. *The South Carolina Review, 48,* 93–109.

Mendenhall, A. P. (2015b). *Oliver Wendell Holmes Jr., Pragmatism and the Jurisprudence of Agon.* Lewisburg: Bucknell University Press.

Midgley, M. (1989). *Wisdom, Information and Wonder.* New York: Routledge.

Milgram, S. (2009). *Obedience to Authority.* New York: Harper Perennial Classics.

Mithen, S. (1996). *The Prehistory of the Mind.* London: Thames and Hudson.

Moreno, J. D. (1995). *Deciding Together.* Oxford: Oxford University Press.

Neville, R. C. (1974). *The Cosmology of Freedom.* New Haven: Yale University Press.

Nietzsche, F. (1886/1972). *Beyond Good and Evil.* Mineola: Dover.

Novak, B. (2007). *Voyages of the Self.* Oxford: Oxford University Press.

Novick, S. M. (1992). Justice Holmes and the Art of Biography. *William and Mary Law Review, 33,* 1219–1248.

Parker, K. (2003). The History of Experience: On the Historical Imagination of Oliver Wendell Holmes Jr. *Political and Legal Anthropology Review, 26,* 60–83.

Poirier, R. (1992). *Poetry and Pragmatism.* Cambridge: Harvard University Press.

Posner, R. A. (1988/2009). *Law and Literature.* Cambridge: Harvard University Press.

Posner, R. A. (2002). Is Pragmatic Adjudication Inescapable? In *How Judges Think.* Cambridge: Harvard University Press.

Posner, R. A. (2003). *Law, Pragmatism and Democracy.* Cambridge: Harvard University Press.

Posner, R. A. (2008). *How Judges Think.* Cambridge: Harvard University Press.

Posner, R. A. (1992). *The Essential Holmes.* Chicago: University of Chicago Press.

Pound, R. (1921). A Theory of Social Interests. *American Sociological Society, 15,* 16–45.

Pratt, S. L. (2002). *Native Pragmatism*. Bloomington: Indiana University Press.
Reid, T. (1764/1997). *An Inquiry into the Human Mind*. Edinburgh: Edinburgh University Press.
Richardson, J. (2007). *A Natural History of Pragmatism*. Cambridge: Cambridge University Press.
Richardson, R. D. (1995). *Emerson: The Mind on Fire*. Berkeley: University of California Press.
Rogat, Y. (1964). The Judge as Spectator. *University of Chicago Law Review, 31,* 213–256.
Rozin, P. (1976). The Evolution of Intelligence and Access to the Cognitive Unconscious. In J. Sprague & A. N. Epstein (Eds.), *Progress in Psychobiology and Physiological Psychology*. New York: Academic Press.
Rozin, P. (1998). Evolution and Development of Brains and Cultures: Some Basic Principles and Interactions. In M. S. Gazzaniga & J. S. Altman (Eds.), *Brain and Mind: Evolutionary Perspectives*. Strasbourg: Human Frontiers Science Program.
Santayana, J. (1967). *Animal Faith and Spiritual Life*. New York: Appleton-Century-Crofts.
Schlag, P. (2002). The Aesthetics of Common Law. *Harvard Law Review, 115,* 1005–1118.
Schulkin, J. (2000). *Roots of Social Sensibility*. Cambridge: MIT Press.
Schulkin, J. (2012). *Naturalism and Pragmatism*. London: Palgrave Macmillan.
Schulkin, J. (2015). *Pragmatism and the Search for Coherence in Neuroscience*. London: Palgrave Macmillan.
Schultz, A., & Luckman, T. (1973). *The Structures of the Life World*. Evanston: Northwestern University Press.
Shusterman, R. (1992). *Pragmatist Aesthetics*. Oxford: Blackwell.
Stuhr, J. (1997). *Geneological Pragmatism*. Albany, NY: SUNY Press.
Sunstein, C. R. (2003). *Why Societies Need Dissent*. Cambridge: Harvard University Press.
Sunstein, C. R. (2005). *Laws of Fear: Beyond the Precautionary Principle*. Cambridge: Cambridge University Press.
Swanson, L. W. (2000a). What Is the brain? *Trends Neuroscience, 23,* 519–527.
Swanson, L. W. (2000b) Cerebral Hemisphere Regulation of Motivated Behavior. *Brain Research, 886,* 113–164.
Tushnet, M. (1977). The Logic of Experience: Oliver Wendell Holmes on the Supreme Court. *Virginia Law Review, 63,* 975–1052.
Wahman, J. (2015). *Narrative Pragmatism: An Alternative Framework for Philosophy of Mind*. London: Rowman and Littlefield.

Warner, R. (1993). Why Pragmatism? The Puzzling Place of Pragmatism in Critical Theory. *University of Illinois Law Review, 3,* 535–550.

White, M. (2002). *A Philosophy of Culture: The Scope of Holistic Pragmatism.* Princeton: Princeton University Press.

Whitehead, A. N. (1938). *Modes of Thought.* New York: Free Press.

Wilshire, B. (2000). *The Primal Roots of American Philosophy.* University Park: Penn State University Press.

Wilson, E. (1994). *Patriotic Gore.* New York: W. W. Norton.

Winter, S. L. (2001). *A Clearing in the Forest: Law, Life and Mind.* Chicago: University of Chicago Press.

Wittgenstein, L. (1953/1958). *Philosophical Investigations.* New York: Macmillan.

Wood, G. S. (1993). *The Radicalism of the American Revolution.* New York: Vintage.

7

Bounded Choice, Human Freedom and Problem-Solving

Introduction

Choice does not exist in a vacuum. Sometimes the choice feels inevitable; like hardly choosing; sometimes it is in conflict with other decisions. We approach choice within a context of available options, capability and context, age, temperamental and situational issues.

Pragmatism situates the individual in a context. Holmes understood this, which is why, like Peirce or Dewey, he couched choice in the larger social milieu. What has come to be known as bounded choice or decision making emphasizes decision making in social context, with available options and within a cultural milieu. Choice is always contextual.

In this chapter, I begin with a discussion of duty and the law, bounded decision making, choice in social context or contact. One adaptive feature in decision making is limiting options in choice. And one feature is the consideration of freedom within the bounds of reason. This issue matters in the context of law; namely human choose within constraints, options, capability, etc.

© The Author(s) 2019
J. Schulkin, *Oliver Wendell Holmes Jr., Pragmatism and Neuroscience*,
https://doi.org/10.1007/978-3-030-23100-2_7

Limiting Choice

The law allows us to pre-commit to avoid or reduce the likelihood of violence, amongst many other things. Of course, pre-commitments bound us to our choices (Elster 1989). We limit choice by our commitments. We limit choice by the laws that bind us. And we limit it by the kind of neurobiological states that underlie us. Rather than unbounded choice, we live with possibilities within constraints in which to understand choice (Schulkin 1992). We expand by the laws that bind us—that bind us against violence, thievery and the whole range of negative susceptibility for which constitutions were heralded to protect us. To persevere was a common theme in a variety of thinkers (e.g., Hobbes, Spinoza). Self-preservation is indeed a dominant motivation. We know that indeed that our biology is oriented toward longer term viability, we are just imperfect and myopic often in staying on track towardsmany ends that matter to us. Social binding is one we do it; limiting choice is also an adaptive feature of neural design (Friston 2010; Schulkin 2015).

Social binding (Elster 1989) is necessary, like sign posts in the direction of human traffic, human wants and ambitions; we are the accountable species. The law in that respect is a sign post; morals are an internal and external sign post of what is expected, or could be expected, or should be.

We limit behavior not just by law or from ethical impositions or goals but by self-binding; one orientation is to find ways to limit our vulnerabilities; it is an important cognitive adaptation, and one way is through diverse forms of recommitments. They include: throwing away the key, imposing costs, creating rewards, creating delays, changing or bolstering beliefs, changing preferences, avoiding exposure to cues, avoiding company, and seeking company.

Limiting choice and binding behaviors is a social adaption. Of course, it is also stoic good sense. To bind to self-regulate is also to withdraw from what is harmful. This act reduces dissonance and removes obstacles from goals. It is an important strategy (Elster 1983). There is nothing sour about it. It is something Holmes saw as a part of duty, removing vulnerabilities to gluttony (a very good feature of humanity in which biology and culture are utterly continuous). Rationality entails knowing what to avoid in addition to knowing how to achieve ends that satisfy our wants: wants that are not gluttonous and manifestly destructive, the endless work in progress. This is where understanding the brain is understanding a good about us.

One view of pre-commitments is that they bind behavior (Elster 2000); they set social precedent and social binding (Currie 1985, 1990). In the constitution, pre-commitments set the conditions for social discourse, social decency and accepted competition in the market place of ideas in action. When it is legal, it is a matter of necessity. When it is merely personal, it is a matter of choice. Both speak to the diverse ways in which social contact is constrained by precedent. The law is a good social context for understanding this fact about us.

As Elster suggested, behavioral adaptations may include the following:

---

1. Reducing options
2. Imposing costs
3. Establishing rewards
4. Creating timelines
5. Tagging preferences
6. Investing in bargaining
7. Reducing knowledge
8. Enhancing certain passions

Human decision making, for pragmatists, is demythologized and placed in the context of precedents of well worked habits, of sanity and common law like interactions of life and the practical side of meaning and coherence in everyday life. Models of decision making are what Herbert Simon called "satisficing." Context matters for good enough explanations. Yet good enough is not the fall from rationality identified with formalism, but practical action and survival and coherence.

Perhaps this orientation takes into account the lived experiences of the individuals trying to adjudicate in compromised real-world contexts. These contexts were and are laden with history and precedent and full of constraints that provide self-binding capabilities.

There is an interesting convergence and intersection of thought between the study of the brain and the cultural evolution of the law. This convergence combines a sense of our biological predilections and our cultural achievements. The brain is a biological achievement, the study of an epistemic wonder. Both are now merging in the law as the advances of our knowledge find home in legal contexts, in contexts where the understanding and measurement devices underlie our understanding of human action.

Choice can be an existential context of despair or delight. Freedom of opportunity, creativity within a pluralism, is an evolving ideal (Neville 1974) structured by battle. Human progress is frail within an evolving set of dangers with our combative tools of war, from protection to destruction which are features of utter continuity from our nature to culture.

Holmes took an original approach toward what he thought were some primitive predilections towards a common law that grows and that represents the body politic and "actual feelings and demands of the community" (Diamond 1935; Baker 1991), in which there are recurring themes that reflect ritual, group formation, social structure, retribution, marriage, and civil inquiry. The metaphor of growth is fundamental for pragmatists such as Dewey. Holmes's is more focused on a task at hand. It was not philosophical or broadly deep and was less optimistic more fatalistic.

Freedom Within the Bounds of Reason: Kant, Holmes and Dewey

Holmes may have flirted with Kant when he was young, as Emerson remembered, but little of Kant remained in Holmes, except for a level of intelligence and an appreciation of ideas, and a broad array of friends within a social group of intellectuals.

What makes morality possible for Kant is free will. The categorical imperative, as Kant formulated in the *Critique of Practical Reason*, is a profound insight and rule in deciding to do something. Anyone and everyone could and should do the same as part of the legitimation process of choice.

Both Kant and Holmes emphasized duty and consistency. They were both rationalists. Holmes though was a thorough experimentalist in principle and a naturalist; Kant a rationalist about the foundation of knowledge. Kant thought that we discover the truths that lay in us (1789, 1792), such as the rational inferences about things and events that lay within us; discovery is internal to the workings of the mind, the structure of thought, something that would be discussed frequently in the next centuries.

For Holmes, duty and clarity defined the social space of decision making, but an evolving space in competition. What we owe each other is something perhaps Kant understood as well as what we owe ourselves. For example, I think about you when I do something broadly. Ought anyone in my position do the same. We can hold onto something like a Kantian categorical imperative and other related moral strenuous suggestions, and render them concrete and normative without being divorced from the continuous transactions of the everyday.

Holmes (like Dewey) is in the everyday world of discerning ends that are viable that became substantial in the battle for ideas and institutions in an evolving democracy of participation. Holmes and Dewey both have their strengths in an idea of broadening human experience. But, unlike Dewey, for Holmes human possibilities were not at the forefront of his mind. Yet, for both men, the moral law is not within the mind as Kant believed (1790), but rather anchored to human action and decision making and the battles of the body politic.

In contrast to Kant and Dewey, Holmes is anchored more to fatality, despair, and duty, a combination of values that gave Holmes meaning when faced with the utter stupidity of human expression. There is also a glimmer of creative expression amongst the brutal competition that is human possibilities (see also Nietzsche 1886/1972). Holmes would say in his Memorial Day speech (1895), "I do not know what is true. I do not know the meaning of the universe." He goes on later to say the there can be little doubt "which leads a soldier to throw away his life in obedience to a blindly accepted duty in a cause that he little understands." There is a fatalistic absurdity in the world of Holmes's experience.

Experience and Reason

Holmes (1881, 1895) makes it clear over and over that experience matters. Jurisprudence judgment is not something to be purified but should be embedded in the mix of the life. In particular, experience matters to a depth in judgment, not a purification toward some foundation of thought. This is part of his pragmatist roots that figures in judicial reasoning.

In contrast, Kant was silent about the development of how experience matters to reason, except in his appreciation of Rousseau and aesthetics. Indeed, as Aristotle noted, "justice is the only virtue which is regarded as benefiting someone else than its possessor" (*Ethics*, 142). Aristotle places the human condition in the naturalistic phrase of combating and developing temperance and good judgment. This is what Holmes strove for. Experience matters. Much more knowledge is "worked into the living texture of the mind" (*Ethics*, 200)—or now what we would call the mind/brain or what I call cephalic to include the meaning of both. Mind is not detached from a body in the flux of problem solving, in the mix of life.

The role of experience and reason underlies pragmatist considerations. Bodily capabilities are not discarded as they were in rationalists (e.g., Descartes, Fodor). They are to be tempered and integrated into an evolving human being with habits that codify behavior—a common theme in law and life. Holmes may not have been a Kantian of any sort, except for being detached in many respects, but the sublime and the beautiful would remain close in his prose.

The battle for coherence required stability, viability, and predictive coherence. They require the codification of behavioral expression. Peirce well understood this and called it "frozen mind." I don't like the frozen metaphor, but codified habits, moral or otherwise, are facts. We know how reason about those facts works in the context of the brain. Regions of the brain such as the basal ganglia are tied to the organization of action. They bind behavioral chunks into meaningful expressions and create syntactically organized forms of behavior (Berridge 2004).

Holmes eschews all certainty about knowledge. In contrast, for Kant, the moral law from within oneself was the imperative of reason's fate. For nature is a thing of beauty and morals a thing of imposition, knowledge the congruence of space and time, and eventually probability a strutting of knowing. We move from certainty, although we could get a lot of agreement with regard to moral sentiments and perhaps some moral rules.

Kant was a rationalist, which in his time meant certain knowledge: Newtonian and Euclidian foundations, not Riemian alternate geometries and Einstein's alternative space-time frameworks. Obedience, duty, and the moral law form a concept of freedom, and its expression is in the pure terms of Kant's great moral prose. After all, Kant (1789, 1792) had at least two battles, to wrestle morality from a Church and make it compatible with knowledge of nature.

For Holmes, there is no separation of nature, body and mind; mind is embodied in action, scarred or not by the contours of experience. The fluidity of an adapting mind searching for coherence, not a purified reasoner, is the life blood of classical pragmatism; Holmes's sense of the fluidity is a piece of this perspective. He believed in the embodied reason adapting to circumstance as a norm. Holmes was keen on jurisprudence of the particular case: the particular experience and the expected particular outcome are what dominates in his sense of reason and experience, problem solving in the law or elsewhere.

Freedom is an achievement, by hard labor within a mind in a body without separation. But a rule that renders fairness possible as a consideration and possibly implemented is another issue.

Holmes's sense of the human condition is "self-preference," but public policy can sacrifice the individual for some social good. Crucially, for Holmes, "the tendency of the law must always be to narrow the field of uncertainty" (Howe 1963) and by "taming chance" by reasonable inferences, embedded in orchestrated habits of behaviors (Peirce 1898/1992; Hacking 1990). As Peirce (1878a, b) put it, "the essence of belief is the establishment of a habit."

Good enough problem-solving means that resources are expended to the unexpected, the less expected, the less familiar. This kind of problem solving aims to acquire a coping mechanism for longer term security amidst forms and pockets of security and insecurity. The law is a key feature in our adaptation, in providing some semblance of security and coherence. Although like everything else, the law is likewise everything wholly fallible. Fallibilism was a philosophical position ushered in by one of Holmes' colleagues and member of the Metaphysical Club, Charles Saunders Peirce, in the later part of the nineteenth century (Menand 2001).

Here we meet again what Holmes was embracing: Fallibilism is the intellectual prophylactic to guard against certainty. Law is part of that struggle for coherence, the fight against our primordial sense of fear, and the evolving and culturally pregnant sense of fairness. But there is a bruteness about life certainly as Holmes well understood.

Law is necessary to lift us from the nasty brutish and short side of our experience—this is a part of our nature. But making peace in the war of nerves, a war on the nervous system, is non-trivial. Ultimately, we seek peace and law provides protection. In order to protect liberty, self-initiative and self-interest converge. Law facilitates rational social contact.

Two forms of sensibility converge. One emphasized in the West is the exaggerated emphasis made so inherent across philosophy on the many senses of the self. The self is not a thing, however. Like "metabolism," the self is a highly theoretical term. The emphasis is obviously not the "me" vs the diverse social milieu that we inhabit. Like our social milieu, the "me" is not a frozen concept. Our world is one of fluidity amid pillars of stability. It is not frozen but solid for us to be viable within the uncertain world in which we are adapting. Law follows us like oxygen, essential for our breathing, for having the space to breathe. The isolated self is an aberration from essentials, and the tie to others. But it is also a fortunate feature of our evolutionary capabilities.

After all, the law is about order in the social space, a space co-inhabited by others and predictable but evolving. Excessiveness was not part of Holmes's intellectual vocabulary: whether in the form of social programs, endless control (communism), or its opposite, anarchy. Balance or some form of mean is a golden ideal in adjudication on principle and rationality. One rational consideration is the impact on others, where the law can step or not. This is perhaps at the heart of Holmes along with a streak of "a very conservative anarchist" (Gilmore 1977, 1999).

The example of harming others with free speech by shouting out something (such as shouting "fire" in a crowded theater) is one example, certainly for Holmes, where the law steps in (e.g., *Schneck vs United States* 1919b). Free thought or free speech is contextualized, despite it being a core value, and certainly a fundamental value for Holmes. But it has limits, and contexts matter. Therefore, adjudication is called for.

Anarchists or communists were something akin to terrorists in the minds of many Americans in the late nineteenth and early twentieth century. Here Holmes was at his independent best, and tied to diverse others (e.g., Felix Frankfurter, Learned Hand, Zechariah Chafee).

But again, law is not ethics, something else reiterated by Holmes. Commenting upon a sensationalist trial of two purported arsonists in Boston, he said, sadly, "we practice law not justice" as he would be lamentable about this (e.g., Leo Frank lynching case). Holmes would not overturn their convictions or listen to pleas to halt execution and pleas for clemency. Holmes did have his doubts about the conviction but not enough to overturn it. But, as in other cases, not getting "a square deal" and degrading state sovereignty and federal encroachment is a core principle. Frankfurter appreciated his book on the trial, despite the fact that he would have acted differently from Holmes (Baker 1991).

Fear, bigotry, ambition and myopia surrounded the conviction of these two individuals. But the context was socially explosive. This case is but one example of endlessly recurring events in human history, not one likely ever to leave the world stage of human expression.

Choice and Social Contact

We live within the confines of choice, of what Herbert Simon (1957, 1962, 1996), the Nobel Prize laureate, called "bounded reason." We experience choice and decision making within more limited domains where our choices are constrained (Elster 1983, 2000). They do not exist within a vacuum or existential flora of unending possibilities. Holmes, being a psychological realist, would easily appreciate these facts about us.

For Holmes or Dewey, values are inherent, not some externality or extra property. They inhere in our very doing, being, and acting. Holmes may not be consistent with regard to issues about values, whether he is a positivist—what has been called "law without values" (Alschuler 2000)—or whether there are properties in our animal nature that predispose an evolution of law in cultures that promote democracy. That is certainly Dewey's (1939) cultured view as expressed in *Theory of Valuation*.

Ends or values turn into and are part of the consideration of means. Both of which are inherent in the appraisals or valuation determinants of human rationality or reasoning (Dewey 1939; see Neville 1974). Freedom to participate in an expanding arena is tied to our sense of inquiry, to political sensibility and an expanding sense of human capability (Sarokin and Schulkin 2016).

Holmes, as I have indicated, was more narrow, more fatalistic, less democratic, and less hopeful, in his political aspiritions for the human condition than Dewey. Holmes's concern is a democracy of "self-devotion" (Weaver 2003). Dewey labored for an evolving democracy (Sullivan and Solove 2013). But Holmes was not in the social optimistic business as Dewey was.

Holmes did not advocate for social change. He was certainly embedded in survival and adaption. He felt that this sense is captured in Dewey's metaphysics of experience—the broad categorical depiction of the sense of the precarious and the stable.

Holmes, as a pragmatist, understood that appraisals of value are inherent in decision making. That is the major difference between earlier versions of pragmatism versus positivism. For example, some things were just inherently better. There are real facts embedded in history and the variation of change, but appraisal of worth is in the ends we pursue and the means we pick to achieve (Weissman 1993).

As Dewey (1939) wanted to make clear, "valuation was not ejaculatory," another non-trivial proposition. Attitudes were not necessarily distant and divorced from experience. The line between theory and experience, theory and evidence and warrantability, was and is quite permeable. It is sort of like the blood brain barrier: the barrier between the peripheral systems and the central nervous system is not absolute and is in fact quite permeable, depending upon context.

Indeed, one adaptation is to self-impose choice limits within what is possible. There is a fundamental link between technical innovation and choice (Elster 1983). Innovation is tied to participation with others and law protects the innovation, the discoveries, the patents, the new horizons. Protection is one thing, as opposed to straight jacketing the activities of others. Therein always lies the bedeviling dilemma between choice and constraint.

De Tocqueville mapped out an early country on a path of democracy, despite the degradations of slavery and the depravity of our behavior towards the American Indian. What America had was an appreciation of equality, with a burgeoning class of merchants and entrepreneurs. Perhaps not surprisingly, the first society of intellectual worth was one founded by Ben Franklin. The devotion was practical: solving problems and invention predominated. Invention is at the heart of this collective battling of individual initiative, something underlying a culture that prized liberty of self-expression and invention and that always underlies innovation and self-worth and reliance.

Local communities and local responsibility were the context of authority, the prevalent predilection of human action and thought. But Franklin was too quick to conclude with "I think that in no country in the civilized world is less attention paid to philosophy than in the United States" (vol. 11, p. 3). Pragmatism emerged as an original philosophical orientation, a term perhaps coined by Kant and made real by Peirce, popularized by James and expanded into the larger culture of democracy and equality of opportunity by Dewey and Holmes.

Competition of Ideas; Choice and Intellectual Freedom

Holmes is the great champion of the freedom to compete in the body politic, but not a big champion of who is allowed in. Holmes glorified intellectual competition with statements like "free competition is worth more to society than it costs" (*Vegelahn vs Guntner* 1896). Intellectual freedom is a core and consistent value of our democracy, Holmes's conception of the living law, within the bounds of a fluid, context-dependent, and historically-bound context. We will see that Holmes is never an absolutist. His pragmatism and his sense of the law are modified more by context (Grey 2000; Horwitz 1992a, b).

Holmes is known and remembered for his defense of free speech, a defense of the first amendment with clarity and dignity. But he was not consistent (cf. *Gitlow vs NY, Schneck, Abrams vs US*). But then Holmes was situational not ideological—rejecting absolutism as Holmes did, was perhaps an intellectual virtue in the open-ended universe in judicial reasoning we inhabit. In *The Path of the Law*, Holmes asserts that "behind the logical form lies a judgment as the relative worth and importance of competiting legislative grounds, of an inarticulate and unconscious judgment."

Our age is one in which free speech is under siege in parts of the world, as it always has been. But like the rest of us Holmes was not consistent on this issue (Healy 2013; White 2015). After all, what is a "clear and present danger" abnegating freedom of speech is sometimes hard to discern. Principles guide, but reason reaches beyond principles of rules or statutes and asks, what is right? What Holmes is suggesting is the pragmatist view about action; what "ideas incite" can lead to action and then of course down the slippery slope into persuasion. As Holmes put it, "eloquence may set fire to reason" (see also Fried 2004).

Holmes waxed and waned on the Sedition Act of 1798, the threat of Bolshevism, and the protection of African Americans and their right to vote, Asian Americans, women's equal rights and voting rights, and the status of American Indian populations. Holmes was distant, often mute. He was no John Dewey when it came to the ideals of the progressive conception of what is possible. Holmes was fatalistic. He tells us that his penultimate test of truth is brute: "I can't help believing so and so" (Grey 1989).

But his fatalism is limiting, perhaps a remnant of what he experienced during the war; the senseless loss of life, though that was a war worth fighting for. Perhaps these are remnants of the impact of the war on his brain.

But Holmes did believe in the competition of ideas, and of people, Holmes would I think would have embraced Thurgood Marshall. Holmes would I think have no problem engaging Ruth Bader Ginsburg. In the end it is about freedom of opportunity, a competition for the resourceful and the fortunate.

Free speech, like free inquiry, is at the heart of pragmatism and the law (Dewey 1928; Posner 2003), guaranteed by the first amendment except under conditions of utter insecurity and danger. There is no absolute predictive condition of adjudication, just the hard slog of good judgment and battle in a market place of competing interests. And, of course, self-corrective inquiry is normative in legal inquiry. The larger view articulated by Dewey is not only reliable coherence and predictive capability, but also settled options and habits that have competed for expression understood within a framework of participatory democratic sensibility (Dewey 1935; Butler 2010). Holmes (see Letters to Pollock 1942/2015) at best is a very modest democrat.

Dewey celebrated Holmes and found he thought in Holmes a kindred spirit, particularly with regard to understanding theory. Theories underlie human practice, as thought is embedded in action. Indeed, Dewey would note that "theory ceases to be remote and otiose and becomes what Holmes called – the most practical of things" (2012, p. 285).

Where he is not ambivalent is about his experience in war, his sense of duty and his adherence to inference. He says law is not logic but experience. Experience is based on the sort of logic of surviving; making sense of our world by adapting, coping, predicting, and expanding experiences embedded in the practices and inferences that inform our life. Of course, the pragmatist notion of experience can be notoriously vague, as many critiques have nicely or not nicely put it (e.g., Hollinger 1977).

As Menand, commenting on Holmes, says, "experience makes everything blurry at the edges; it reduces knowledge to a prediction of what should be the case most of the time…" (2001, p. 45). Holmes is understood here very much like other well-known pragmatists (Dewey 1925/1989; James 1890; see also Quine 1969)—the epistemic and spacious sense of experience with regard to the future and to the outer fringe of the knowing process. The center is anchored, and the fringe is the continued expansion of our present epistemic states.

Holmes (1881) also had a bodily feel for which practices and well worked adaptations to our surroundings inform and guide us—not the mere logic of a pure thinker, untainted as Kant (1789) imagined the process of judgement.

But Holmes, like many pragmatists, dilutes the dualism of thought and action (White 1947). That is not to say that they are not separate. But reasoning does not stop when we move nor when we sit. Action and thought run together like a stream of diverse forms of adaptation, of reasoning, of coping, enjoying, building, creating, etc.

The life of the law Holmes (1881) professes is experience. It is the pragmatist notion of experience; active not passive, rich in the expression of problem-solving proclivities. Judges like Holmes, indeed like all judges, were informed and enriched by experiences. There was no purity of separation of reason from the contours of experience, the sculptured sensory/motor experiences.

Holmes straddled these worlds. For he probably saw emotion in a negative light as many still do. Excessive-compulsive thought, sheer logic, is as deadening and irrational as sheer exaggerated emotion: that is probably what Holmes was referring to, with regard to experience versus logic. His was a rhetoric that acknowledges that "when we study law we are not studying a mystery but a well-known profession" (Holmes 1897)—a profession in which experience matters. Reason is not divorced from experience, logic from context. Ultimately, judges like Holmes, of which there are probably many, arrive at some form of pragmatism in their legal reasoning (Grey 1989; Posner 2002; Kellogg 2018).

"The law of fashion is a way of life," Holmes asserts in his paper on "Law in Science and Science in Law." He also notes in many places that logic runs through experience, because inferences and probabilities and computational systems run through experience like oxygen. They are common place.

Holmes, no philosopher of logic or of reason, was propelled not to accept the idea of disembodied logic or reason as opposed to experience being what matters. Thus in practice, Holmes was no purist. He understands that in law, as in war, the practice of judges is bound to experience. So is science, so is art, so are all meaningful activities, both for the good or not. Indeed, the idea of being faithful to experience is a key pragmatist orientation and phrase, though it is not always clear what that means.

While there is deductive certainty in matters of jurisprudence, or elsewhere as Holmes would argue, "views of policy are taught by experience of the interests in life. Those interests are in the fields of battle" (Holmes 1894). But what sort of battle? Holmes understood, like James (1890), that brain expression underlies judgment, and neural incapacities matter. However, "the tests for liability take no account of incapacities, unless is so marked as to fall into the well-known exceptions, such as infancy and madness. They assume that every man is as able as every other to behave as they command" (Holmes 1881).

Constitutional Possibilities and Constraints: Bounded Choices

Holmes was the champion of free competition of ideas and expression. Freedom, like choice, is considered in context: competition for resources, competition of expression, experience, capability and fortune. While Holmes would assert that "an ideal system of law should draw its postulates and its legislative justification from science" (1895 address), he was sanguine enough to know there was no such idealized system. While he idealized science into perhaps more than what it could deliver under the best conditions, he knew that it might not actually be practicable. Science, Holmes imagined, is embedded in statistical inference (Peirce 1898/1992; Hacking 1965, 1990).

In our social context, diverse laws constrain us. For those of us for whom the Constitution is an evolving precedent tied to context, instrumental reason, and not pure principle, rationality is preserved by longer term viability and not rational certainty. Holmes set a context for a precedent that enhanced the long-term viability of our species, such as inclusion and fairness. The ratification of the Constitution is a wonderful example of a society deciding something together (Moreno 1995) and blending differences within arenas of consciousness, difference, and diffidence. A root metaphor for our Constitution, a blueprint for government, for rationality and adjudication is a version of "American Scripture" (Maier 2012).

Neuroscience Considerations

One dominant way in which to understand neural function are the ways we limit choice; we reduce our options to promote reliable outcomes. In other words, we limit options, relax neural activation, by relying on habits that work (Peters et al. 2017; Schulkin 2000; Sterling 2004). This is a basic core pragmatist orientation to hypothesis formation, learning, inquiry and habit formation. The basic interplay is between exploring new options and relying on old reliable ones; the pragmatist orientation is to explore when there is breakdown of reliable habit formation; it preserves neural tissue from over use and neural detioration. Bounded choice aids decision making; bonded choice does not mean neural frozen tissue.

The possibilities and constraints that figure on neural expression and that are reflective of choice is the exploration essential in ontogeny is the expansion of choice and neural innervation and expression across the neural axis (Schulkin 2004). Human choices reflect our evolutionary and cultural options; devolution of neural function is the converse, limiting human choice and perhaps accountability (Jackson 1884; James 1890).

As we take into account choice and responsibility in the adolescent brain, we now consider what we know about cortical inhibition or lack of it in some individuals. Inhibition is a primary feature of neural control over behavioral choices (Sapolsky 2017; Moll and Schulkin 2009). Choice for pragmatists, in some context is experienced. And it is active not passive. The blurring boundaries of our epistemic experiences are testimonials for the transitions and transactions with others that mark our experiences.

Indeed, the experiences of choice are tied to regions of the brain vital for the organization of action, for responding to what is needed, in the search for solutions of adaption. The motor systems in this regard are not by stangers to thought but essential to it (Smith and Graybiel 2013; James 1890; Lakoff and Johnson 1999; Schulkin 2007); exploration is part of epistemic foraging. Movement pervades thought and thought pervades movement, which is why the dualism of thought and action was always spurious. All of this is compromised with diverse forms of brain damage; brain damage that lessens agency, choice and the ability to forage (Sapolsky 2004). All of this impacts the extent to which we attribute punishment when we transgress from expectations and the law (Hirstein et al. 2018). And these are critical features of judicial decision making.

Conclusion

Our sense of freedom exists within constraints; neural and cultural constraints and the utter permeability of the two through our experiences (Schulkin 1992; David Weissman, unpublished manuscript). Enriched experience of both our sense of nature and culture are not bifurcated; the separation of nature and culture is spurious and misleading.

Law figures in limiting choice, protection, and patterning human viability and enhancement. It fosters ideals of social coherence in democracies that promote reduction of violence, and human capability through pedagogy and opportunity. Choice is also a piece of cultural evolution (Dennett 2003; Neville 1974).

Law is part of that human growth and pedagogy. Holmes, in *The Common Law*, noted that our urge "to codify the law into seemingly self-sufficient propositions will only be a phase in a continuous growth" (p. 25). Ours is a long pedagogy and exploration, getting longer as society grows more complicated.

Law and our cultural evolution is placed in this context of the utter continuity of biological and cultural components of human nature. Culture is a natural outgrowth of our capabilities. What so offends the spirit of inquiry was the narrow reduction of culture to biology. That is not to say that there is not a lot to learn in the study of ants, fish, rabbits, and monkeys. But the diversity is relatively fixed. The notion of progress and adapting and creating diverse contexts is a human expression.

Our cognitive capability reflects the anticipatory side of adaptation. We are not just reactive to events. We predict, prepare and organize. These events impact biological tissue. Anticipatory, cognitive architecture is reflected in the epigenetic changes that permeate our experiences in life. These events start early. Holmes liked the idea of statistics, what is likely to happen, but he rejected claims of certainty.

Law is in the ideal to lift us from the insensitive side of our experience. Of course, this was a part of our nature. What goes on in the brain is a reflection of everything about us, including that of law. Holmes entered a world in which ideas competed in a market place, and competition was essential, as we shall see throughout the text, for Holmes.

What we can discern is that the philosophical debate about knowing was in dispute, between rationalist and empiricist, instrumentalist and naturalist. Understanding the Heavens, as Kant or Peirce would intimate, was to be understood by theory and evidence.

Theory is about orientation. Sense data and the logic of sense data do not compete against a foundation for knowing. For knowing is too multi-varied. The law is a good example. Holmes's universe is porous. He was endlessly allusive and literary (Grey 1989, 2014) as well as inconsistent (White 2000, 2016). After all, he was evolving too. He lived a long public life, a life in the law, a life of initial privileges and excitement of a continent spurring adventure.

The many senses of the law are something that permeates the intellectual air of Holmes's world. So too is his link to modern science and naturalism. I think it is fairly clear that Holmes grew up in a rarified world of broadly understood intellectual acumen. Debate, engagement, and the enlargement of experience were valued at the Holmes's family breakfast table. The discussion at the table was surely, palpably rich. His family was an elite, intellectual family.

The law is an extension to group formation and coherence. Restraint and self-discipline are the major ingredients that inhere in a life worth living, or at least the one that Holmes valued. Indeed, at 90 years of age, in a radio broadcast on March 8, 1931, he would say, "for to live is to function. That is all there is to living."

Law, in part, comes in to reduce our options. Likewise, we reduce our options through self-regulation and self-binding, which are forms of pre-commitments that bind us to certain paths. The issue is how to determine the rationality of these options (Elster 2000), the achievement of the goals, and the worthiness of those goals (Dewey 1925/1989). Reduced options, if they enhance the quality of our lives, are an achievement. For example, it is a benefit if a person or a society avoids addictions. Reduced options serve as strategies in the regulation of reason and behavior through self-management, something the stoics would have appreciated. Understanding this, we better understand Holmes's commitment to the notion of duty.

Law is as basic to our cultural evolution as breathing is to our biological evolution. Both rest on an expanded brain and a penchant for group formation and social contact. Law pervades all features of our lives, our property, freedom, rights, economics, and relationships (Friedman 1973, 1993). Law, normatively, is tied to the expansion of order and coherence. The list is endless, but one feature stands out: regulation within freedom to compete and thrive and life worth living. Pragmatism is rightly characterized as a philosophy that is forward looking (Peirce 1898/1992) and tied to problem solving and social coherence.

References

Alschuler, A. W. (2000). *Law Without Values: The Life, Work and Legacy of Justice Holmes*. Chicago: University of Chicago Press.

Baker, L. (1991). *The Justice from Beacon Hill: The Life and Times of Oliver Wendell Holmes Jr*. New York: Harper & Row.

Berridge, K. C. (2004). Motivation Concepts in Behavioral Neuroscience. *Physiology & Behavior, 81*, 179–209.

Butler, B. E. (2010). Democracy and Law: Situating Law Within John Dewey's Democratic Vision. *Ethics and Politics, 12*(1), 256–280.

Currie, D. P. (1985). *The Constitution in the Supreme Court: The First Hundred Years*. Chicago: University of Chicago Press.

Currie, D. P. (1990). *The Constitution in the Supreme Court: The Second Century*. Chicago: University of Chicago Press.

Dennett, D. C. (2003). *Freedom Evolves*. New York: Penguin Books.

Dewey, J. (1925/1989). *Experience and Nature*. LaSalle, IL: Open Court Press.

Dewey, J. (1928). Justice Holmes and the Liberal Mind. *New Republic, 53*, 210. In J. A. Boydston (Ed.), *The Later Works of John Dewey, Volume 3, 1927–1938* (pp. 177–183). Carbondale: Southern Illinois University Press.

Dewey, J. (1935/1963). *Liberalism and Social Action*. New York: Capricorn Books.

Dewey, J. (1939). *A Theory of Valuation*. Chicago: University of Chicago Press Press.

Dewey, J. (2012). *Unmodern and Modern Philosophy*. Carbondale: Southern Illinois University Press.

Diamond, A. S. (1935). *Primitive Law*. London: Longmans.

Elster, J. (1983). *Sour Grapes*. Cambridge: Cambridge University Press.

Elster, J. (1989). *Solomonic Judgments*. Cambridge: Cambridge University Press.

Elster, J. (2000). *Ulysses Unbounded*. Cambridge: Cambridge University Press.

Fried, C. (2004). *Saying What the Law Is: The Constitution in the Supreme Court*. Cambridge: Harvard University Press.

Friedman, L. M. (1973/1985). *A History of American Law*. New York: Simon & Schuster.

Friedman, L. M. (1993). *Crime and Punishment in American History*. New York: Basic Books.

Friston, K. (2010). The Free Energy Principle: A Unified Brain Theory? *Nature Reviews, 11,* 127–136.

Gilmore, G. (1977). *The Ages of American Law*. New Haven: Yale University Press.

Gilmore, G. (1999). Some Reflections on Oliver Wendell Holmes Jr. *Green Bag, 12,* 374–379.

Grey, T. C. (1989). Holmes and Legal Pragmatism. *Stanford Law Review, 41,* 787–870.

Grey, T. C. (1991a). *The Wallace Stevens Case: Law and the Practice of Poetry*. Cambridge: Harvard University Press.

Grey, T. C. (1991b). What Good Is Legal Pragmatism. In M. Brint & W. Weaver (Eds.), *Pragmatism in Law and Society*. Boulder, CO: Westview Press.

Grey, T. C. (2000). Holmes on the Logic of the Law. In S. J. Burton (Ed.), *The Path of the Law and Its Influence: The Legacy of Oliver Wendell Holmes Jr.* Cambridge: Cambridge University Press.

Grey, T. C. (2014). *Formalism and Pragmatism in American Law*. Boston: Brill Press.

Hacking, I. (1965/1979). *Logic of Statistical Inference*. Cambridge: Cambridge University Press.

Hacking, I. (1990). *The Taming of Chance*. Cambridge: Cambridge University Press.

Healy, T. (2013). *The Great Dissent: How Oliver Wendell Holmes Jr. Changes His Mind and Changes History of Free Speech in America.* New York: Henry Holt.

Hirstein, W., Stifford, K. L., & Fagan, T. K. (2018). *Responsible Brains: Neuroscience, Law and Human Culpability.* Cambridge: MIT Press.

Hollinger, D. A. (1977). Review: The Culture of Experience: Philosophical Essays in the American Grain. *Transactions of the C. S. Peirce Society, 13,* 312–315.

Holmes, O. W., Jr. (1881/1952). *The Common Law.* New York: Dover.

Holmes, O. W., Jr. (1894). Privilege, Malice and Intent. *Harvard Law Review, 8,* 1–14.

Holmes, O. W., Jr. (1895). The Soldier's Faith. Graduating Class of Harvard University.

Holmes, O. W., Jr. (1896). *Vegelahn vs Gunther.*

Holmes, O. W., Jr. (1897). The Path of the Law. *Harvard Law Review, 10*(8), 45.

Holmes, O. W., Jr. (1919a). *Abrams vs United States.*

Holmes, O. W., Jr. (1919b). *Schneck vs United States.*

Holmes, O. W., Jr. & Pollock, F. (1942/2015). *Letters, Volumes 1 and 2* (M. DeWolfe Howe, Ed.). Cambridge: Cambridge University Press.

Horwitz, M. J. (1992a). The Place of Holmes in American Legal Thought. In R. Gordon (Ed.), *Legacy of Oliver Wendell Holmes Jr.* Palo Alto: Stanford University Press.

Horwitz, M. J. (1992b). *The Transformation of American Law.* Oxford: Oxford University Press.

Howe, M. D. (1957/1963). *Justice Oliver Wendell Holmes: Volumes 1 and Volumes 2—The Proving Years.* Cambridge: Harvard University Press.

Jackson, J. H. (1884/1958). Evolution and Dissolution of the Nervous System. In *Selected Writings of John Hughlings Jackson.* London: Staples Press.

James, W. (1890/1952). *The Principles of Psychology.* New York: Dover Press.

Kant, I. (1789/1997). *Critique of Practical Reason.* Cambridge: Cambridge University Press.

Kant, I. (1792/1914). *Critique of Judgment.* London: Macmillan.

Kellogg, F. R. (2018). *Oliver Wendell Holmes Jr. and Legal Logic.* Chicago: University of Chicago Press.

Lakoff, G., & Johnson, M. (1999). *Philosophy in the Flesh.* New York: Basic Books.

Maier, P. (2012). *American Scripture.* New York: Vintage Books.

Menand, L. (2001). *The Metaphysical Club.* New York: Farrar, Straus, and Giroux.

Moll, J., & Schulkin, J. (2009). Social Attachment and Aversion: On the Humble Origins of Human Morality. *Neuroscience and Biobehavioral Reviews, 33,* 456–465.

Moreno, J. D. (1995). *Deciding Together.* Oxford: Oxford University Press.

Neville, R. C. (1974). *The Cosmology of Freedom.* New Haven: Yale University Press.

Nietzsche, F. (1886/1972). *Beyond Good and Evil.* Mineola: Dover.

Peirce, C. S. (1878a). Deduction, Induction and Hypothesis. *Popular Science Monthly, 13,* 470–482.

Peirce, C. S. (1878b). Doctrine of Chances. *Popular Scientific Monthly, 12,* 604–615.

Peirce, C. S. (1898/1992). *Reasoning and the Logic of Things: The Cambridge Conferences Lectures of 1898 (Harvard Historical Studies)* (K. L. Ketner & H. Putnam, Eds.). Cambridge, MA: Harvard University Press.

Peters, A., McEwen, B. S., & Friston, K. (2017). Uncertainty and Stress: Why It Causes Diseases and How It Is Mastered by the Brain. *Progress in Neurobiology, 156,* 164–188.

Posner, R. A. (2002). Is Pragmatic Adjudication Inescapable? In *How Judges Think.* Cambridge: Harvard University Press.

Posner, R. A. (2003). *Law, Pragmatism and Democracy.* Cambridge: Harvard University Press.

Quine, W. V. O. (1969). Epistemology Naturalized. In *Ontological Relativity and Other Essays.* New York, NY: Columbia University Press.

Sapolsky, R. M. (2004). The Frontal Cortex and the Criminal Justice System. *Philosophical Transactions of the Royal Society of London. Series B, 359,* 1787–1796.

Sapolsky, R. M. (2017). *Behave.* New York: Penguin.

Sarokin, D. J., & Schulkin, J. (2016). *Missed Information.* Cambridge: MIT Press.

Schulkin J. (1992). *The Puruit of Inquiry.* New York: SUNY Press

Schulkin, J. (2000). *Roots of Social Sensibility.* Cambridge: MIT Press.

Schulkin, J. (2004). *Bodily Sensibility: Intelligent Action.* Oxford: Oxford University Press.

Schulkin, J. (2007). *Effort: A Behavioral Neuroscience Perspective on the Will.* Mahway: Erlbaum Press.

Schulkin, J. (2015). *Pragmatism and the Search for Coherence in Neuroscience.* London: Palgrave Macmillan.

Simon, H. A. (1957). *Models of Man, Social and Rational: Mathematical Essays on Rational Human Behavior in a Social Setting.* New York: Wiley.

Simon, H. A. (1962). The Architecture of Complexity. *Proceedings of the American Philosophical Society, 106,* 470–473.

Simon, H. A. (1996). *The Sciences of the Artificial.* Cambridge: MIT Press.

Smith, K. S., & Graybiel, A. M. (2013). A Dual Operator View of Habitual Behaviors Reflecting Cortical and Striatal Dynamics. *Neuron, 2,* 361–374.

Sterling, P. (2004). Principles of Allostasis: Optimal Design, Predictive Regulation, Psychopathology and Rational Therapeutics. In J. Schulkin (Ed.), *Allostasis, Homeostasis and the Costs of Physiological Adaptation.* Cambridge: Cambridge University Press.

Sullivan, M., & Solove, D. J. (2013). Radical Pragmatism. In A. Malachowski (Ed.), *The Cambridge Companion to Pragmatism.* Cambridge: Cambridge University Press.

Weaver, W. G. (2003). The Democracy of Self Devotion: Oliver Wendell Holmes Jr. and Pragmatism. In A. Morales (Ed.), *Renascent Pragmatism: Studies in Law and Social Science.* Burlington, VA: Ashgate.

Weissman, D. (1993). *Truth's Debt to Value.* New Haven: Yale University Press.

White, G. E. (2000). *Oliver Wendell Holmes Sage of the Supreme Court.* Oxford: Oxford University Press.

White, G. E. (2016). *Law in American History.* Oxford: Oxford University Press.

White, J. W. (2015). *Lincoln on Law, Leadership and Life.* Naperville, IL: Source Books.

White, M. (1947). *Social Thought in America: The Revolt Against Formalism.* Boston: Beacon Press.

8

Naturalizing Decision Making: Heuristics and Concerns

The naturalization of reasoning looks at diverse forms of capabilities and certainly to the dissolution of diverse forms of dualism, so endemic to diverse pragmatists, including Holmes. For Dewey (1910, 1925), there is no separation between cultural capability and our biological inclinations; the empiricism is in structuring the best sort of outcome. Dewey perhaps was naïve and over exaggerated the relationship between neural engineering and cultural moral expression. But the training and the means for those ends is part of the continuing evolving perspective in the progressive agenda, Holmes's agenda. Holmes was less optimistic than Dewey.

Holmes noted in *Science of the Common Law* that law "is forever adapting new principles from life at one end and it always retains old ones from history at the other… It will come entirely consistent only when it ceases to grow." Holmes sounds almost like Dewey here with the organic growth and decay conception of law, growth, and human action, which is why he would see in the expression of what he thought important in a philosophy of nature within a cultural democratic milieu (Grey 1992).

© The Author(s) 2019
J. Schulkin, *Oliver Wendell Holmes Jr., Pragmatism and Neuroscience*,
https://doi.org/10.1007/978-3-030-23100-2_8

Metaphors are not on one side and reason on the other. As I indicated at the outset of the book in the context of neuroscience, metaphor figures in our understanding. Metaphor is just a tool in reasoning and in doubt (Boyd 1993; Johnson 1993, 2007), solid but not frozen into eternity—hard fought through experience and won for its value in life—in getting along, in thriving, in sustaining the good things in our life and in the consideration of future generations.

Theory in law is hitting on roughly the right theory, the right heuristic, or roughly the right orientation; and this is empirical. Whatever is sort of good enough reasoning, generates the right heuristics (see below). While about heuristics, pragmatism is also about human interactions, determination of consequences. It is not anti-formalism, in the sense of formalism as analytic mathematics or logic.

The chapter starts with a sense of "good enough" reasoning—not mythologized perfect reason—heuristics, and anchoring realism about reason. These are familiar pragmatist themes. There is then consideration of Holmes's anchoring legal reasoning to more general problem-solving capabilities to niche capability, assessment of prior probabilities and self-corrective inquiry and are at the basis for reasoning and a pragmatic sense for reasoning about law and neuroscience.

Demythologized Reason or Good Enough Problem Solving

Holmes, like other pragmatist (e.g., Peirce) understood the fallibility and limitations of human reason. He exalted the competition of ideas, but not reason. He supported the rough and ready language of heuristics (Simon 1962, 1996; Gigerenzer 2000). They permeate the law, science, and all forms of human decision-making.

All forms of human interactions, including that of law, grow out of basic capabilities in problem solving. Tool use and its extensions, such as telescopes and microscopes during the seventeenth century, are but examples. These tools are extensions of ourselves, metaphors of limbs that expand our horizons for seeing. Metaphors run through legal reasoning or neuroscience, as in "I see what you mean" or "the analysis of is illuminating" or "your argument is clear." Metaphors aid our reasoning legal or neuroscientific; and they can be helpful or not. Nevertheless, they run through our reasoning and understanding of events.

The tools get integrated with common exploration to the examination of less common and novel. Such events have been continuous in our cultural evolution, with the law one example amongst others. Inferences in mathematics and tears are embodied in action, in foraging for coherence.

Fast heuristics are embedded in longer-term or reflective capabilities. Heuristics is just a phrase for a form of problem solving—something that works and something that transfers easily, the common ways in which we take short cuts and use "mother's helpers" (Gigerenzer 2000). Heuristics are not bad things. They are just something to notice and reflect on so that we are not over-vulnerable to making mistakes. Reason though is reflective; you have to make a case. Moreover, even the most banal moments in the law have to refer to a context of reason. So reflection figures—in a context of giving reasons, making a case. That is, in part, what we mean by being rational. Making a case is one thing, however; being roughly right is quite another.

Holmes did not mythologize human reason. He understood something about the limitations of reason, pure reason, analytic inferences. Reason is about good enough problem solving. Heuristics is good enough in adapting to problematic contexts.

The fact that heuristics (orientations that tend to work and are ecologically reliable or contextualized) or biases (a specific orientation not dependent upon context) exist is not a surprise. We want know how they operate in human decisions, legal (Sunstein 2005) and more generally (Kahneman et al. 1982; Gigerenzer 2000). After all, legal decision making is just a part of the larger class of decision making in general, not a special part of the brain.

One context for legal heuristics (Sunstein 2005) is the regulation of fear, precautions against danger, hazards. Diverse forms of biases emerge: orientations, such as the availability heuristic, a vulnerability which appears easily in decision making, or the most salient, reliability bias, probability bias, and neglect bias. Managing fear is pregnant in the law and manifests endlessly in human experience. Low probability catastrophic events are but one instance; and so one's tendency is to over-regulate. Sunstein makes clear "fear and liberty" are twin conflicts (Sunstein 2005). Holmes understood, following Lincoln, that in terms of war and other grave dangers to the foundation of the state, we compromise the law.

Good Enough Reasoning

A standard of legal reasoning that Holmes referred to, namely the so-called "reasonable person" standard, of course is misleading. For it depends upon what you mean by that term, particularly when understood in the context of biases and heuristics (Dahan-Katz 2013). However, there are some common ways we reason and are vulnerable.

Steven Toulmin, a student of Ludwig Wittgenstein, the great twentieth century philosopher, would argue in his book, *The Uses of Argument*, that reasoning was more akin to making a case in the law than in formal logic. Argument is in a context of knowledge for which we use diverse tools: probability, context inference, and knowledge (see Toulmin 1950, 1958, 1972).

Toulmin was dismissive of formalism because like Wittgenstein or Holmes he was in rebellion against the search for logical foundations of thought. Wittgenstein left the *Tractatus* for the *Investigations* (see Wittgenstein 1922, 1953). He left formalism for life forms of activity, social context, and history. Wittgenstein came to appreciate diverse forms of experience.

Dewey was not silent on this subject and pointed out what we have at stake in the commons: the ground of objective and hard-fought values, rights, and duties that we owe each other. It was what Holmes was referring to as experience, or what Dewey called "lived experience" (see McDermott 2007).

Stephen Toulmin, in *Reason and Ethics* and *the Uses of Argument*, would couch reasoning in terms of two perspectives: "we have before us two rival models, one mathematical, the other jurisprudential" (p. 95). Perhaps a better way to put this is that we use both formal and non-formal modes of reasoning. Indeed, for pragmatists, they fuel each other for reasoning and are rich in expectations about future events and consequences.

For Toulmin (1950, 1972), "Logic is generalized jurisprudence" (p. 7). Well, that might be a bit much: we use many tools in reasoning, including formal reasoning. Toulmin at that period was not aware of the pragmatists (see his later work). What he rightly was responding to was divorced from detached reason, i.e. formalism in a vacuum. Toulmin comes close to Holmes and Dewey when he asserts that "an argument is like an organism" (p. 94). Dewey used the phrase "warranted assertion" to dilute the formalization of truth, and the utterly abstract divorced notion. Nonetheless, mathematization of propositions is a thing of beauty. Idealization of data into such forms is just one tool amongst others, and in some contexts more important and more vital than in others.

Toulmin was certainly not against logic; just not the idealized logic that still permeated the epistemic space of giving reasons for events, making a case. Formal logic is just one amongst our many epistemic tools. Holmes understood that perhaps. Formalism runs amok when simply foundational and not simply viable, validated, rooted, justified (Toulmin 1950, 1972). For Holmes, law is not just logic. Logic is inherent in reason, something Aristotle understood.

Holmes's sense of the law is knotted to what Toulmin called "legal ecology," a loose confederation of precedent and argument in touch with history and context. Concepts are developed, cultivated, and expanded if deemed worthy in our democratized culture and the battle for ideas. Our reason for rationality lies in these relationships. Indeed, the rational justification is pragmatic.

Like many recurrent trends, the drive for formalism or foundationalism exists in our epistemic endeavors along with the common-sense property of striving for coherence, good enough problem-solving capabilities—interesting and informative theories that have traction and perspective and insight. Law is no different from other knowledge pursuits in this respect (Stone 2002).

That is not to say that law is like neuroscience, for instance. For some, law is biblical, rationalistic, or mathematical and foundationalist.

Law is again this hodge-podge of events in which foundationalistic and piecemeal epistemics coexist and are therefore akin to other epistemic investigations. Holmes understood this. When commenting on the Constitution he said, "the provisions of the constitution are not mathematical formulas... they are organic living institutions" (Holmes 1914 in Gompers decision).

Law has this added property as representing the way we reason about events, i.e., the language of justification for our claims for our views. Formalism is not an alternative to piecemeal bootstrapping and problem solving in the law or in a science. Formalism is continuous with this boot strapping and problem solving. Both are important. It is just that foundations are more about stability or viability, etc.

Importantly, the Constitution is historical, as Holmes certainly understood, and not geometrically given. When Holmes (1881/1952) asserted that the law was not just about logic, but about making the right decision, the right judgment, he was noticing that it was the larger social context that also mattered: judgment in context with an eye for the larger good. Societal depictions of law (as shown in below figure in a statue representing justice) show the fight for balance and fairness in law and justice. This is law in the service of social consequences and the acknowledgement of facts. Logic is one feature in reasoning, consistency within a social milieu that matters and is evolving. Biology feeds right into culture. This was a running theme for the younger John Dewey, for which the later Holmes felt a great affinity.

As Holmes noted, judges do what judges do; the variation is endless. What Holmes, like other pragmatists, also noted is that reason is rooted in biology, and common sense is an outgrowth of these capabilities instantiated in our practices, because they have adaptive capability, they are tied to ecological and social environments, and they reflect well-worn practices that are common place; the vast commons of social participation whose air we breathe.

Good judgment in law or elsewhere as a norm is something we often, in different circumstances, call optimization or optimal judgment. It is hard and is not the only standard but an important goal. The emphasis is on an external measure, a predicted outcome. But there was also another fact; something endemic in the classical pragmatist, the enhanced sense and very active notion of experience. And, of course, another palpable sentiment is learning by doing—getting a foothold by learning from older lawyers with experience (LaPiana 1994).

Law in part legislates the possibilities of human action that limit human expression, to hold individuals responsible in torts, contracts, duties (Austin 1832; Holmes 1881/1952; Hart 1961; Epstein 2000)—accountability within a political milieu of laws. Harm and fear are non-trivial considerations in the context of interests and the larger sense of conflicts in which the legal is infused with the ambiance of the larger social milieu in which Holmes understood the law (Gordon 1982, 2017). Holmes, looking for a common thread in our experience, would sardonically comment in a case, *Baltimore and Ohio Railroad vs Goodman*, in 1927 that we still hold one responsible for a risk from a train "when he knows he must stop for the train and not the train stop for him" at an intersection. Context and circumstance matter in judgment.

Many of Holmes's critics lament his inconsistent judicial judgments and his lack of framework (e.g., Epstein 1998; Treanor 1998; Winter 2001). In this regard, Holmes (1881/1952, 1917) is an experimentalist, historically minded, situational and open ended. Moreover, Holmes is no ideologue. He is within a pragmatist sensibility. His sense of inquiry is close to Peirce and Dewey (Kelogg 2007; Butler 2017), despite his distaste for the one and his ambivalence about the other.

Crime and punishments are social and economic concepts (Friedman 1993; Hart 1968). Holmes inherited the common law concept from England. What our culture has evolved is inherent due process for all, at least as the normative goal. The participants have increased under this normative goal of due process. Our legal system is one of the best examples of our culture. Holmes understood this as well as its imperfections across race and class.

Rationality is in context and grounds human reasoning and information processing within the confines of the setting of what is possible, desirable, time warranted, etc. Inferences are understood as grounded in possible action (Cherniak 1986), where tracking events is part of the adaptive capability.

Naturalizing epistemology recognizes the limitations as well as our capabilities. Menial rationality amounts to inferences justified in context that make sense in the context of the ends desired and the means or tools at hand. Holmes placed the evolution of the law in this very human context—a common law like a common faith from a naturalist who displayed piety and understood something about what we owe each other.

Holmes, a pragmatist by inclination, understood the value of heuristics, making just enough of a good case, something sound that hopefully will stand. But foundations are permeable and mostly not everlasting in law and much else.

Heuristics represents an adaptive set of tools (Tables 8.1, 8.2, 8.3, adapted from Gigerenzer [2000]) that are grounded in context and ecological and social boundary conditions for which it is suitable, much like adaptation in biology.

But our world has never been like that. In nature, we are bound to search for security and to drive down uncertainty, to placate our fears by the search for predictive order and preemptive control and coherence. The point of view here is not giving up rationality for eternal law, but adaptive rules that enhance human experience. Goals and action are fueled into mobilization, what Herbert Simon called "satisficing." Instead of being beyond human rationality in action, it is an action tied to human wants and the need for security and predictive capability. This is the stuff of biological cognitive adaptation. This is perhaps something instinctively understood. One of the exciting features about Herbert Simon was the utter permeability of the created or artificial and the natural—the infusions of our creation, computing machines embedded in adaptive design, rationality in context, rationality by selection and competition, rationality with design principles and suited for certain tasks.

Indeed, predictive capability is a core feature of neural function in brains like ours, with an expansive sense of experience, tied to situation and core survival (Clark 2013). This is not the stuff of abstraction, but of the precarious and the search for the stable (Dewey 1925/1989).

One possible way the brain computes some forms of probability assessment is via Bayes' theorem (Clark 2013, 2017). Bayes' theorem is a fundamental theorem of probability that states that, for any two

Table 8.1 Examples of cognitive biases

Bias	Explanation/example
Time category	
Hindsight bias	When decision makers with outcome knowledge exaggerate the chances that they would have predicted the outcome in advance
Sunk cost fallacy	Persisting in a negative expected value activity because a significant investment has already been made
Projection bias	Projecting onto the future not only affective states but any state that influences preference
Ignore category	
Omission bias	The tendency to choose not to do something when doing something might cause harm
Attribution bias	Incorrectly determining who or what was responsible for an event or action
Base rate neglect	Ignoring empirical statistics when making a probability judgement
Confirmation neglect	Seeking information that, if consistent with the current hypothesis, would yield positive feedback and to interpret evidence as consistent with the hypothesis
Egocentric bias	Subjects will over report their contribution and underreport their group members' contributions
Paternalistic category	
Anchoring	Different starting points yield different judgements which are biased toward the initial values
Framing effects	Variations in framing information yield systematically different preferences
Diversification bias	More variety is chosen when choices are bracketed together than when they are bracketed individually
Unit bias	The tendency for people to eat less when serving sizes are smaller and more when serving sizes are larger
Risk/loss category	
Ambiguity/avoidance	People avoid gambles with an unknown distribution of possible outcomes
Loss aversion	Losses loom larger than gains
Regret avoidance	Averting a feeling that a decision will have an undesirable consequence
Status quo bias	Preference to remain in the current state

Adapted from Gigerenzer (2000)

Table 8.2 Examples of heuristics

Heuristic	Explanation/example
Recognition heuristic	If one of two alternatives is recognised, infer that it has the higher value on the criterion
Take-the-best	To infer which of two alternatives has the higher value: (a) search through cues in order of validity, (b) stop search as soon as a cue discriminates, and (c) choose the alternative this cue favours
Availability heuristic	Probability of an event is estimated by the ease with which instances or occurrences can be brought to mind

Adapted from Gigerenzer (2000)

Table 8.3 Examples of intellectual foibles and heuristics

Intellectual foibles	
Ambiguity bias	Stay with what you do for lack of certainty
Dominance bias	Narrow scope, single mindedness, ignore facts and observations
Status quo bias	Stay with what is already the case
Recency bias	Unduly influenced by recent occurrences
Omission bias	Resisting action
Naturalistic bias	What is natural is right
Intertemporal bias	Inability to choose a reward and outcome
Heuristic	**Definition**
Recognition heuristic	If one of two alternatives is recognized, infer that one has the higher value on the criterion
Fluency heuristic	If one alternative is recognized faster than another, infer that it has the higher value on the criterion
Take the best	Infer which of two alternatives has the higher value by (a) searching through cues in order of validity, (b) stopping the search as soon as a cue discriminates, (c) choosing the alternatives this cue favors
Tallying (unit-weight linear model)	To estimate a criterion, do not estimate weights but simply count the number of favoring cues
Satisfying	Search through alternatives, and choose the first one that exceeds your aspiration leve

Adapted from Gigerenzer (2000)

events A and B, the probability of A given B can be computed from the probability of B given A, as well as the overall probabilities (known as the "prior probabilities") of A and B.

$$p(A|B) = \frac{p(B|A)p(A)}{p(B)}$$

Statistical reasoning, like many forms of cognitive adaptation, is not conscious. A critical issue in the logic of statistical reasoning is what constitutes support for a hypothesis. A vast array of ratios and likelihoods is required (Bayes and Price 1763). Bayes' theorem is one formal tool to scrutinize how frequency information is related to probability judgments in attempting to link relative frequencies and sample size. Bayes' theorem penalizes, amongst other things, unnecessary model parameters and thus encourages simplicity. We arrive at some ideas rather quirkily, a fact that Peirce linked to abduction (genesis of an idea) and instinctive responses, rapid and heuristic. There is an experimental sensibility that pervades human experience and human inquiry that are constants across cultures and throughout the world.

The predictive brain works in the organization of action and perception in everyday life events, and perhaps functions something like Bayes' theorem (Clarke 2013; Schulkin 2004), which is one mechanism amongst others used to give the epistemic value in tracking events in the organization of action. Importantly, the structural systems are inherent to the perception and action in the appraisals.

But, for many of us, there is no separation; there are just forms of information systems. And this is a piece of how we understanding neural function (Schulkin 2015). Indeed, diverse regions underlie predictive capability; they range from the neocortex to the subcortical regions (e.g., basal ganglia). The brain, perhaps to Holmes's delight, is readily tied to predictive inference, and in an adaptive context is drawn to the suitable best explanation in resolving problematic contexts that warrant resolution or attempt resolution. Good enough problem solving is the phrase, not perfectionistic. This is the stuff of adaptation and neural design (Schulkin 2004).

The evolution of problem solving reveals many species-specific capabilities. What we have in part are a diverse set of problem-solving specific capabilities, including problem solving about space, time and probability, as well as problem solving about language, forms of logic and algebra, people (faces and bodily posture), etc. Bounded rationality

(Simon 1962, 1996), as this has been called, just sets ecological or social capabilities. Naturalism and social evolution run continuously, not as a reduction of one to the other, something Holmes would appreciate, as it runs from Montesquieu through anthropology to biological discoveries.

Gigerenzer (2000, 2007) and his colleagues have shown that heuristics underlie decision making; they are bound by context. Of course, pre-adaptive capabilities are a feature of evolution, including that of human problem solving, which has expanded from initial adaptive use (Rozin 1976). The evolution of complexity noted by Rozin (1976, 1998) or Simon (1957) is just this expansion of problem-solving capabilities, no trivial feature in the context of law (Table 8.4).

Gigerenzer (2000) talks about an adaptive tool box, ecological or social boundary conditions, and the heuristics of good enough problem solving to fit the niche or design capabilities and solutions. Reason is demythologized. Indeed, we readily see movement and transitions as endemic to problem solving. Purposeful relations pervade the practice of inference and the practices of life.

They are less abstract, since they are embedded in the machinery of action itself. This holds whether the decisions or actions are fast and unreflective or more reflective and slower. Meaning is found in ingesting the larger body politic that we live in that fuels or sustains the meanings we derive in action. Meaning is not strictly in the head (Putnam 1990, 2014); it is spread across the worlds we live in, ways in which we communicate with one another (Heelan and Schulkin 1998).

The brain materializes into well-expressed habits, codified into the fabric of expectations of others. These are not abstract issues, but the common allure of the everyday. It is just that we know more about the brain, the codification in two senses, regions that orchestrate habits, and the social milieu in which it takes place. The science is still young, the knowledge base not what one would like. Still, knowing which parts of the brain underlie habits, statistical inferences and attention are important to know but only crudely understood.

Of course, these capabilities, and language being a critical one in addition to statistical capabilities, are in the context of our social evolution (Dunbar 1996). The point to realize in the consideration of cognitive capability is how much of it is rooted in action, something dear to pragmatists such as Dewey (Table 8.5).

Table 8.4   Table (patial) of heuristics

Less is more effects	The phenomenon of proportion dominance can result in people preferring less to more in a between-subjects design
1/N: Equality heuristic	Allocate resources equally to each N alternatives
Affect heuristic	Every stimulus evokes an affective evaluation that influences judgments, and that this evaluation can occur outside of awareness
Ambiguity avoidance	People avoid gambles with an unknown distribution of possible outcomes
Anchoring	Different starting points yield different judgments which are biased toward the initial values. Related to primacy effect (tendency to remember the first item)
Availability heuristic	Probability of an event by the ease with which instances or occurrences can be brought to mind
Fluency heuristic	If one alternative is recognized faster than another, infer that it has the higher value on the criterion
Prototype heuristic	A prototype is substituted for its category, but in which representativeness is not necessarily the heuristic attribute
Recognition heuristic	If one of two objects is recognized and the other is not, then infer that the recognized object has the higher value
Representative heuristic	Probabilities are evaluated by the degree to which A is representative of B. By the degree to which A resembles B
Simulation	A mental operation that is used to study the biases in the construction
Heuristic	Class of availability heuristic (constructing examples or scenarios vs. other mental operations used to study the availability heuristic such as recall or retrieval)
Take the best	Infer which of two alternatives has the higher value by (a) searching through cues in order of validity, (b) stopping the search as a soon as a cue discriminates, (c) choosing the alternatives this cue favors
Tallying (unit-weight linear model)	To estimate a criterion, do not estimate weights but simply count the number of favoring cues

Adapted from Gigerenzer (2000)

Even mathematical inference is built into the very fabric of cognitive everyday inference (Lakoff and Nunez 2000). Numbers, to be sure, are inherent in the everyday, in something like experience and nature, and the culture we are constructing (Table 8.6).

Table 8.5 Relationships that underlie cognition and action

Thinking as perceiving
Imaging as moving
Knowing as seeing and responding
Attempting insight as searching
Representing as doing
Becoming aware as noticing
Communicating as showing
Knowing from a "perspective"
Listening as detecting, knowing

Adapted from Lakoff and Johnson (1999)

Table 8.6 Cognitive inferences

Universality	Just as external objects tend to be the same for everyone, so basic mathematics is, by and large, the same across cultures. Two plus two is always four, regardless of culture
Precision	In the world of physical subitizable objects, two objects are two objects, not three or one. As an extension of this, given a sack of gold coins there is a precise answer to the question of how many there are in the sack
Consistency	For any given subject matter, the physical world as we normally experience it is consistent. A given book is not both on the desk and not on the desk at a given time
Stability	Basic physical facts-that is, particular occurrences at a given time and place- don't change. They are stable over time. If there was a book on your desk at 10 a.m. this morning, it will always be the case that on this day in history there was a book on your desk at 10 am
Generalizability	There are basic properties of trees that generalize to new trees we have never encountered, properties of birds that generalize to birds yet unborn, and so on
Discoverability	Facts about objects in the world can be discovered. If there is an apple on the tree in the backyard, you can discover that the apple is there

Adapted from Lakoff and Nunez (2000)

Mathematics is in everything we do, the way we are anchored to objects. Echoing Aristotle, Pico, the great Italian humanist, claimed "the science of numbering is supreme" (p. 52). Many before and after have held this view. Holmes, the empiricist, had some sort of sense of what

Dewey was driving at in *Experience and Nature*. Our being thrust into a world of adaptation and making sense of things—what I have called "foraging for coherence."

It is a feature of our experience to live amidst the endless disorder and patchwork in nature, in the law (Grey 1991; Rorty 1990a, b; Posner 2008), and, more generally, in human understanding. Disunity runs though biology (Dupre 1993); disunity runs through the law. This is very much the Holmesian experience. There is no forced integration.

Problem Solving, Inference, and Concern

Inference to a good enough explanation is the normative goal of a reflective brain. Good enough inference is surviving, achieving goals, given weighted values in the balance of life. Indeed, the reflexive machine has given way with regard to the economic considerations of inferences about weighted values to probabilities, and that is something that Holmes or Green and particularly Peirce wanted to see materalize.

What Holmes realized is that there is no overarching logic, which is why he emphasized in his early work the importance of human experience. It is no fall into utter irrationality in science or the law to realize that number logic is not a safeguard for the truth or warranted assertions. It is better than the alternative, something without logic—though I'm not sure what that would mean anyway.

Inferences and tracking them are part of what we mean by logic. The logic of practice pervades most features of life. Like law, there are contexts to draw inferences, rules, codes, statutes, etc. It is just that, like most of life, it floats on softer foundations than the phantasies one wants in life. Law is a reminder of the frailty of the fabric, but also its ultimate durability—like Belgian lace—in the contours of life. Law is also a reminder of the absurd and the stupid in life.

Our species is rooted in diverse forms of problem solving; problem-solving has its roots in primitive forms of adaptation, such as seeking safety or shelter, being with others, reduction of uncertainty under conditions of want and fear, etc.

What we call common sense is something that seems obvious, a set of problem solving or assessments. Common sense grows out of common contact with others. Whatever works well becomes integrated into everyday affairs. After all, common sense is a social feature based in part of the social affairs of everyday life. But common sense is wholly fallible. At various points in history (and sadly still today) around the world, it was "common sense" to commit what we know now to be horrible, inhumane acts of slavery, sterilization, and genocide. We have a natural inclination to attack and alienate what is other, to characterize them as different and, thus, as dangerous.

What should we care about? Well, cooperative behavior is a key factor in our evolution. Promoting cooperative behaviors is in our interest. Justice Hoffman (2004) insightfully noted that Holmes did not take seriously enough the social cooperative side of our evolutionary success. Inferences and statistical literacy are what Holmes saw as the future; and indeed he was right. In fact, statistical literacy is a normative goal for law and certainly for medicine (Gigerenzer 2000; Anderson and Schulkin 2014).

Law is one critical feature in our cultural evolution, or not. There is no panacea with regard to whether we will treat each other in a fair and equitable manner, whether we will minimize the pain and discomfort of others. Law figures in impositions of sanctions. Laws can keep some from doing harm, but not others. It is the percent of others that matter, and, of course, what we do about it as it reflects the larger culture as Holmes so well understood.

Naturalizing decision-making was a preference for Holmes. Though, Holmes eschewed the psychological and had a very limited notion, with his emphasis to eschew or eliminate diverse ways of thinking about what associated with internal events (willing, intending). This was and still is common. He was, however, oriented to the social for epistemic clarity and viability, and predictive events—like Peirce but unlike James. Pointing out the many forms of vulnerabilities in human decision making or the heuristics that guide decision making would have been something that Holmes might have appreciated. The emphasis in these traditions is about statistical inferences, what we do and do not get right, including in the law.

The brain is prepared to drawing rapid inferences; and we are prepared to correct our mistakes, at least in principle, and in cultures where it is emphasized. Pragmatist view of inquiry emphasizes the formation of habits guiding beliefs that are reliable, and that reflect a tendency to rely on statistical inference. The inference generating neural systems are pregnant across diverse regions and that are also tied to minimizing disturbance and excess neural expenditure.

Conclusion

Holmes was sanguine about human problem solving. Problem solving is at the heart of legal appraisals. What Dewey and Peirce emphasized was not a "blooming buzzing confusion" but adaptive capability. Sentiments run through rules in selective capabilities that underlie social cohesion. The empirical issue is how and when they manifest themselves.

Holmes's world is about reasoning, problem solving and surviving. Holmes understood the move from considerations of perfect reason with sound problem solving, good enough problem solving, formidable problem solving. Law in this sense is no different from neuroscience; some things are clear and others not. We search for ideas that are searchlights for discovery within tools for exploration and a common knowledge to build from (Simon 1996).

Practice and precedent are dominant in a common social space, a space for participating with others. Common sense is collective, and common law is collective. Both grow from an ability to problem solve that underlies what we call common sense. Common sense does not necessarily mean just and right. These sets of concepts are part of the fabric of our social milieu: common law becomes a feature of pragmatism based on worked out practices of value (Gordon 1992, 2017; Kelley 1989).

For Holmes, law was not about ethics (1897). While he was eager to separate law from a moral theory, he would also highlight the following in his "The Bar as a Profession": "A system of law at any time is the resultant of present needs and present notions of what is wise and right....and of rules handed from earlier states of society"

(Collected Papers, p. 156). The ideal of law, he would suggest, is that "when properly taught jurisprudence means simply the broadest generalization and the deepest analysis of the ideas at the bottom of an actual system" (p. 157).

Richard Posner, in a review of two books on Holmes and who perhaps comes very close to capturing the best sense of Holmes, notes that Jeremy Bentham was "the first great debunker of the idea of law as a sacred mystery… Holmes was the second and so far if the word is great is taken seriously the last." Holmes's remedies are modest. Posner like Holmes is a realist about expectations. There are three aspects about Holmes's intellectual legacy he notes that stand out:

1. Bad man theory of law;
2. Predictive theory of judge related law; and
3. An objective sense of criminal liability.

There are no mythologies here. There are other things as well, but rights talk and a theory of morality and about what matters except for debate and competitions and the continued influx of science. But the bad man theory is only a feature of us (Alschuler 2000).

Holmes gravitates in law in the mix of human adaptation, in a flux of experience (Margolis 2002). Holmes's ambiguity is the good enough clarity and the deep sense of the indeterminacy that surrounds human experience and adaptation. The stuff of experience is often inter-determinate. We preserve, adapt, cope, forage for forms of coherence and viability.

Holmes understood the problem-solving rootedness in the indeterminate, a key feature since problem solving and his philosophy of law reflected that fact (cf., Kellogg 2007, 2017; Leiter 2007). Holmes is a naturalist (White 1947) and like other pragmatists taken by the burgeoning theories in evolution (Wiener 1949). Holmes is biologically oriented under the guise of his mentor Chauncey Wright, not an ideologue like Spencer. Indeed, both thinkers lamented Spencer.

Between biology and history, there is a rich tapestry of human expression that underlies the law; the continuity of nature and culture are seamless and the boundaries porous. Such distinctions were diluted and are diluted in the vernacular of everyday reasoning and life, the stuff the pragmatists were interested in understanding. The same in the law and what we presuppose matters in the logical inferences that we impose. Holmes was part of the larger "revolt against formalism" (White 1947). The emphasis on experience is about the transaction in battling for knowledge—battle scarred and weary but still alive.

Holmes was neither a conservative nor a twentieth century liberal or progressive. He grew up in a rich ambiance of intellectual conversation. What many have noticed is that "Holmes's conception of experience is the key to understanding almost everything that is distinctive about his view of the law" (Menand 2002, p. 38). That statement might be a tad strong. But Holmes emphasized the vulnerability of what judges actually do, the way they decide.

Moreover, like Peirce, Holmes emphasized the social or the collective sensibility more than James, the personal and the individual. Risk and negligence, retribution and punishment reflect the larger collective assembled thought—what Peirce would think of as settled thought within a "community of inquirers " (Wells-Hantzis 1988). "The aim of the law," Holmes suggests to us, "is not to punish sins, but to prevent certain external results" (*Commonwealth vs Kennedy*, Massachusetts Supreme Court, 1897). He is always looking for an external anchor to lean on (Wells-Hantzis 1988; Rosenberg 1995).

There is just one important difference: Peirce believed in the advance of knowledge through inquiry and the community of inquiry. Whereas Holmes always seemed less sanguine about all possibilities of advance (Wells-Hantzis 1988). One, though, had the brain scars of a war (Rogat 1964; Grey 1989); the other avoided the war (except the self-destructive scars of personal and professional failure [see Brent 1993]). Of course, diverse regions of the brain underlie the remembered risk, including the regions of the neocortex, limbic regions, such as the amygdala, and a range of circuitry vital for animal/human survival (Reyna and Zayas 2014).

Continued use and common satisfaction underlie this sense of legal inquiry and jurisprudence. Demythologized reason is practical, statistical, and reliable in settling disputes. But it is not a dispute about individuals as such. It is across individuals into what binds collectively.

Experience is social for Holmes, not unlike George Herbert Mead. But it is wartime experience that appears the most social for Holmes, not the collective sense of others across many other domains of human contact.

Of course, this will not do, not in the hands of Peirce or Holmes. Individuals, as James understood, are essential for understanding human experience. These two interesting men undermined this fact in their attempt to make all things objective and outside individual minds, not unlike in many regards the later Wittgenstein (1953).

References

Alschuler, A. W. (2000). *Law Without Values: The Life, Work and Legacy of Justice Holmes*. Chicago: University of Chicago Press.

Anderson, B. L., & Schulkin, J. (Eds.). (2014). *Numerical Reasoning in Judgments and Decision Making About Health*. Cambridge: Cambridge University Press.

Austin, J. (1832). *The Province of Jurisprudence Determined*. London: John Murray.

Bayes, T., & Price, R. (1763). An Essay Towards Solving the Problem in the Doctrine of Chance. *Philosophy of the Royal Society, 43*, 370–418.

Boyd, R. (1993). Metaphor and Theory Change. In A. Ortony (Ed.), *Metaphor and Thought*. Cambridge: Cambridge University Press.

Brent, J. (1993). *Charles Sanders Peirce*. Bloomington: Indiana University Press.

Butler, B. E. (2017). *The Democratic Constitution*. Chicago: University of Chicago Press.

Cherniak, C. (1986). *Minimal Rationality*. MIT: Cambridge University Press.

Clark, A. (2013). Whatever Next? Predictive Brains, Situated Agents and the Future of Cognitive Science. *Behavioral and Brain Sciences, 36*(3), 181–204.

Clark, A. (2017). *Surfing Uncertainty*. Oxford: Oxford University Press.

Dahan-Katz, L. (2013). The Implications of Heuristics and Biases: Research on Moral and Legal Responsibility: A Case Against the Reasonable Person Standard. In N. A. Vincent (Ed.), *Neuroscience and Legal Responsibility*. Oxford: Oxford University Press.

Dewey, J. (1910/1965). *The Influence of Darwin on Philosophy*. Bloomington: Indiana University Press.

Dewey, J. (1925/1989). *Experience and Nature*. LaSalle, IL: Open Court Press.

Dunbar, R. I. M. (1996). *Grooming, Gossip and the Evolution of Language*. Cambridge, MA: Harvard University Press.

Dupre, J. (1993). *The Disorder of Things*. Cambridge: Harvard University Press.

Epstein, R. A. (1998). Pennsylvania Coal vs Mahon: The Erratic Taking Jurisprudence of Justice Holmes. *Georgetown Law Review, 86*, 875–905.

Epstein, R. A. (2000). *Cases and Material on Torts*. New York: Wolters Kluwer.

Friedman, L. M. (1993). *Crime and Punishment in American History*. New York: Basic Books.

Gigerenzer, G. (2000). *Adaptive Thinking, Rationality in the Real World*. New York: Oxford University Press.

Gigerenzer, G. (2007). *Gut Feelings*. New York: Viking Press.

Gordon, R. W. (1982). Holmes's Common Law as Legal and Social Science. *Hofstra Law Review, 10*, 719–740.

Gordon, R. W. (Ed.). (1992). *The Legacy of Oliver Wendell Holmes Jr*. Palo Alto: Stanford University Press.

Gordon, R. W. (2017). *Taming the Past: Essays on Law in History and History in Law*. Cambridge: Cambridge University Press.

Grey, T. C. (1989). Holmes and Legal Pragmatism. *Stanford Law Review, 41*, 787–870.

Grey, T. C. (1991). What Good Is Legal Pragmatism. In M. Brint & W. Weaver (Eds.), *Pragmatism in Law and Society*. Boulder, CO: Westview Press.

Grey, T. C. (1992). Holmes, Pragmatism and Democracy. *Oregon Law, 71*, 521–542.

Hart, H. L. A. (1961). *The Concept of Law*. Oxford: Clarendon.

Hart, H. L. A. (1968). *Punishment and Responsibility*. Oxford: Oxford University Press.

Heelan, P. A., & Schulkin, J. (1998). Hermeneutical Philosophy and Pragmatism: A Philosophy of Science. *Synthese, 115*, 269–302.

Hoffman, M. B. (2004). The Neuroeconomic Path of the Law. *Philosophical Transactions of the Royal Society of London. Series B, 359*, 1667–1676.

Holmes, O. W., Jr. (1881/1952). *The Common Law*. New York: Dover.

Holmes, O. W., Jr. (1897). The Path of the Law. *Harvard Law Review, 10*(8), 45.

Holmes, O. W., Jr. (1914). *Gompers v. United States*.

Holmes, O. W., Jr. (1917). *Southern Pacific Co. v. Jensen*.

Johnson, M. (1993). *Moral Imagination*. Chicago: University of Chicago Press.

Johnson, M. (2007). Mind, Metaphor, Law. *Mercer Law Review, 58*, 845–868.

Kahneman, D., Slovic, P., & Tversky, A. (1982). *Judgment Under Uncertainty: Heuristics and Biases*. Cambridge, UK: Cambridge University Press.

Kelley, P. J. (1989–1990). Was Holmes a Pragmatist? Reflections on a New Twist to an Old Argument. *Southern Illinois University Law Review, 14*, 427–467.

Kellogg, F. R. (2007). *Oliver Wendell Holmes: The Legal Theory as Judicial Restraint*. Cambridge: Cambridge University Press.

Kellogg, F. R. (2017). Take the Trolley Problem… Please! Pragmatism, Moral Particularism and the Continuum of Normative Inquiry. *Contemporary Pragmatism, 12*, 8–18.

Lakoff, G., & Johnson, M. (1999). *Philosophy in the Flesh*. New York: Basic Books.

Lakoff, G., & Nunez, R. E. (2000). *Where Mathematics Comes From*. New York: Basic Books.

LaPiana, W. P. (1994). *Logic and Experience*. Oxford: Oxford University Press.

Leiter, B. (2007). *Naturalizing Jurisprudence*. Oxford: Oxford University Press.

Margolis, J. (2002). *Reinventing Pragmatism*. Ithaca: Cornell University Press.

Menand, L. (2002). *American Studies*. New York: Farrar, Straus and Giroux.

McDermott, J. J. (2007). *The Drama of Possibility: Experience as Philosophy of Culture*. Bronx: Fordham University Press.

Posner, R. A. (2008). *How Judges Think*. Cambridge: Harvard University Press.

Putnam, H. (1990). *Realism with a Human Face*. Cambridge: Harvard University Press.

Putnam, H. (2014). *Philosophy in the Age of Science*. Cambridge: Harvard University Press.

Reyna, V. F., & Zayas, V. (2014). *Neuroscience of Risky Decision Making*. Washington, DC: American Psychological Association.

Rogat, Y. (1964). The Judge as Spectator. *University of Chicago Law Review, 31*, 213–256.

Rorty, R. (1990a). *Philosophy and Social Hope*. New York: Penguin.

Rorty, R. (1990b). The Banality of Pragmatism and the Poetry of Justice. *Southern California Law Review, 63*, 1811–1819.

Rosenberg, D. (1995). *The Hidden Holmes*. Cambridge: Harvard University Press.

Rozin, P. (1976). The Evolution of Intelligence and Access to the Cognitive Unconscious. In J. Sprague & A. N. Epstein (Eds.), *Progress in Psychobiology and Physiological Psychology*. New York: Academic Press.

Rozin, P. (1998). Evolution and Development of Brains and Cultures: Some Basic Principles and Interactions. In M. S. Gazzaniga & J. S. Altman (Eds.), *Brain and Mind: Evolutionary Perspectives*. Strasbourg: Human Frontiers Science Program.

Schulkin, J. (2004). *Bodily Sensibility: Intelligent Action*. Oxford: Oxford University Press.

Schulkin, J. (2015). *Pragmatism and the Search for Coherence in Neuroscience*. London: Palgrave Macmillan.

Simon, H. A. (1957). *Models of Man, Social and Rational: Mathematical Essays on Rational Human Behavior in a Social Setting*. New York: Wiley.

Simon, H. A. (1962). The Architecture of Complexity. *Proceedings of the American Philosophical Society, 106,* 470–473.

Simon, H. A. (1996). *The Sciences of the Artificial*. Cambridge: MIT Press.

Stone, M. (2002). Formalism. In J. Coleman & S. Shapiro (Eds.), *Jurisprudence and the Philosophy of Law*. Oxford: Oxford University Press.

Sunstein, C. R. (2005). *Laws of Fear: Beyond the Precautionary Principle*. Cambridge: Cambridge University Press.

Toulmin, S. (1950). *Reason in Ethics*. Cambridge: Cambridge University Press.

Toulmin, S. (1958). *The Uses of Argument*. Cambridge: Cambridge University Press.

Toulmin, S. (1972). *Human Understanding: The Collective Use and Evolution of Concepts*. Princeton: Princeton University Press.

Treanor, W. M. (1998). Jam for Justice Holmes. *Georgetown Law Review, 86,* 813–874.

Wells-Hantzis, C. (1988). Legal Innovation Within the Wider Intellectual Tradition: The Pragmatism of Oliver Wendell Holmes Jr. *Northwestern Law Review, 82,* 541–595.

White, M. (1947). *Social Thought in America: The Revolt Against Formalism*. Boston: Beacon Press.

Wiener, P. P. (1949). *Evolution and the Foundations of Pragmatism*. Cambridge: Harvard University Press.

Winter, S. L. (2001). *A Clearing in the Forrest: Law, Life and Mind*. Chicago: University of Chicago Press.

Wittgenstein, L. (1922). *Tractatus Logico-Philosophicus*. London: Routledge and Kegan Paul.

Wittgenstein, L. (1953/1958). *Philosophical Investigations*. New York: Macmillan.

9

Ethics, Body Politic, and Neuroscience

Law is a reflection of continuity with both our evolutionary past and our cultural history. As a reflection of the larger body politic, law is a little too much for a simple consideration of the brain. As Holmes noted, "the law is the witness and external deposit of our moral life. Its history is the history of the moral development of the race" (*The Path of the Law*, p. 170).

Dewey (1893), commenting on Holmes, would say, "on the legal side I shall draw my material almost entirely from a book of equal interest to the student of law in its professional phases of history, of anthropology, of ethics – I refer to The Common Law of Oliver Wendell Holmes Jr." (Flower and Murphy 1977).

It is in the Common Law that Holmes pontificated that "the great body of the law is found in the principles of liability." For Holmes, the principles of liability were tied to basic human social origins, basic human contact, and virtual connectivity between anthropologists and psychologists (Kagan 2002). Holmes's Common Law is within the larger culture.

© The Author(s) 2019
J. Schulkin, *Oliver Wendell Holmes Jr., Pragmatism and Neuroscience*,
https://doi.org/10.1007/978-3-030-23100-2_9

The juxtaposition of the individual and the social is diluted by pragmatists. Certainly, Dewey and Holmes viewed purposes and ends as continuous with adjudication methods, and clarity in both as an imperative for intelligent action.

Moral judgment is an expression of the brain; diverse regions of the brain underlie ethical assessment (Moreno and Schulkin 2020). Mostly, these are also tied to social assessment. Prosocial capabilities, something Dewey emphasized, is what is fundamental probably for moral sentiments, moral judgments and socially cooperative behaviors. I have suggested, like others, that there is no special region of the brain for morality (Schulkin 2015). There is actually a variety of neural capabilities expressed in suitable social contexts. Dominant preconditions lie in the diverse regions of the brain that underlie social contact.

Prodding morality in individuals starts early; what are the moral conditions required to promote character? The issue is empirical, and not just conceptual; what do we value in a moral individual? Human moral behavior takes place against the backdrop of almost everything we do. Understanding who we are is in the backdrop of the life world in which we participate (Thompson 2015), in which we have meaning and transactions with one another.

This chapter is about law, ethics, and neuroscience. Consistent with the pragmatism that Holmes might have been comfortable with, as we are social animals. Ethical judgment is pervasive and pervades the social milieu. But there are no special parts of the brain devoted to just ethics, or the law (Moll and Schulkin 2009). There are however, diverse regions of the brain that underlie pro-social behaviors which underlie ethical judgment, and considerations of the law.

Ethics, Brain, and Body Politics

Holmes went to great lengths to separate law from diverse ethical traditions, particularly natural law (Horwitz 1992). Political or legal positivism is what we were comfortable with during periods of his long life. But still for him when something is wrong, it is wrong. While I don't think there is a special area in the brain exclusively for ethics, I also don't think there is for law or for that matter music (Schulkin 2013). There are regions of the brain tied to reasoning and what many of us would consider roughly the right thing, given the context and the social milieu.

The brain that underlies ethical judgment rests in the larger subset of social discernment. Key regions are the temporal lobe and amygdala, regions tied to the meaning of social events. Prosocial capability as Darwin noted and Holmes probably understood underlies our cultural evolution, of which the law is an essential feature in providing protection from danger of self and property.

Holmes (1897) is not anti-moral as some critics have suggested. He is trying to clarify when something is moral and when something is strictly legal. He is worried about the confusion of legal reasoning and moral stance; for they are not the same. He is not abandoning ethics, but housing ethics in the larger body politic.

Both Dewey and Holmes held out for a morality without foundations (Johnson 2014). For demythologized moral appraisals that permeate in a community in which science and the human condition are not separated or reduced to one another but continuous with one another. It is rich in a sense of human fallibility, a feature of our historical record and our existential expressions. As pragmatists they understood that moral codes are embedded in social practice, in development, and in social policy. The brain may not have a separate region for ethical assessment, but valuation is a basic property of diverse regions of the brain, in which ethical considerations are a feature of valuation.

Issues in neuroethics (Levy 2007, 2011) overlap with traditional issues in the philosophy of mind (Roskies and Morse 2013). Indeed, information processing requires that informational content is beyond the material encoded in the head; we indeed abbreviate informational content as a neural adaptation essential for problem solving. We know a little about, at least, the core regions of the brain that are tied to diverse forms of problem solving. Law is another problem-solving vehicle for us, an unbelievably important one.

Of course, there are changes in experience, changes in mind; neuroethics also deals with that (Levy 2007; Moreno and Farah 2012). But that is nothing special to ethics and the changes will not be specific,

I do not think, at least at this time, to specific features of an ethical circuit. It will reflect a variety of events, and we want whatever insight comes from understanding neural events, or the enactment of events or experience. Ethics, of course, is tied to this, so are most kinds of events, such as issues about causation, symptoms vs causes, memory, resilience, issues of the self, choice, responsibility, deception (Levy 2007, 2011). But these are both related to ethics and broader than ethics. These issues are at the heart of what we are and the way in which we understand the brain (Racine 2010).

As much as we are ever going to know about the brain is not going to tell us what is worth pursuing because it leads to the good life, a moral life (Gert 2012). Neuroscience does not replace ethics, law, economics; it just contributes to the discussion. Talk of foundations is replaced with usefulness, at least, this is the pragmatist perspective (Illes and Bird 2006).

Interestingly and not surprisingly, neuroethics also entail something of bioethics (Illes et al. 2010). While ethics and the law are separate, they obviously intersect across human judgment. They do so palpably in neuroscience, for example, in drug enhancement of performance, enhancing memory, attention, endurance (Farah 2005). Ethics as an account of moral behavior runs through the synaptic connectivity of the human experience. Law is one fundamental feature in that binding behavior, holding obligation against the backdrop of what Erich Fromm called the "Sane Society." Our ethics and our laws are depicted in our social institutions, something, again, that Holmes understood.

Jonathan Moreno (1999, 2003) has suggested that bioethics is a feature of naturalism, a feature of problem solving, coping with problems, solving problems within the larger body politic. Bioethics is placed in something close to cooperative critical reflection and inquiry, a fundamental feature of a pragmatist point of view (Seigfried 1996). But the law is endlessly adverserial; perhaps too much as Judge Posner suggests (2008) but also exaggerates. Posner view is narrow, Holmes was narrow, but held out the possibility of a Deweyian view about Experience and Nature.

Holmes separated law from ethics and further separated law into methodological maneuvers within a larger body politic of insight, coherence, and stability. He understood the endless interactions where ideas battle for coherence; the larger culture for Holmes is less the absolute separation of facts from values (Weber 1903/1949) than a social science separate from moral impositions, or being clear where moral notions and the law collide. But he is not consistent.

Pragmatists like Dewey (1939) and many others (Neville 1974; Putnam 2014; Kitcher 2014) understand that all pervasive sense that appraisals of every kind are laden with valuation, and so the separation of fact and value is less absolute and much more permeable. Of course, appraisal systems are facts and they value events. They provide ethics and ethics in habits for most of human social contact and performance.

Moral Inclinations

Ethics turns to law when social and moral control is not enough or is dangerous in itself. We do not impose on each other in a participatory democracy legally when it causes harm and goes beyond the bounds of reason, as the bounds of reason is a rather abstract concept. Metaphor sizes morals down to human proportions (Johnson 2014), something Holmes appreciated. Indeed, metaphor is embedded in thought, in action, in understanding. The question is about the constraints and the empirical link in discovery, understanding, and capturing the phenomenon at hand.

Moral imagination is part of imagining others; getting to think about others, their experiences, why they do what they do. Such imagination is critical in understanding others, sharing experiences. Metaphors are aids to thinking and underlie most events in our lives. They underlie what is right and what is just. They underlie inquiry and the diverse cognitive systems in adapting to events, in understanding events, in extending one event with another.

The issue is not getting overwhelmed; we limit the moral context of experience. It is difficult to deal with too much information. Cognitive dissonance is an adaptation to limit exposure and thought; it is also an easy aberration. I am robbed from considering you and what I might owe you (Scanlon 1998, 2014).

To the contrary of Kant or Plato, the spoils of bodily contact are vital for the enhancement of cognitive events. Bodily contact leads to the expansion of horizons into the wider moral space of interacting with others and understanding moral reproach for offense and moral attributions that underlie envy, procreation, betrayal, and loyalty (Sabini and Silver 1982). The moral fabric is facing the facts of everyday moral transgressions without being demoralized, beaten down by plain endless setbacks.

Adam Smith in his great book on *The Theory of the Moral Sentiments* suggested that there were sets of inclinations that move human beings in diverse ways toward morally relevant behaviors. Depending upon who writes it, however, the list can vary. But almost always it includes the following items (Table 9.1).

Sympathy and empathy are easy to understand. Fairness, of course, means a number of things, including equal access. The list expands into self-restraint, reliability, and the more traditional social virtues identified with stoicism and duty (Wilson 1993).

Duty is one of the moral sentiments, a sentiment close to the heart of Holmes. Smith went on to capture the larger culture in which participation, innovation, and self-initiative within minimal constraints are at the heart of human development. His context was partly stoical, with the language of self-control, moderation, and balanced appetite. The evolution of virtues for stoics was essential to human happiness. Benevolence is a natural outcome of the right kind of balance in one's life. These are broadly construed and embody the thoughts of Smith and Hume, as well as Bentham and Mill. Indeed, the balance, of course, is a prized goal of morality and well-being across diverse cultures (Stoicism, Taoism).

Table 9.1 A partial list of moral sentiments

Sympathy/empathy
Fairness
Self-regulation
Discipline
Duty
Loyalty

Adam Smith and others, such as Darwin, also asserted that such moral sentiments underlie the evolution of social behaviors, including the solidification of the group. The consideration of the greatest good as a rule might be a consideration of doing the greatest good from sentiments about sympathy towards others. Empathy is perhaps a greater expansion of sympathy (Churchland 2011).

Moral disgust is perhaps easily rooted in transgressions towards others cheating, enforced by visceral disgust. Of course, this muddies the waters of moral purity. But what purity? The sentiments are not pure; they are culturally housed in a social climate. Disgust, a primordial appraisal tied to gastrointestinal distress (Rozin 1976, 1998), is superimposed in our cultural evolution or devolution. It is a very powerful moral sentiment towards bad behavior. While some may believe that it is tied to moral norms or natural moral sentiments, I think it is more opportunistic and more neutral; disgust is just a mechanism for withdrawal and condemnation (Haidt 2007; Nussbaum 2004).

Indeed, we now know the regions of the brain that are linked to disgust. Disgust emanating from foods is, to be sure, its primary motivator. These sources of disgust are embedded in moral disgust. They include regions of the neocortex, basal ganglia, limbic regions such as the amygdala, hypothalamus, and brainstem regions such as the parabrachial region. Indeed, looking at spoiled or rotten food elicits a similar pattern of neural activity as responses to the taste of the food (Phillips et al. 1997).

Disgust reactions are ancient responses, initially tied to food resources, tagging others about what they like or not. Disgust is a key moral sentiment. Moral sentiments run through moral rules like sensations run through cognitive orientation. Some are closer to the sensory side and some closer to the cognitive side; but there is no radical separation. What we have are orientations to events. Behavioral response is embedded in the larger culture and biologically endowed with capabilities, one of which is the moral sentiment we call sympathy (Nussbaum 1997).

But we are always vulnerable for getting things wrong, morally or other wise (Hirstein et al. 2005). Of course, Holmes understood. Our endless vulnerability palpable, our mistakes transparent. We don't need neuroscience, as Holmes would probably say to know this sort of thing. But know we have a science about human decision making, our vulnerability to mistakes of diverse sorts and the neural systems that figure in the decision making.

Of course, we easily endow our visceral responses with notions of who is one of us and who is not—those are the ones that are putrid, who foul the scent of the air with their moral transgressions. Moral disgust allows us to kill more easily and indeed to commit endless expressions of barbarity towards others (Jews, Gypsies, Kurds). We are repulsed so easily by their sight; we exterminate theologically, ideologically, racially, without constraints. These are built in brain/body responses.

Darwin himself, who like Smith thought that diverse moral sentiments underlie our social contact, described facial and bodily expressions linked to disgust. Their origins may emerge from basic revulsion to roasting food. Visceral sickness is one dominant result. The belongingness of the food with the visceral response is primary learning, important and long lasting (Rozin 1976). People come to be repulsive or attractive, through what others like or do not like, and by what comes to make them viscerally disturbing.

Revulsion to what is morally repugnant is neutral. It has endlessly been abused, but it is a basic and primary feature of approach and avoidance behaviors. The neural systems that traverse the brain, from forebrain to brainstem, including primary taste/visceral cranial nerves (7th, 9th and 10th nerves), underlie this revulsion (Norgren 1995).

Disgust also plays a central role in social aversion. Distaste, nausea, and vomiting occur following exposure to potentially toxic or contaminated foods. Odors have a clear adaptive function. In humans, disgust and its close relative, contempt, play a clear role in interpersonal settings (Jones 2004). In contrast to anger, disgust and contempt are slower to fade out; they tend to "stick" or to become a property of the object, intensely devaluing it (Rozin 1998). Thus, in the same way that neural systems underlying primitive forms of pleasure and social bonding operate in highly complex social situations associated with human cooperation, neural systems underlying aversive responses related to physical properties of odors and foods seem to have been adapted to sustain social disapproval.

Indeed, while morality often promotes cooperation and helping, it can also steer hostility among individuals and social groups (Moll et al. 2005; Moll and Tomasello 2007; Jones 2004). The close affiliations between disgust, detection, and experience might help explain the similarities among the neural substrates of disgust and perception (Phillips et al. 1997).

Brain regions involved with basic forms of disgust and with emotionally mediated social disapproval, such as moral disgust, appear to be largely shared (Moll and Schulkin 2009). Accordingly, decisions to oppose charities linked to societal causes, whether at a personal cost or at no cost, were associated with activity in the lateral OFC and anterior insula, in keeping with the suggestion that basic aversive mechanisms have been adapted to enable forms of disapproval (Moll et al. 2006), which is at the heart of what Holmes called "primitive law" in the Common Law (e.g., vengeance).

Appraisal runs through the nervous system activation. It is the breathing space of judgment. Appraisals of moral disgust or worthiness are basic features of our assessment tied to a phylogenetic and vibrant central nervous systeml. When coupled to law and a cultural evolution of protecting the most vulnerable as a moral imperative or normative goal, it serves one cultural view: pragmatist view of democracy, participation, and human possibilities (Nussbaum 1997, 2004). Dewey's philosophy of law and his pragmatism embrace such possibilities and their social origins, notably in the development of warranted practices and customs, a common and shared social space (Westbrook 1993; Butler 2010).

Moral Actions

Moral sentiments are intrinsically linked to daily social interactions. Anticipated or actual violations of one's own principles and beliefs trigger aversive feelings such as guilt and shame (Moll and Schulkin 2009). Standing up for one's core values, however, will tend to trigger positive feelings such as pride and joy. Moral sentiments are thus strong motivators for human action in social contexts.

Social pressure capitalizes on guilt-ridden propensities, something Nietzsche rallied against and Holmes seemed to consume. Making sense of guilty minds is not easy but is pursued (Shen et al. 2011; Hoffman 2014) and even appreciated; consider, for example, Judge Hoffman of Colorado.

Neuroscience provides some insight into this (Green 2008), but nothing overwhelming for a court of law. Nonetheless, evidence with some statistical accuracy of a tendency to sort individuals into categories of purposeful and negligent harms (Shen et al. 2011). Such appraisal systems depend on the engagement of several cognitive processes, including action and conceptual knowledge, emotion and motivation, requiring a tight integration among human neocortical and limbic circuits (Moll and Schulkin 2009).

As such, moral sentiments should not be considered to be either purely cognitive or emotional, nor a simple sum of "cognition" and "emotion." It has been proposed that specific moral sentiments will be elicited depending on the precise recruitment of component neural and psychological events (Moll et al. 2005, 2006), which include: basic forms of affiliation, anger, anxiety, hedonic states, agency, intentionality, and prospective thinking and outcome prediction.

This "prosocial" cluster includes guilt, embarrassment, compassion, and gratitude, which promote cooperation, helping, reciprocity, reparative actions, and social conformity. A subclass of those, the so-called empathic moral sentiments (guilt, gratitude, and compassion), putatively share the attachment component, and play a central role in behaviors linked to empathy (Nussbaum 2004; Moll and Schulkin 2009).

Moral sentiments underlie behavior are highly motivational. Darwin believed that moral sentiments are pro-social and are fundamental for moral development (Kagan 1984). Indeed, as Judge Hoffman has articulated in his book *The Punishers' Brain*, we come prepared to attribute blame, to distribute harm in the context of group formation. As we are "predisposed to cooperate" as the judge reminds us, we are also "born cheaters" (p. 6). Truth is all too often faint and in competition with other motivational states.

However, sentiments linked to interpersonal aversion—the other-critical sentiments (disgust, contempt, and anger/indignation)—are experienced when others violate norms or one's rights, and endorse aggression, punishment, group dissolution, and social reorganization (Moll and Schulkin 2009). Still, acting in accord with one's own values often triggers self-praising sentiments—among which pride is the prototype, while witnessing praiseworthy actions of others will lead to the experience of other-praising sentiments—gratitude when one is the recipient of such actions, and admiration when praiseworthy actions are directed to a third-party.

Indeed, moral theory tied to practical outcomes is one of the ideals of a working democracy in which economic adjudication is and should be a major factor in the law (Posner 1992)—where statistics, as Holmes envisioned, functions underlying epistemic inferences, but where, I submit, moral theory is still the valuable and arguable end (Nussbaum 2000). Frankly, I don't know really what it would be like to discard moral theory, despite various forms of limitations.

Indeed, moral theorists such as Dewey noted the need to be rooted in understanding human psychology and neuroscience (Greene 2014). One view about ethical appraisal is a fast set of appraisals and a slower more reflective set. Such dual factors systems are represented across human decision making (Kahneman 2011). Another way to characterize these dual systems is that one is more automatic than the other; the other more reflective. One involves settled habits, as Peirce or Dewey might say, orchestrated by basal ganglia in the brain. Recruitment, not surprisingly when measuring consequences and being less reflexive, is compromised with frontal cortical damage (Greene 2014).

Empathy might require more, like who is this person to whom I am sympathetic? What do I know about her? What would it mean for me to understand this individual? Why would that matter anyway? Empathy is the larger consideration, at least for me, for the wider context; it entails more understanding.

The moral sentiments also entail duty—duty towards others, towards what we could owe each other under certain conditions. Moral sentiments for Adam Smith were active responses that help bind us together.

What matters is that we have evolved something Smith and many others have identified as social instincts towards the plight of others. It is not a particularly strong motivation in diverse contexts that matter, as we can all attest; but it is there. I, at least, think so.

Engaging the person, the other, in our calculus of magnifying goods, opens access in more impartial ways. Providing opportunities is a good moral rule tied to sentiments that matter. Conflict between the personal and the impersonal is the stuff of life (Nagel 1991). It is part of what we owe each other to consider others, under a moral fabric (Scanlon 1998). The law as Dewey understood it, helps to promote this (Butler 2010).

This conflict is the life-blood of social space with other conflicts. It is part of being an adult; namely not to deny conflicts between the personal and the impersonal. But also our interests lie with others. It is in our social nature. It happens in the context of both war and peace. We self-sacrifice or at least as Nagel (1970) put it, we consider others without larger based "ulterior motives."

Our biology includes others; human meaning is anchored to others. Solipsism is an aberration. Idiocy, as the Greeks understood, is the loss of a world in which to be, in which to participate. Social attachment is a prime feature of the human condition. But so are, as Rousseau noted, solitary walks and the sense of wonder in the universe in solitude and awe.

But it is a matter of degree between me and others; the range of behaviors are vast in expression and culturally expressed. Ethics is part of the empirical sciences. What we know is the interdeterminacy of expression and is reflective of the larger social and historical architecture.

An evolving constitution with formative and foundational principles exists within a larger body of the competition of ideas and ethics, or set of ethical inclinations. Ethics is the study of moral judgment or behavior. Moral sentiments and moral rules are embedded in this class of events.

For some thinkers who eschew what G. E. Moore called "that naturalistic fallacy," nothing particularly is revelatory in our nature, in our metaphysics or rules or sentiments. The sense of the good is just that: it is. As Moore put it, "the good is a simple notion, just as yellow is a simple notion" (*Principia*, p. 7). Wittgenstein later critiques Moore in the *Investigations* and elsewhere, belittling this demonstration as no demonstration at all.

Moore was out to dismiss and dispel any influence of what he called "the modern vogue of evolution," which as he put it "is chiefly owing to Darwin's investigation as to the origins of the species" (p. 47). The naturalistic fallacy is anchored to a distaste for the naturalistic turn in all matters, and indeed crude forms of reductionism have always appeared. When biological capabilities are just factored in the context of the cultural milieu, there is less of a reduction to bees or abstract genes and selfish motivations. That is not to say that reproductive motives are not ultimate causes along with more proximal causes (Tinbergen 1951/1969) in the expression of behavioral options and the competition of expression in motivation. But morals and the study of ethics go far beyond that, though biology always matters and the separation of ethics from the larger categories of human biology is to be eschewed. These are continuous events.

Empathy is a state in part in which we can place ourselves close to the other, what the other might experience. Sometimes that is the very thing we do not want to do; it makes it harder to carry out nefarious deeds against others. Demonizing others is a common adaptation and may be necessary for all sorts of barbaric behaviors. ISIS is perhaps the latest example, as they cut off the heads of the infidels.

A sense of fairness, concern for others, and observance of cultural norms permeates human social existence (Smith 1759/1974; Kant 1789/1997; Sabini and Schulkin 1994). This social sensibility is the essence of human morality, which are shaped through cultural exposure. Morality is thus a product of our cultural and biologic evolutionary history and represents an important adaptive element for social cohesion and cooperation (Darwin 1871/1982).

Prosocial behaviors are at the heart of morality; what we do for each other, owe each other, how we get along with each other. But grounding morality in evolution (Joyce 2000) is not the same as reducing morality to evolution; it is just the utter continuity of both our biological and cultural evolution, which underlies certainly Holmes's understanding of these events.

The allure of social affiliation Holmes ironically found in combat and war. A natural social affiliation facilitated by common bounds however, as he knew, is facilitated in diverse social contexts. Zone morality (Weissman 2014) involves diverse contests in which moral relationships conflict. They are pervasive across human expression and are about as omnipresent and necessary as breathing in our social evolution.

Naturalistic ethics ties social contact or prosocial behaviors to the basic forms of innate and learned behaviors. We come prepared to innately organize social contact as a fundamental preadaptation and condition along with diverse forms of social and moral learning. The degree to which we learn moral codes essential for functioning is empirical. Moreover, the interaction of moral rules that are learned or easily expressed such that one might consider them innate is also empirical. I think what it is not is an utterly abstract set of rules, analogous to Chomsky's syntax; morals are grounded in human contact, action, and pedagogy (Johnson 2014).

Linking moral intuitions, namely fast and immediate with more deliberate reasoning, is something that develops and, indeed, underlies most forms of human reasoning (Haidt 2007). A combination of exploration and inhibition underlies moral judgment as it does most forms of human social reasoning. This is something Holmes understood. What Dewey would have celebrated were the prosocial features.

Neuroscience of Prosocial Sentiments

The neuroscience of moral judgment goes directly to the frontal cortex. Individuals with prefrontal damage are vulnerable to autonomic diminution. Such diminution is thought to increase the susceptibility for sociopathic behavior (Damasio 2007). One example is decreased activation of skin conduction and visceral responses to facial and bodily expressions. Interest in this issue has been reawakened more recently by systematic studies of acquired personality changes due to brain damage, mostly to the frontal cortex (Damasio 1994). Of course, this large region encompasses a great bulk of the neocortex. But importantly brain imaging studies have pulled a part, regions of the brain tied to alarm and autonomic responses (e.g., amygdala) from moral judgment and more cortical regions of the brain (ventromedial prefrontal cortex, Shenhav and Greene 2014).

Given the similarities with developmental psychopathology such impairments in social conduct have been dubbed "acquired sociopathy" (Damasio 2007). A review of lesion studies of patients with acquired sociopathy and preserved general cognitive abilities showed, however, that current models of normal social conduct have emphasized the prefrontal cortex (PFC) more than other brain regions (Moll et al. 2005; Moll and Tomasello 2007; Greene et al. 2001).

Greene's experiments highlight contact. For instance, Greene and colleagues (Greene et al. 2001) probed another important aspect of moral judgment using fMRI. Normal subjects were exposed to moral and non-moral dilemmas that were imposing a higher load of reasoning and conflict. Moral dilemmas were divided into moral-personal (the agent directly inflicts an injury to another person to avoid a worse disaster) and moral-impersonal (the agent does it in indirect ways, such as by pressing a button in this sacrificing one person to save five). This is the famous trolley dilemma: pushing one person in front of a train to save multiple lives.

In one experimental context, the dilemma is up close and personal; in another it is simply pushing a button with no personal context. The dilemma, not surprisingly, is more difficult for subjects up close and personal in this experimental context.

The brain regions measured in the above experiment included the frontal gyrus, posterior cingulate cortex, and regions of the temporal and parietal lobe in addition to the amygdala. That is a lot of brain. These same regions are active under other conditions, which is why it is not a signature terribly significant to morality. But it is to a larger class of social contexts. It does show that ordinary contact matters, viz., the stuff of embodied experiences, considering others, impacted by others, etc.

Now a pragmatist would assert that such events are more continuous with each other than they are so radically separate: actively or more passively allowing one to die from the trolley strikes a pragmatist as continuous on a theme of moral problem solving (Kellogg 2017). Moral sensibility pervades our breathing space, our habits of life as Peirce, Dewey or Holmes would have understood such events. There are more connections or continuity here. There is something about touch and contact that matters.

In war as Holmes knew, and also in everyday life, we dehumanize the other. It makes it easier to kill: the less familiar, the easier under diverse conditions. It is not axiomatic, but it is something we have learned in the common lure of experience. Hannah Arendt (1963) points this out about Eichmann and the final solution of the Nazi extermination of the Jews: first dehumanize your enemies as much as possible.

Other studies have addressed additional key issues in moral judgments, including the contribution of general emotional arousal, presence of bodily harm, response times, semantic content, cognitive load, conflict, intention, consequences vs. means, emotional regulation, and justice vs. care-based judgments (Greene et al. 2004).

One interesting study from Greene et al. (2008) has shown that the greater the cognitive load the more inference with utilitarian judgments, with more to consider. Perhaps time and consideration are interfered with due to the complexity of context and the demand of cognitive control, cognitive control being the primary consideration in this study. Studies like the above have extended our knowledge of the neural substrates of moral conflict emphasizing the consistent involvement of lateral and medial sectors of the orbitofrontal cortex (OFC), prefrontal cortex, amygdala.

What makes moral judgments translate into real-life action? Moral sentiments, which have long been recognized as powerful motivational forces (Hume 1738; Smith 1759/1974), stand out as strong candidates. Personal commitment to values also plays a crucial role in moral behavior, though their psychological and biological underpinnings are even less well understood. Moral sensitivity, through the experience of specific moral sentiments, allows humans to quickly apprehend the moral implications in a social situation depending on context, agency, and the consequences of one's choices.

A critical step of human evolution might have involved the functional adaptation of basal forebrain-limbic circuits linked to social attachment and aversion, which are readily identified in other species, and their integration with greatly expanded isocortical regions. Intertwining of social attachment/aversion and complex social knowledge through cultural learning became thus a unique feature of our nature (Moll and Schulkin 2009).

Prosocial values, such as friendship and loyalty and the dispositions they evoke in the agent, could emerge by connecting culturally shaped information represented in cortical structures (e.g., conceptual and action knowledge related to "friendliness" or "friendly manners," represented in frontal and temporal association cortex) with affiliative motivations arising from limbic circuits (Moll and Schulkin 2009). This interaction of affiliative experience with social concepts, social perceptual features, and prediction of action outcomes thus provides the basis for sentiments such as compassion, guilt, gratitude, and empathy.

Moral character, like temperament, figures in judgment in neural expression (Kagan 2002). Indeed, the perception of moral character impacts neural activation in rewards in games about trust (Delgado et al. 2003). Both monetary rewards and decisions to donate activated the mesolimbic reward system, in agreement with the warm glow hypothesis: it feels good to do good. In addition, in Moll et al.'s study, a direct comparison of decisions to donate to the pure monetary reward condition revealed that donations selectively activated the subgenual-septal area, which is intimately related to social attachment in other species. These findings extend the role of frontal-limbic networks in social cooperation from interpersonal economic interactions, as addressed by a number of studies, to the realm of internalized values and preferences shaped by culture (Moll and Schulkin 2009; Greene 2014).

Greene et al. (2004), noted in situations involving decision conflicts, reason and emotion may conflict with each other. In such situations, a cognitive control mechanism must be invoked to overcome emotional biases so that a rational decision can be made. In contrast, another view proposes that emotion and reason normally operate in an integrative fashion in moral decisions, including in situations of conflict (Moll et al. 2005).

There is a long tradition in philosophy and psychology supporting the division of mental processes into two broadly separable types: (1) rational, effortful, and explicit, and (2) intuitive, emotional, and quick (Kahneman 2011). Such a view is expressed in several modern psychological accounts which pit rational or cognitive processes against intuitive and emotional ones in the context of decision conflict (Sunstein 2005). Of course, in judgment no such distinction actually exists, as Holmes and other pragmatists assert. Control is a lofty term as it is anchored to external events, and not measured by internal measures. The same holds for the measurement of neural events.

According to the cognitive control and conflict theory of moral judgment (Greene et al. 2004), cognition will compete for behavioral output when pre-potent responses arising from "emotional brain regions" favor one outcome while "cognitive brain regions" favor another. This dual-process view is particularly intuitive. When faced with difficult choices—say, admitting a mistake or getting away unnoticed after misbehaving—we vividly experience a feeling of conflict. Fast and slow decision making is a feature of not just moral decision making but most kinds of decisions (Haidt 2007; Kahneman 2011). Of course, it is not that emotion is without cognition. And that is a pragmatist view about human action and judgment.

Such a separation of emotion from cognition might have been an anathema for Holmes, definitely for Dewey, and, certainly, this is the case for pragmatists (Schulkin 2007; Parrott and Schulkin 1993). The issue of control would not have been. Failing to admit a mistake can thus be naturally interpreted as a failure of this control mechanism over our emotional responses. In fact, it has been consistently demonstrated that prefrontal cortex lesions can lead to poor judgment, "disinhibition," and impaired decision-making (Moll and Schulkin 2009).

The massive prefrontal cortex is known to be particularly important when we face unexpected and novel situations, especially when different behavioral options can be made. Frontal damage results in more reflexive reactions. Utilitarian attributions, perhaps, are easily reached because they are more formulated. Losing such a large part of your brain, particularly regions tied to judgment, should result in devolving to what might be easier (Greene et al. 2008). But it does not tell us much about the brain. It does perhaps say something about decision making and brain damage. The findings are not ready for conclusive findings for the court.

Sometimes behaviorally salient situations are straightforward, so that actions driven by automatic motivational-emotional mechanisms can be taken swiftly. In many occasions, however, complex contextual demands make behavioral choices difficult, and basic emotional responses can be insufficient to ensure appropriate behavioral choices. In the context of moral decision making, moral dilemmas are the prototypical example. Moral dilemmas invoke dissonant choices of comparable motivational relevance, giving rise to a slow and effortful process that is linked to moral calculus (Moll and Schulkin 2009; Greene 2014).

Moral dilemmas require a careful analysis of available choices according to outcomes and side effects, and how they relate to personal preferences and values. Such considerations depend on a host of higher-order cognitive abilities, which include prospective evaluations, cognitive flexibility, and priority judgments. These rational processes, however, work in the service of goals which are, in essence, motivationally relevant. This is the subtle but essential element that distinguishes this integrative view from the cognitive control approach: true behavioral choices cannot be split into cognitive vs. emotional. In the context of classical moral dilemmas, choosing to kill or not to kill one innocent to save five other lives represents the struggle between suffering the angst of becoming a murderer or, instead, bearing the responsibility of letting five people die because of an act of omission.

This interpretation is in agreement with the finding that the prefrontal cortex and other brain regions are engaged whether or not decisions or behavioral outputs are required in moral scenarios, suggesting that the prefrontal cortex does not merely manipulate information stored elsewhere in the brain, but in fact represents certain aspects of social knowledge, motivation, or action. Recent neuroimaging studies showed that the prefrontal cortex is involved in both voluntary enhancement and reduction of emotional experience. One hypothesis is that the prefrontal cortex may play a central role in enabling the experience of moral sentiments and values. Indeed, this region of the brain underlies most human activities, learned actions, movements of every sort, and anticipation of events (Passingham 1993; Greene 2014).

Two independent studies demonstrated that patients with ventromedial prefrontal damage tend to opt, more often than control subjects, for "utilitarian" choices in highly conflicting moral dilemmas. One possible interpretation would be that the emotional blunting observed in these patients might render them less sensitive to immediate emotional reactions. In other words, reduced aversive emotional experience when choosing to sacrifice an innocent person in order to save five other ones would favor "rational" or utilitarian choices (e.g., killing one instead of five). Intriguingly, another study in which patients with VMPFC damage played the Ultimatum game, an economic decision-making task, demonstrated the opposite behavioral pattern. When faced with the choice of accepting (or not) unfair but financially rewarding monetary offers from an anonymous player, patients tended to reject such offers, thereby punishing the non-cooperators and losing money (Greene 2014).

One possible explanation would be that these patients suffer from a selective deficit in experiencing prosocial sentiments such as guilt and compassion, while showing a relative preservation of other-critical sentiments, such as indignation and contempt—i.e., a dissociation within the moral sentiment domain. In sum, these findings point more strongly to functional segregation between prosocial sentiments (associated with affiliative components, such as guilt and compassion) and socially aversive sentiments (indignation, contempt; Moll and Schulkin 2009) than to dissociation between cognition and emotion in guiding moral evaluations. The hypothesis that an intact ventral medial prefrontal cortex is more critical for the experience of prosocial behavior, whereas dorsal and orbito-lateral sectors of the PFC are more relevant for socially aversive sentiments, needs to be tested in future lesion and neuroimaging studies (Greene 2014).

These data support the notion that moral judgments and sentiments both rely on a close interplay of cortical and limbic circuits. Although some brain regions are intimately tied to motivational/regulatory mechanisms and others are less so, this does not imply clear anatomical boundaries or hierarchical top-down relationships between cognition and emotion. Competition between behavioral options, instead, can only occur when available choices are emotionally salient (Moll and Schulkin 2009). None of this is not helpful in the courtroom but is part of the larger body of epistemic understanding of ourselves. This is how knowledge and the law are woven from the same continuity of science and an evolving culture in which knowledge and not revealed knowledge is the heralded task and end.

But these, the brief examples above, are isolated contexts in the laboratory. Like John Rawls's (1971) depiction of the original condition for the consideration of a fair decision, these examples are abstracted away from the bustle of life. It is hard to link the laboratory consideration of neural discovery and the rationalist consideration of what might promote fairness into what underlies everyday decision making. And which individuals (psychopaths) for which fairness is not remotely on the horizon of possibilities or considerations. We don't yet have a handle on the brains of psychopaths that is court room ready. But the diverse ways in which we understand developmental disorders, devolution of function and the diverse forms of what James would characterize as the "sick soul" is the mindset of available considerations with neural considerations.

Beyond the Brain: Pragmatism and Expanded Social Polity

Sentiments and judgments tie us to others. They are essential for getting a foothold and surviving. For pragmatists, and perhaps for Holmes, when he moved from epistemic positivism to pragmatism as well, cognition runs through emotions; there are no pure emotions divorced from cognition (Parrott and Schulkin 1993). There are not pure moral sentiments for which appraisal systems are operative (Haidt 2007). Some emotions or moral sentiments seem more viscerally related; but all moral sentiments are embedded in bodily sensibility (Schulkin 2000; Johnson 2014). Both neuroscience and ethics look at many factors, one of which is cognitive capability (Greene 2014; Moll and Schulkin 2009; Solymosi and Shook 2014).

Moral sensibility underlies social interaction. Moral inquiry reflects diverse traditions, many of which are posed in terms of opposition to one another. Perhaps they can also complement each other. A historical (genealogical) and moral development and even formal rules and systems can be linked (MacIntyre 1990). There are nevertheless diverse forms of understanding who we are as moral agents; Holmes was a behaviorist and externalist; the whole concept of agency is probably too Kantian for him.

Body Politic: Rights

Dewey emphasized the expansion of human rights. Holmes seemed to loathe it, or at least its impact. "All rights" are "limited by the neighborhood of principles of policy" or "what a crowd will fight for" (1908 Letter to Laski). Dewey is my kind of pragmatist, for whom expansion of the possibility of the polity and participation is a normative goal in a democracy.

Both Dewey and Holmes understood that rights were fought out in the competition of history and ideas in a democratic culture—a culture of experimental sensibility (Butler 2017). Rights are part of the larger body politic in a culture of democracy. Rights are embedded in the customs of the larger culture (see Menand 2001). Rights are fought for: there is no inherent trajectory in history for the expression of rights.

The first American Revolution was about the protection of individual freedoms, freedom from impositions; the Civil War can be regarded as a second American revolution focused more on the rights of others, towards freedom of participation in the body of polity (McPherson 1991; Berlin 1976)—a second revolution of not just negative freedom, freedom from imposition, but positive freedom, enlargement of participation and equality of opportunity. Freedom in a market accessible to the larger expanded citizenry is the larger normative goal in an evolving imperfect democracy. Pragmatist view of neuroscience and the law, of the sort that Dewey (1935) labored for grounds ideals of societal contribution to individual development. We exist with others, social by nature, individual by constitution, the mixture the endless fluidity in the battle for living and living better. The normative goal is to materialize human ends that are worthy and to use our intelligence in experimental platforms. Easy to say, hard to do, but we know it is possible. The issues is always context, availability with a mixture of furtuna, capability—all things frail.

Thus, rights co-evolved in a culture of moral considerations of the law. They are not axiomatic or necessary but are more co-evolved with our sense of democracy and adjudication and the law. One thing that matters is, amongst other things, integrity in the law (Dworkin 1977). Integrity is a powerful concept that can mean many things; here perhaps it is tied to Kantian principles (Rorty 1990).

Having a sense of the evolving sense of rights talk is part of thinking about integrity in the law, participation in practice, and our evolving sense of democracy. I think we mean a variety of things when we speak about rights (Dworkin 1977). I do think and have argued elsewhere that they are historical and co-evolved within certain historical contexts one of which was the Enlightenment (Schulkin 1991). Laws in many countries represent what might be construed as basic rights, which varies with cultures.

Rights are intimately connected to the goals of a culture. Our sense of justice is a reflection of the acculturation process and the sense of justice that pervades. We have a large sense of justice. The right to equality and liberty (Dworkin 1977) is one that has been embedded in a constitution and a sense of jurisprudence, which gives moral oxygen to everyday behavior. It is external. The fluidity between mind and the larger world is continuous and not so different from law and science, as we live in the larger body politic.

Rights exist in a social political world. We also have a right in our culture not to participate, to stay out of the polity (Dworkin 1977, 1986)—a right to retreat to our personal Walden and find solace in solitary walks, and the sense of being with oneself. Of course, one is embedded in the larger culture too.

Dworkin wants a theory of law rich in integrity. He worries about "cheap" pragmatism, an old worry about the anti-intellectual feature of the philosophy (West 1984–1985) and its perceived lack of depth. For Posner, that is no worry, in fact depth is not needed. What is are ways of prediction and the further integration of an economic view about the law and consequences of human decision making, including that of legal decision making.

Pragmatists like Dewey refer to worthy ends (democracy, human well-being, education, etc.). So, Dworkin's characterization of cheap pragmatism is not a core feature of classical pragmatism, although it has often been characterized that way (e.g., West 1989). In fact, most judges are pragmatic if not outright pragmatists, as there is enough theory within the pragmatist perspective (Grey 1992). What dominates judicial normative considerations are contexts, practice, instrumentalism, and effective decisions in action—of course, that is a norm (Luban 1998).

For most pragmatists, as Holmes (1901) noted, "all thought is social" (talk on John Marshall). Certainly, "law so conceived is a set of practical reason for cooperative social life" (Grey 1989, 1992). At least, that is an underlying theme among diverse pragmatists, particularly Dewey.

Dewey, citing Keats, dreamed of a "grand democracy of forest trees." He had a poetic sense of a balanced nature, and a normative fantasy of a rich democracy which acknowledged the experiences of others—a democracy that Whitman alluded to in a similar normative fantasy.

This is the fantasy of possibilities. The reality is hard fought and endlessly imperfect. Holmes really understood that, and so did Dewey. There is no easy panacea of method. Despite the best intentions of educational ideals, the reality, even under the best of conditions, is something quite different. It is easy to be a fatalist as Holmes appears.

Holmes returns to his recurring theme; the scars of war. "We are sacrificed without a scruple… but when the interest of society, that is the dominant power in the community is thought to demand it".

Thomas Paine asserts, in *The American Crisis*, the common sentiment, "the natural right of the continent to independence" and the call for revolution. He appealed to guilds of associations of laborers and entrepreneurs in Philadelphia. Loose associations of creative men at the heart of a making and doing class of folks fueled Paines's sensibility for a revolutionary America (Foner 1976). These same sorts of associations de Tocqueville noted more generally as contributing towards the larger sense of America. What became "the rights of man" were the basic sentiments of an evolving culture embedded in individuals who were already self-reliant, self-expressed, and unbounded within possibilities of human expression, a view held by many.

Law, then, is placed in the consideration of the diverse rights that we have. The constant tug that Holmes understood is the balance of federal and state power, including where the individual fits in the mix. There is no one overall theory, except to maximize, where possible, freedom to choose and to create, without harm to others.

And then there is massive variation of titles, contracts, land, and uneven state variation about rights and punishments, taxation, liability (White 2015). Americana literary realism reaps from the pen of Holmes (Thomas 1997; Schoenbach 2012).

Indeed, the very concepts of rights is a cultural evolutionary trajectory. There is no one linear relationship to be found but a hodge-podge of events loosely connected, fomented, and accentuated by revolutionary movement, resembling in fact those moments in science that expand the horizon. One feature is the right to know; the right to know is key. It is at the heart of the pragmatist sense of inquiry, and it lies at the heart of Holmes's sensibility.

The intellectual drama is about power, states, and the role of a federal government. Freedom is to be balanced within constraints. The battle formulated clearly in the struggling states was made transparent in both the intellectual and personal animosity between the likes of Jefferson and Adams, Hamilton, Jay, and Madison, and the formulation of the Articles of Confederation, the Federalist Documents, and ultimately the Constitution.

Madison represented balance between diverse competitive interests (Rakove 1990,1996, 2010) and perhaps recognized, as Holmes would later, the historical nature of the event: ages of cumulative reasoning and testing reaching a precipice of discovery and now statement. Our Constitution represented cumulative reasoning about governing and participation (e.g., Montesquieu, Locke). Herein lies one essential feature of the law and the Constitution: fair laws and a maximum sense of social and political liberty (Montesquieu, Locke).

The battle continued; a broad array of conflicting views were expressed in formulating the Constitution, and still are, both within and across individuals. Some are better at it than others. Having principles is virtuous, and compromising, when it counts, is saintly. Democracy depends upon the ability to be and do both.

While frail, embedded in the struggle, is the evolution of "the civilization of experience" (Whitehead 1933; Hall 1973). The deepening of human experience and human contact is the larger consideration despite the predilection and reinforcement of a narrow enclosure. Narrowness always offers the endless allure of comfort and safety. This is part of the pragmatist emphasis on experience: our considering others, the piety we have towards others, the kindness we present towards, and the civilizing features of culture. But whose culture? Our culture is just one in a world and is inhabited by many. It is just as Holmes understood, always fraught with danger, with the apocalypse of degradation, destruction, and moral vulnerabilities always a possibility.

In any case, the battle lines of the Civil War were drawn in the American Revolution. The Federalists demanded solid unity across states; the Jeffersonians were drawn to a romantic view of the individual, its glory, agrarian sensibility, the fundamental tie to nature, and looked for less union, fewer federal constraints. Gains for opportunity would evolve on the discourse of inequality. For Rousseau, it remained romantic, operatic, with a sense of the collective social will and an active choice of participation.

Non-negotiable rights—*inalienable* rights—were the slogan for the consideration of rights. The moral landscape like all landscapes was not quite as architecturally sound or as foundational in the way Jefferson perhaps understood. But still he used the language of rights, and that was an advance for coherence. An Enlightenment perspective about inquiry was established. Fundamental protections were legislated that set the conditions for a broader consideration of human participation, albeit in the hands of Jefferson, as opposed to Paine, much more narrowly posited.

The Bill of Rights is a less than perfect document, but it was part of a cultural progression towards a sense of freedom and accountability. It was an expandable document that would take us later through the emancipation of slaves and women's suffrage. A large part of the pragmatism associated with Emerson, James or Dewey is anchored here and perhaps the strain of pragmatism tied to progressivism (Seigfried 1996; Koopman 2009, 2011).

As noted, Holmes disdained and disliked rights talk. But Lincoln, much admired by Holmes, initiated an expansion of rights. Here Holmes departs from a person he admired and one he did not: Franklin Delano Rooselvelt.

The Second Bill of Rights, as Sunstein (2004) calls them, is about human possibilities. But possibilities are always within constraints: cultural, biological, and the continuum of these two. Franklin Delano Roosevelt spoke of the four freedoms (see Kennedy 2001):

1. Freedom of speech
2. Freedom of worship
3. Freedom from want
4. Freedom from fear.

The last two are the sort of thing that made Holmes cringe, but that Dewey embraced: the idea of equality in addition to opportunity is one reading of our constitution (Allen 2014).

Holmes, somewhat brutally, would describe Roosevelt as a "second class intellect" but with a great temperament, which is not so different from Keynes's description of him as well. Of course, you only have to be good enough, and Roosevelt was certainly smart enough.

While Holmes would not be comfortable with an expansion of rights, he did think property and value are inherently tied to the law. Sunstein quotes Holmes: "property is a creation of law, does not arise from value, although exchangeable—a matter of fact" (Sunstein 2003). Of course, value is embedded in the transactions of exchange. Holmes believed in literary expression about the fundamental role of law, and not absolute libertarian freedom.

This expansion of rights, associated with Roosevelt's New Deal, was building on a Bill of Rights ironically associated with the antifederalist movement (e.g., Jefferson 1787/1982; Sunstein 2003) and then eventually federalism (Rakove 1990). Of course, the constitution as a living and not a frozen document is riddled with the out-dated conceptions of our historical past (e.g., Danson 1980).

Dworkin reads into a breathing constitution and an expanding set of rights, a moral reading, as he puts it, of the constitution: *Roe vs Wade*, the right to die, and affirmative action being three outcomes. Expanding freedoms is the use of the law to expand individual rights. Such rights are, of course, "Matters of Principle" and evolve within rules, contexts, and precedents (Dworkin 1986).

I am one of those investigators for whom the expansion of human rights and human participation is part of the expansion of democracy, literacy, and vested interests. This expansion brings more individuals to the decision-making table, where we all have things at stakek, which, of course, is an idealization. The reality is something else. But this idea has been battle-tested and it exists, circulating in quarters of the world, a world perilous.

Rawls (1971), for instance, placed an emphasis on setting contexts in which people can thrive and where consideration of others is part of the fabric of everyday decision making (see also Lyons 1993, 2010). This consideration takes into account how to insure fairness with what we might owe each other, in a context where our self-interests are covered by a hypothetical veil of ignorance. But everyday life is not like this. We are thrown into a world where we try to forage for coherence. Coherence is an essential primary heuristic in decision making, as Holmes preferred.

Fairness for all is the normative goal in a theory of justice. But fairness in communities where Jim Crow dominated does two things. It acknowledges history and rights wrongs (Woodward 1955/1966). It does so by providing contexts of access. Fairness is illusory if it goes with wealth, privilege, and availability. Holmes was privileged to have endless access with position, social grace, talent and good fortune, and he knew that.

There are no inevitable rights, but there are historical hard-battled rights. And therein lies a bulk of human worth, a hodge-podge of issues loosely woven together in pockets of meaning. A neuroscience perspective gives some of the conditions that deliver decisions, that offer another vehicle to couch legal adjudication and human organization and meaning.

Conclusion: Ethics, the Brain, and the Body Politic

Neuroscience will not replace ethics or the law (cf., Scott 2012; Meyen 2014). They are, nevertheless, continuous disciplines, drawing connections between diverse avenues of inquiry that underlie our behavior and our experience. We want to know about the neuroimaging of the genetics for violence, deception, or lying (Markowitsch and Staniloiu 2011). Still, the social context is mostly the deciding feature.

The continuing epigenetic modulation of gene expression over the course of one's life is where one finds some sense of who is vulnerable and who is not to lie, cheat, etc. Who is free or less free will continue to be a matter of what we think is morally worthy, or legally accountable. We hold people accountable and that forms the basis of our social order. Depletions of dopamine or serotonin which degrade behavioral competence and decision making, are the factors we want to grapple with, while we dwell in gray areas of accountability.

Neuroscientists are trying to capture just that (Gazzaniga 2008). But the gray is still endless, and I am talking about both gray matter in the brain and the gray areas in what we know—let alone the historical precedent of promising more than can be delivered for the courts, or misleading the courts with bad science (Judge Rakoff 2016).

An unfair advantage, of course, can be legal but not moral. Morality is a consideration with regard to most features of the human condition. Moral appraisals are endemic to social problem solving in the social milieu, or, for that matter, nature. The resources, the appraisals of proportionality and sustainability, of course, turn to law when there is undue instability and danger.

Moral fallibility and moral inquiry is limited by human capability in addition to social context, which was an endless limiting factor. Dim hope riddled Holmes's sensibility of the human condition, a far cry from his parents or Emerson. But then he did experience war. Perhaps that sculptured a brain destined, motivated by greatness with literary and critical capabilities, but also detachment.

Both Holmes and Dewey understood, following Darwin, that morality developed for social cohesion and for social cohabitation—in our case with diverse individuals. Morality is the oxygen that runs through the corpus of the larger body politic of which we are part. The brain is just a piece of what makes this possible in individuals. Morality is naturalized, like everything else about us. But culture is just continuous with what we come prepared with, and what we build upon from natural inclinations or capabilities.

It is easy to see Holmes bedeviled by aristocratic detachment from the plights of others, and elitism that runs rampant through eugenics to fatalism and utter power politics (e.g., Alschuler 2000). Holmes is appalled by rights talk, and by future possibilities of the good through social engineering. The existential plights of others is anchored by the futility that pervaded the lives lost of his memory of a war he never forgot. That famous aloofness of Holmes might reflect some of the degraded tissue of social hope lost in wanton destruction. Holmes, to be sure, is limited, and I have made that clear throughout. He is far from consistent.

Cautious and rational, Holmes (1913) would assert that "I have no belief in panaceas." Fundamentally entrenched rights, including economic rights (Brandeis 1916; Frank 1932), were not sacrosanct under all conditions for him. Individual rights emboldened and fought for, over a cultural historical evolution, are still subject to context and necessity (McPherson 1988). In *Hudson Water Co. v. McCarter* (1908), Holmes says that "all rights tend to declare themselves absolute to their logical extreme. Yet all in fact are limited by the neighborhood of principles of policy which are other than those on which the particular right is founded."

The law is designed to enhance human safety and entrepreneurial activity. Laski, who edited one of Holmes's first collections of representative writings, compared Holmes to Spinoza and Jeremy Bentham (Laski 1931). Holmes understood the historical context but was less sanguine about "rights" talk. He (1921) would say "rights are one of the most deceptive of pitfalls; it is easy to slip from the qualified meaning in the premise to an unqualified one in the conclusion. Most rights are qualified" (*American Bank Trust Co. v. Federal Reserve Bank*). For

Holmes, there was a vulnerability to the exaggeration of rights talk into immobility. He believed that the balance of rights is not pre-given or axiomatic but historical. In fact, the point is the non-axiomatic, culturally rich context in which these discussions should take place, and how frail and vulnerable these events are. Regardless, Holmes understood that the discussions occur.

Naturalistic moral capabilities, cephalic capabilities of problem solving with group capabilities, and biological capabilities combine together into moral adaptability. Nature and culture fuel endlessly and porously in an evolving species, with no necessary trajectory, and with frail possibilities amidst human want and destruction—an endless bad dream and drama, eternally recurrent in fatalists like Holmes or Nietzsche.

Yet the diverse moral sentiments, or the moral sense and its many variants, deliver a palpable sense of others, a moral realism that under diverse conditions can be cultivated, expanded, sustained. This realism has its roots in biology, in development, in culture, and, of course, in history (Flanagan 1991). Underlying the moral sentiments are our easily expressed moral rules. They vary depending upon the culture and who is or is not participating in that culture.

Survival skills are essential in a moral life. We discern events, form collaborations of meaning and effectiveness rooted in self-sacrifice and self-survival. The world, when matured into a moral life, moves beyond the "blooming buzzing" confusion of self-involvement. In fact, part of our survival is getting a foothold in the larger social world from the start as early as possible.

Holmes, like many others, found Dewey's writings troubling and confusing sometimes. Holmes writes to Laski (June 14, 1922) that he found "John Dewey's human nature" "… hard to characterize…" He later states that Dewey "of the exploitation of man by man…which always rather gets my hair up" (Posner, *The Essential Holmes*, p. 71). Dewey was radical; Holmes was not. He was neither a conservative nor a radical in anyone's vernacular (Grey 1989; Horwitz 1992; Touster 1982).

Holmes was to some extent a "public intellectual," particularly in his letters and papers. There he is like Dewey. But he is mostly moderate, independent, and not idealistic.

The fact that we are social animals runs through the study of ethics, from Aristotle to the present. The degree of social contact is tied to our survival; the greater the contact in social development, the greater the viability for longer-term survival. Self-preservation, so much noted for Hobbes, is tied to others. Law is the grounding force in groups; social laws, self-protection, facilitation of social growth are set within rules for engagement.

The particular sensory field, though minimal with perspective, is still perspectival. What we owe each other highlights the fact that we have a predilection towards others, faint and frail though it be. The ego's central envelopment, not to be denied, co-exists with a moral social cohesion tied to the existence of others.

The child comes ready with a set of problem-solving skills aiding moral judgment, assessment filled in by experience (Kagan 1984). We express fairly uniform development across diverse cultural contexts in the modern era; no blank slate but diverse forms of plasticity. From the point of view of the brain, we can trace somewhat the neural plasticity that underlies human development but not with great rigor and exactness in the brain. After all, in comparison to animals, we are quite limited, even if we were to look at the brain of Bonobos, dolphins, etc.

The issue about our cultural evolution within our biological evolution is the interplay of specialization with expansion. In a culture like ours, like Holmes's, free expression is a premium within bounds of context and capability. Emerson was a hero of Holmes; but clarity of expression is critical in the law. Tools for reasoning and persuasion can take, however, many forms, including the poetic—perhaps the poetic grounded in pragmatism—of the sort that came from Emerson (see Chapter 5).

Freedom exists within constraints; war is one such constraint, as Lincoln asserted and Holmes appreciated. Holmes perhaps did not consider enough what we owe each other (Scanlon 1998, 2014), or people different from himself (women, minorities, etc.). A culture of experts tempered by nuance and literary grace is where Holmes is at home. A culture of experts is an important goal in an educated electorate. Yet Holmes was skeptical about the human condition.

Holmes also did not appreciate sufficiently the new experimental science tied to psychology. Behavioral science, and eventually behavioral economics, was something Holmes adumbrated; its importance would materialize. This is a milieu in which valuation is embedded and acknowledged in reasoning, so that facts blur into the contours of adaptation and viability with the promise of good-enough rationality. Rationality may be faint as Holmes understood it, but it does occur; just as deciding together (Moreno 2003) and cooperative behaviors towards common ends occur. In Holmes's day, the concept of informed consent, a context for deciding together, barely existed. Vulnerable populations were the ones most marked for neuroscience treatments in the forms of surgery and medication.

What did occur is the interesting link of regions tied to the larger categories of human preference, choice and inference. It is part of the larger body of coherent choice and predictive capabilities. Something very much of interest for how Holmes understood the law and its continuation with the body politic and the larger culture of inquiry—a culture of democracy and the law (Butler 2014; Talisse 2014; Vannatta 2014).

Now law and its many variants come intact in some instances in modern neuroscience. It comes in quite dramatically in understanding the regulation of behavior and violations of the law, in understanding violence—in promoting protection from violence of predatory behaviors that undermine human well-being and social comfort.

References

Allen, D. (2014). *Our Declaration*. New York: Norton.

Alschuler, A. W. (2000). *Law Without Values: The Life, Work and Legacy of Justice Holmes*. Chicago: University of Chicago Press.

Arendt, H. (1963/1965). *Eichmann in Jerusalem: A Report of the Banality of Evil*. New York: Penguin Press.

Berlin, I. (1976). *Vico and Herder: Two Studies in the History of Ideas*. London: Chatto & Windu.

Brandeis, L. D. (1916). The Living Law. *Illinois Law Review, 10*, 4–22.

Butler, B. E. (2010). Democracy and Law: Situating Law Within John Dewey's Democratic Vision. *Ethics and Politics, 12*(1), 256–280.

Butler, B. E. (2014). Pragmatism, Democratic Experimentalism and the Law. In G. Hubbs & D. Lid (Eds.), *Pragmatism, Law and Language*. London: Routledge.

Butler, B. E. (2017). *The Democratic Constitution*. Chicago: University of Chicago Press.

Churchland, P. S. (2011). *Brain Trusts*. Princeton: Princeton University Press.

Damasio, A. (2007). Neuroscience and Ethics: Intersections. *American Journal of Bioethics, 7,* 3–7.

Damasio, A. R. (1994). *Descartes Error*. New York: Grossett/Putnam.

Danson, N. L. (1980). *Brandeis, Frankfurter and the New Deal*. New York: Anchor Books.

Darwin, C. (1871/1874). *The Descent of Man and Selection in Relation to Sex*. Chicago: Rand, McNally & Co.

Delgado, M. R., Frank, R. H., & Phelps, E. A. (2003). Perceptions of Moral Character Modulate the Neural Systems of Reward During the Trust Game. *Nature Neuroscience, 8,* 1611–1616.

Dewey, J. (1893). Anthropology and the Law. *Islander, 3,* 305–308.

Dewey, J. (1935/1963). *Liberalism and Social Action*. New York: Capricorn Books.

Dewey, J. (1939). *A Theory of Valuation*. Chicago: University of Chicago Press Press.

Dworkin, R. (1977). *Taking Rights Seriously*. Cambridge: Harvard University Press.

Dworkin, R. (1986). *Law's Empire*. Cambridge: Cambridge University Press.

Farah, M. J. (2005). Neuroethics: Trends in Cognitive. *Science, 9,* 334–340.

Flanagan, O. (1991). *Varieties of Moral Personality: Ethics and Psychological Realism*. Cambridge: Harvard University Press.

Flower, E., & Murphy, M. G. (1977). *A History of Philosophy in America* (Vols. 1 & 2). New York: Capricorn Press.

Foner, E. (1976). *Tom Paine and Revolutionary America*. Oxford: Oxford University Press.

Frank, J. N. (1932). Mr. Justice Holmes and Non-Euclidean Legal Thinking. *Cornell Law Review, 17,* 568–588.

Gazzaniga, M. S. (2008). The Law and Neuroscience. *Neuron, 60,* 412–415.

Gert, B. (2012, May/June). Neuroscience and Morality. *Hastings Center Report, 42*(3), 22–28.

Green, J. M. (2008). *Pragmatism and Social Hope*. New York: Columbia University Press.

Greene, J. D. (2014). Beyond Point and Shoot Morality: Why Cognitive Neuroscience Matters for Ethics. *Ethics, 124,* 695–726.

Greene, J. D., Morelli, S. A., Lowenberg, K., Nystrom, L. E., & Cohen, J. D. (2008). Cognitive Load Selectively Interferes with Utilitarian Moral Judgment. *Cognition, 107,* 1144–1154.

Greene, J. D., Nystrom, L. E., Engell, A. D., Darley, J. M., & Cohen, J. D. (2004). The Neural Bases of Cognitive Conflict and Control in Moral Judgment. *Neuron, 44,* 389–400.

Greene, J. D., Sommerville, R. B., Nystrom, L. E., Darley, J. M., & Cohen, J. D. (2001). An fMRI Investigation of Emotional Engagement in Moral Judgment. *Science, 293,* 2105–2108.

Grey, T. C. (1989). Holmes and Legal Pragmatism. *Stanford Law Review, 41,* 787–870.

Grey, T. C. (1992). Holmes, Pragmatism and Democracy. *Oregon Law, 71,* 521–542.

Haidt, J. (2007). The New Synthesis in Moral Psychology. *Science, 316,* 998–1002.

Hall, D. L. (1973). *The Civilization of Experience.* New York: Fordham University Press.

Hirstein, W., Poland, J., & Radden, J. (2005). *Brain Fiction: Deception and the Riddle of Confabulation.* Cambridge: MIT Press.

Hoffman, D. A. (2014). *The Punishers's Brain: The Evolution of Judge and Jury.* Cambridge: Cambridge University Press.

Holmes, O. W., Jr. (1897). The Path of the Law. *Harvard Law Review, 10*(8), 45.

Holmes, O. W., Jr. (1901). One Hundredth Anniversary of the day on which Marshall took his seat on the bench.

Holmes, O. W., Jr. (1908). *Hudson Water co. v. McCarter.*

Holmes, O. W., Jr. (1913). *Law and the Court.* Speech at a dinner at the Harvard Law School.

Holmes, O. W., Jr. (1921). *American Banks and Trust Co. vs Federal Reserve Banks.*

Horwitz, M. J. (1992). The Place of Holmes in American Legal Thought. In R. Gordon (Ed.), *Legacy of Oliver Wendell Holmes Jr.* Palo Alto: Stanford University Press.

Hume, D. (1738). *A Treatise of Human Nature.* Start Publishing LLC.

Illes, J., & Bird, S. J. (2006). Neuroethics: A Modern Context for Ethics in Neuroscience. *Trends in Neuroscience, 29,* 511–517.

Illes, J., Tairuan, K., Federico, C. A., Tabet, A., & Glover, G. H. (2010). Reducing Barriers to Ethics in Neuroscience. *Frontiers in Human Neuroscience, 4,* 167.

Jefferson. T. (1787/1982) *Notes on the State of Virginia.* New York: Norton.

Johnson, M. (2014). *Morality for Humans*. Chicago: University of Chicago Press.

Jones, O. D. (2004). Law, Evolution and the Brain. *Philosophical Transactions of the Royal Society of London. Series B: Biological Sciences, 29,* 1697–1707.

Joyce, R. (2000). *The Evolution of Morality*. Cambridge: MIT Press.

Kagan, J. (1984). *The Nature of the Child*. New York: Basic Books.

Kagan, J. (2002). *Surprise, Uncertainty and Mental Structure*. Cambridge: Harvard University Press.

Kahneman, D. (2011). *Thinking Fast and Slow*. New York: Farrar, Straux, and Giroux.

Kant, I. (1789/1997). *Critique of Practical Reason*. Cambridge: Cambridge University Press.

Kellogg, F. R. (2017). Take the Trolley Problem… Please! Pragmatism, Moral Particularism and the Continuum of Normative Inquiry. *Contemporary Pragmatism, 12,* 8–18.

Kennedy, D. M. (2001). *Freedom from Fear*. Oxford: Oxford University Press.

Kitcher, P. (2014). *Life After Faith*. New Haven: Yale University Press.

Koopman, C. (2009). *Pragmatism as Transition*. New York: Columbia University Press.

Koopman, C. (2011). Genealogical Pragmatism. *Journal of the Philosophy of History, 5,* 533–561.

Laski, H. J. (1931). The Political Philosophy of Mr. Justice Holmes. *Yale Law Review, 40,* 683–695.

Levy, N. (2007). *Neuoethics*. Cambridge: Cambridge University Press.

Levy, N. (2011). *Hard Luck: How Luck Undermines Free Will and Responsibility*. Oxford: Oxford University Press.

Luban, D. (1998). What's Pragmatic About Legal Pragmatism. In M. Dickstein (Ed.), *The Revival of Pragmatism*. Durham: Duke University Press.

Lyons, D. (1993). *Moral Aspects of Legal Theory*. Cambridge: Cambridge University Press.

Lyons, D. (2010). *Ethics and the Rule of Law*. Cambridge: Cambridge University Press.

MacIntyre, A. (1990) *Three Rival Versions of Moral Inquiry*. Notre Dame: Indiana: University of Notre Dame Press.

Markowitsch, H. J., & Staniloiu, A. (2011). Neuroscience, Neuroimaging and the Law. *Corte, 47,* 1248–1251.

McPherson, J. M. (1988). *Battle Cry of Freedom*. New York: Ballantine Books.

McPherson, J. M. (1991). *Abraham Lincoln and the Second American Revolution*. Oxford: Oxford University Press.

Menand, L. (2001). *The Metaphysical Club*. New York: Farrar, Straus and Giroux.

Meyen, G. (2014). Neurolaw. Ethics Theory Moral. *Practice, 17,* 8119–8129.

Moll, H., & Tomasello, M. (2007). Cooperation and Human Cognition: The Vygotskian Intelligence Hypothesis. *Philosophical Transactions of the Royal Society of London. Series B: Biological Sciences, 362,* 639–648.

Moll, J., de Oliveira-Souza, R., Moll, F. T., Ignácio, F. A., Bramati, I. E., Caparelli-Dáquer, E. M., et al. (2005). The Moral Affiliations of Disgust: A Functional MRI Study. *Cognitive and Behavioral Neurology, 18,* 68–78.

Moll, J., Krueger, F., Zahn, R., Pardini, M., de Oliveira-Souza, R., & Grafman, J. (2006). Human Fronto-Mesolimbic Networks Guide Decisions About Charitable Donation. *Proceedings of the National Academy of Sciences of the United States of America, 103,* 15623–15628.

Moll, J., & Schulkin, J. (2009). Social Attachment and Aversion: On the Humble Origins of Human Morality. *Neuroscience and Biobehavioral Reviews, 33,* 456–465.

Moreno, J. D. (1999/2003). Bioethics Is a Naturalism. In G. McGee (Ed.), *Pragmatic Bioethics*. Cambridge: MIT Press.

Moreno, J. D. (2003). Neuroethics: An Agenda for Neuroscience and Society. *Nature Reviews, 4,* 149–153.

Moreno, J. D., & Farah, M. J. (2012). Neuroethics. *Science Progress, 3,* 13–18.

Moreno, J.D., & Schulkin, J. (2020, in press). *The Brain in Context: A Pragmatic Guide to Neuroscience*. New York: Columbia University Press.

Nagel, T. (1970). *The Possibility of Altruism*. Princeton: Princeton University Press.

Nagel, T. (1991). *Equality and Partiality*. Oxford: Oxford University Press.

Neville, R. C. (1974). *The Cosmology of Freedom*. New Haven: Yale University Press.

Norgren, R. (1995). Gustatory System. In *The Nervous System*. San Diego: Academic Press.

Nussbaum, M. C. (1997). *Cultivating Humanity*. Princeton: Princeton University Press.

Nussbaum, M. C. (2000). Why Practice Needs Ethical Theory. In S. J. Burton (Ed.), *The Path of the Law and Its Influence: The Legacy of Oliver Wendell Holmes Jr.* Cambridge: Cambridge University Press.

Nussbaum, M. C. (2004). *Hiding from Humanity: Disgust, Shame and the Law*. Princeton: Princeton Law Review.

Parrott, W. G., & Schulkin, J. (1993). Neuropsychology and the Cognitive Nature of Emotions. *Cognition and Emotion, 7,* 43–59.

Passingham, R. (1993). *The Frontal Lobes and Voluntary Action*. Oxford: Oxford University Press.

Phillips, M. L., Young, A. W., Senior, C., Brammer, M., Andrew, C., Calder, A. J., et al. (1997). A Specific Neural Substrate for Perceiving Facial Expression of Disgust. *Nature, 389*, 495–498.

Posner, R. A. (1992). *The Essential Holmes*. Chicago: University of Chicago Press.

Posner, R. A. (2008). *How Judges Think*. Cambridge: Harvard University Press.

Putnam, H. (2014). *Philosophy in the Age of Science*. Cambridge: Harvard University Press.

Racine, E. (2010). *Pragmatic Neuroethics: Improving Treatment and Understanding of the Mind/Brain*. Cambridge: MIT Press.

Rakoff, J. S. (2016, May). Neuroscience and the Law: Don't Rush In. *The New York Review of Books*.

Rakove, J. N. (1990). *James Madison and the Creation of the American Republic*. New York: HarperCollins.

Rakove, J. N. (1996). *Original Meanings*. New York: Knopf.

Rakove, J. N. (2010). *The Revolutionaries*. New York: Houghton.

Rawls, J. (1971). *A Theory of Justice*. Cambridge: Harvard University Press.

Rorty, R. (1990). The Banality of Pragmatism and the Poetry of Justice. *Southern California Law Review, 63*, 1811–1819.

Roskies, A. L., & Morse, S. J. (2013). Neuroscience and the Law: Looking Forward. In S. J. Morse & A. L. Roskies (Eds.), *A Primer on Criminal Law and Neuroscience*. Oxford: Oxford University Press.

Rozin, P. (1976). The Evolution of Intelligence and Access to the Cognitive Unconscious. In J. Sprague & A. N. Epstein (Eds.), *Progress in Psychobiology and Physiological Psychology*. New York: Academic Press.

Rozin, P. (1998). Evolution and Development of Brains and Cultures: Some Basic Principles and Interactions. In M. S. Gazzaniga & J. S. Altman (Eds.), *Brain and Mind: Evolutionary Perspectives*. Strasbourg: Human Frontiers Science Program.

Sabini, J., & Silver, M. (1982). *Moralities of Everyday Life*. Oxford: Oxford University Press.

Sabini, J., & Schulkin, J. (1994). Reconciling Social Constructions and Biological Realism. *Journal for the Theory of Social Behavior, 24*, 207–217.

Scanlon, T. M. (1998). *What We Owe Each Other*. Cambridge: Harvard University Press.

Scanlon, T. M. (2014). *Being Realistic About Reasons*. Oxford: Oxford University Press.

Schoenbach, L. (2012). *Pragmatic Modernism*. Oxford: Oxford University Press.

Schulkin, J. (1991). Science and Human Rights. *The Journal of General Evolution, 32*, 243–253.

Schulkin, J. (2000). *Roots of Social Sensibility*. Cambridge: MIT Press.

Schulkin, J. (2007). *Effort: A Behavioral Neuroscience Perspective on the Will*. Mahway: Erlbaum Press.

Schulkin, J. (2013). *Reflections on the Musical Mind*. Princeton: Princeton University Press.

Schulkin, J. (2015). *Pragmatism and the Search for Coherence in Neuroscience*. London: Palgrave Macmillan.

Scott, T. R. (2012). Neuroscience May Supersede Ethics and Law. *Science and Engineering Ethics, 18*, 433–437.

Seigfried, C. H. (1996). *Pragmatism and Feminism*. Chicago: University of Chicago Press.

Shen, F. X., Hoffman, M. B., Jones, O. D., Greeme, J. D., & Marois, R. (2011). Sorting Guilty Mind. *New York University Law Review, 86*, 1306.

Shenhav, A., & Greene, J. D. (2014). Integrative Moral Judgment: Dissociating the Roles of the Amygdala and Ventromedial Prefrontal Cortex. *Journal of Neuroscience 34*, 4741–4749.

Smith, A. (1759/1974). *The Theory of Moral Sentiments*. Indianapolis: Liberty Classics.

Solymosi, T., & Shook, J. R. (2014). *Neuroscience, Neurophilosophy and Pragmatism: Brains at Work with the World*. London: Palgrave Macmillan.

Sunstein, C. R. (2003). *Why Societies Need Dissent*. Cambridge: Harvard University Press.

Sunstein, C. R. (2004). *The Second Bill of Rights*. New York: Basic Books.

Sunstein, C. R. (2005). *Laws of Fear: Beyond the Precautionary Principle*. Cambridge: Cambridge University Press.

Talisse, R. (2014). Deweyan Pragmatism. In G. Hubbs & D. Lind (Eds.), *Pragmatism, Law and Language*. London: Routledge.

Tinbergen, N. (1951/1969). *The Study of Instinct*. New York: Oxford University Press.

Thomas, B. (1997). *American Literary Realism and the Failed Promise of Contract*. Berkeley: University of California Press.

Thompson, E. (2015). *Waking, Dreaming and Being*. New York: Columbia University Press.

Touster, S. (1982). Holmes a hundred Years Ago: The Common Law and Legal Theory. *Hofstra Law Review, 10,* 673–708.

Vannatta, S. (2014). Pragmatism Without the Fighting Tag: Functional Realism in Holmes's Jurisprudence and Moral Philosophy. In G. Hubbs & D. Lind (Eds.), *Pragmatism, Law and Language.* London: Routledge.

Weber, M. (1903/1917/1949). *The Methodology of the Social Sciences.* New York: Free Press.

Weissman, D. (2014). *Zone Morality.* Berlin: Walter de Gruyter.

West, C. (1989). *American Evasion of Philosophy.* Madison: University of Wisconsin Press.

Westbrook, R. B. (1993). *John Dewey and American Democracy.* Ithaca: Cornell University Press.

White, R. (2015). *The Hidden God: Pragmatism and Posthumanism in American Thought.* New York: Columbia University Press.

Whitehead, A. N. (1933). *Adventures of Ideas.* New York: Free Press.

Wilson, J. (1993). *The Moral Sense.* New York: Free Press.

Woodward, C. (1955/1966). *The Strange Career of Jim Crow.* Oxford: Oxford University Press.

10

Neuroscientific Considerations and the Law

Hypothesis formation, the cornerstone of inquiry, erupts when well-worked problem solving no longer works effectively and the search for new solutions emerges. This everyday link between the practices that predominate and the search for new solutions pervades the practice of science and the law, as well as most parts of the "life world" (Schultz and Luckman 1973).

Law, like neuroscience or other human activities, requires taking interdeterminate contexts and rendering them more stable. The stuff of experiences is adapting to this sense of change and habit formation; that is making the interdeterminate more stable (Dewey 1925; Margolis 2002). Legal interdeterminancy is a feature of jurisprudence, the inquiry that goes into cases not easily covered by law (Kellogg 2004, 2007, 2018). The unsettled features that can appear in legal reasoning are the essential thrusts of most forms of reasoning, particularly in science.

In this chapter, I call attention to some considerations about the brain, neural enhancement and considerations of fairness and safety, issues that impact the fluidity about reasoning about the brain, the law

© The Author(s) 2019
J. Schulkin, *Oliver Wendell Holmes Jr., Pragmatism and Neuroscience*,
https://doi.org/10.1007/978-3-030-23100-2_10

and, and the larger social polity. Neuroscience brings to pragmatism the continued understanding about problem solving, and pragmatism brings to neuroscience a conception of worthy ideals, but not overly theoretical of what is worth investigating for the human condition. One result of this is that normative sensibilities glide into ongoing neuroscientific inquiry.

Epistemic Trajectories, Neural Engine and Neural Design: Discovery

Brain size is a good barometer of these evolutionary trends; the diversity of hominoid expression is quite remarkable and varied. Neanderthals, who had very large brains, perhaps also had severe constraints that led to their eventual downfall. One factor in this downfall may have been that they had less language facility than *Homo sapiens* (Mithen 2006), but is only speculative. Of course, it is the kind of speculation that runs through the naturalist theorizing stories that explain why something happened. That is why our evolutionary trend was so successful; tool use is non-trivial. But with language, the doors to the universe were opened in manifold ways, leading it to be regarded as the feature that made all of the difference.

As human beings, we are prepared to recognize or categorize diverse kinds of objects while regions of the brain run simultaneously in parallel and in unison, tagging and remembering objects in both short and long-term memory. William James, wrote about this process in *The Principles of Psychology* (1890) and was quite prescient about the adaptive side of memory and the vulnerability to what we remember (Schacter 1996). Linking or tracking events are the basic constituents of neural computational capabilities at their most elementary levels. The considerations of evolution underlie most aspects of neural function. Even before there was an explicit theory of evolution, there was much speculation about this relationship. The first recorded depiction of neural function and consideration of neural design dates back to the Egyptians (Gross 1998).

The brain is designed with survival information systems to survive, to discern events; it does not separate emotion from reason as emotions inform and are expressed. From an engineering perspective, emotions are no less computational than other forms of brain function. "*Computational*" is just another more technical sounding term for information processing, which runs through the brain like water from the basic well-spring of expression. This is a biological perspective and perhaps one Holmes would have appreciated. From a naturalist perspective, there is no separation between emotion and reason; naturalism is about adaptation and "good enough" problem solving. Nineteen hundred years later something like that idea became a bit realized. But empiricism is continuously difficult and exists within a cultural milieu in which the idea that science and engineering are continuous is demythologized and with good enough theory, sized up and down. Informed theory is grounded in a culture of invention and experiment, and "good enough" is sufficient for most things human. In fact, adaptive responses are not about perfection. They only require solutions that are good enough for the problem at hand as well as for anticipating problems and instantiating habits in a coherent manner.

Not surprisingly, a Renaissance and reappraisal of the ancients took place and revealed a world in which naturalism and humanism could find splendid expression. In this world, battles of ignorance and greed would bring the constant and universal drama of human expression to light and compel the mind of a Shakespeare, Holmes or Hobbes to put pen to paper and give it voice.

From Galen to Vesalius, scientists spawned an on-going parade of drama—the only constant being the way in which neural function was discerned and neural design depicted. Both were addressed with a realistic appreciation for human flaws and desires and with an acute awareness that arrogance and avarice can exist amidst pockets of virtue. Envy is no small human vulnerability.

Discovery and invention reached a punctuated moment in the period defined as the Renaissance. Although this moment occurred as just one part of a more general cultural renaissance, it was the only part linked to brain anatomy and which put an emphasis on neural discovery.

Of course, this also took place in a period that gave birth to modern experimentation (Shappin and Schaffer 1985) that strove to answer questions like *how indeed does an organ work?* Harvey focused on the heart and his experimentation led to an understanding of its mechanism that gained traction and fluidity. Over the centuries, many scientists, including Harvey, Bernard, Beckterov, Golgi, Helmholtz, Cajal, Sherrington and Pavlov, introduced, as well as advanced, an age of experimentation. They also established the design principles for basic units of the brain and the idea of experimentation in a socially accountable context.

First established were some basic doctrines about the neuron, synapse, reflexive design, sensory and sympathetic and parasympathetic motor systems as well as chemical and electrical signaling systems. These were followed, later, by the development of computational design (Swanson 2000a, b; Sterling and Laughlin 2015) and an understanding of vertebrate design principles and embryologically common development.

Utilization of resources and maximizing local capability are not common occurrences. Neuronal firing patterns and physical diameter underlie capabilities and are reflected in neural design (Sterling and Laughlin 2015). Moreover, cortical design for decision making vertically across six layers, imagined by Cajal, was realized 75 years ago across different systems in the brain (Rakic 2000). There is a realistic sense of the human condition which fits easily under the purview of the law and also brings in neuroscience.

The magnificent advances of this period in neural anatomy set a standard in which depictions could be drawn from the same pallet, or at least understood as continuous (Catani and Sandrone 2015). Perhaps from subsets of human experimentation and categorization schemes, we are able to link to diverse categories of coherence that underlie predictive coherence (Clark 2013, 2017). We infer in a context of expectations, embedded in weighted variables of predictive coherence and meaning.

Science is cultural. Integral to any investigation are the allotment of resources and the availability of minds to see it through. It does not mean, however, that everything is a mere cultural construction. Pragmatists might say that much like good enough parenting, there are good enough facts on the matter—with just enough theory in practice to warrant assertion.

In an age of computational design, the idea of fusing machines with biological matter is, in my view, a real advance in what we are capable of achieving—particularly with regard to neuroscience. The resources are being lined up in *that* direction, and there is interest *there*. Neuroscience is a cultural achievement, a worldwide endeavor that links nearly 50,000–100,000 scientists worldwide—maybe more than that when you include the diverse field of engineering. It is amazing. With broad-based funding, both public and private, we are in a neuroscientific renaissance.

But the interests reflect the culture. If the interests have no empirical grounding, over time the pursuit is rejected or dies out. There is the view, much proselytized by Peirce, that settled opinion and continued or accepted forms of findings tend to win out, reaching some settled acceptance in the community of inquirers.

We are social animals. Solitary delight, creativity and death are also prevailing features of our experience, but we are inherently social and culturally minded as biological entities. The transmission of information that pervades is not just in the genes (Boyd and Richerson 2005). An unlimited array of culture makes the activity of excessive reductionism seem ludicrous yet it still dominates, if perhaps a little less so. It is enough to be biological broadly construed.

Knowing now that our evolution is not linear and unidimensional as Darwin had thought, we still seek to understand ourselves through the understanding of our primate history. In truth, diverse hominoids competed at various times dating back at least 5 million years in our evolutionary past—even before the separation from close primate relatives (e.g., chimpanzees). So, we are now able to reconstruct our past using both morphological archaeological findings and molecular biology and neuroscience.

Neuroscience and Metaphor

In earlier chapters I mentioned the importance of metaphors in reasoning and certainly Holmes was the master of the use of metaphors. Metaphors underlie diverse features of scientific investigations (e.g., physics, Gallison 1988; Lakoff and Johnson 1980/2003) including neuroscience (Schulkin 2015). Metaphors are a foothold into the world of

Table 10.1 Metaphors in neuroscience

The mind/brain as a *computer*
The brain as a *map* (motor/sensory homunculus)
The brain as *thinking* versus the heart as *feeling*
Neurons *communicate*
Moods as *up* (happy) and *down* (sad)
Brain dead
Mirror neurons
Thinking machines
Information molecules
The mind's eye

others and objects; metaphor underlies morality and the law, they provide cognitive contact, they provide extensions vital in the organization of thought and action (Johnson 2007; Winter 2001).

Try to talk or write for more than a sentence without using a metaphor; it's almost impossible to sustain. Other metaphors are also common in neuroscience, such as the human mind as a machine, the Helmholtzian phrase of the telegraphic, Galen's basic temperamental expression in the four humors (Kagan 2002), reverberating circuits and memory, and hydraulic metaphors for the buildup of motivation (Tinbergen 1951/1969) (Table 10.1).

Metaphor expands our horizons as we puzzle about objects and widen our understanding; we use metaphor to keep track of the expansion of the familiar to the less familiar. The diverse categories we use, for instance, have various degrees of clarity. Metaphor pervades all of human inquiry. The issue is tying it to self-corrective inquiry.

Brain and Culture

Neuroscience is also part of the larger body politic and thus what is considered within features of jurisprudence—the primary consideration being that of neuroscience and the law within the bounds of reason (Greely 2009; Morse 2013a, b). References to the state of mind and the brain are often alluded to in the context of responsibility, accountability and punishment; now we have a burgeoning knowledge of neuroscience to contribute to an already growing body of knowledge within the body politic.

Neuroscience thus is another potential epistemic anchor. The broader culture is rapidly forging additional links between neuroscience and law, addressing central states of intention, deceit, and vulnerability to violence (Garland and Glimcher 2006). The developing brain matters in the consideration of the legalities around, for example, death sentences when adolescence or mental competency is involved (*Graham vs Florida*). Neuroscience research is a consideration in the larger understanding of human development and accountability, an understanding of behavioral inhibition is a non-trivial feature in the developing brain, or neural deterioration in considering diverse forms of human expression.

Neuroscience research is a consideration in the larger understanding of human development and accountability. An understanding of behavioral inhibition is a non-trivial feature in the developing brain, nor neural deterioration in considering diverse forms of human expression development or devolution of frontal cortex function for arguing for not knowing right from wrong across a wide array of cases in insanity defenses (Sapolsky 2004).

Of course, the issues in this defense became less about informed thought and more about informed will or effort. And then the endless breadbasket of ambiguous mental terms was opened up. For instance, when considering psychopaths, the smaller the brain volume, perhaps the greater the tendency for such behavior; less behavioral inhibition may be equated to lack of cortical tissue (Bechara 2003). But, in fact, the link between size and neural pathology and such behaviors is highly underdetermined (e.g., Brower and Price 2002).

Neuroscience is undoubtedly a factor when perhaps taking into consideration the background of individuals, and the expression of the brain. But neuroscience should not be oversold in the context of the law, leading to what a University of Pennsylvania professor called "irrational neurolaw exuberance" (Morse 2011). Neurolaw is the impact of neuroscientific findings on legal judgments, with specific regard to culpability, capability of individuals in adjudication.

Neurolaw is easily placed in the context of neuroethics, and more generally decision making (legally speaking, adjudication). Neuroethics is "concerned with the ethical, legal and social impact of neuroscience" (Moreno and Farah 2012; Illes 2006). Neuroscience blends with the epistemic orientations in our considerations of ethics and the law. It is the fluidity and integration and not the reduction of one to another that is important.

Our age is one in which psychologists, biologists, and psychiatrists intersect with the law, both criminal and civil, more than ever (Salvato et al. 2014; Taylor et al. 1991; Markowitsch and Staniloiu 2011; De Kogel et al. 2014). All issues of human experience and problem solving emerge in the context of neuroethics (e.g., Illes 2006), which is a useful consolidating or re-focusing of ethics. Knowing the brain, as I have indicated, however, will not generate a moral theory, something that Holmes would have understood, and Dewey worked to show the naturalization of human reasoning in law, in education and the cultivation of the human condition.

Knowing the brain will not resolve or adjudicate most issues about contracts and reason, or what we ought to do, what we value, etc. (Gauthier 1990; Kamm 1993). Getting a good sense of Oliver Wendell Holmes Jr., probably the most cited jurist in the US (Lerner 1943; Posner 1992) and also a naturalist and pragmatist, is important for getting a sense of how neuroscience is part of the larger fabric of understanding the human condition, without, I suggest, excessive fantasy of reductionism in neuroscience. Ethics and the law are part of the larger body politic, not reduced to brain function. To be sure, the brain and its functioning is essential for ethics and the law and everything else we do.

No Neuroscience Epistemic Magic

There are no epistemic magic bullets here, just the work of neuroscience and behavioral biology and the settled integration in the growing understanding of ourselves as having an evolutionary past, a past with neural design, and neural features that underlie our interactions. One does not have to be a narrow evolutionary determinist. Biology runs continuous with social activities (Jones and Goldsmith 2005).

However, neuroscience is not going to change the fact that we hold people accountable (cf. Morse 2008; Kolber 2014; Alces 2018). It is, however, important to understand the diverse ways in which delusions matter in our sense of agency, an anthropology of the mind (Atran 2002). The concept of an agent is critical in our understanding of ourselves (Holmes 1891; Neville 1974; Weissman 1993). But it is a complicated concept. And agency though embedded in our understanding needs to be critically engaged, warts and all. We may not know each other very well without the concept of agency, or indeed a variety of attributes; but then that is true of many features in our understanding of the world in which we live.

The epistemic pockets of knowledge in neuroscience are relatively new, and they will expand without replacing many core issues about us. Neuroscience is one thing; neuro-mythology is another; and neural excitement quite another. Understanding the brain is a cultural expression of our development. We build and theorize, and we come to a growing body of knowledge of which law is continuous with these events (Jones et al. 2013, 2014a). At least, that is what I suggest.

Neural Enhancement: A Cultural Evolutionary Trend: Fairness, Safety and Access

Today, as in the past, organs like brains are eaten for enhancement. Whether that helps or not is a different issue. There are other things we eat to enhance the brain although most just sustain neural function. The desired ability to facilitate our engagement with what matters is a natural outgrowth of a feature that is endemic. While I am not sure it's what the consumption of brains represents—some cultures consume them on a regular basis—it is an object of study and ingestion. A history of science reveals that fact. Our present cultural evolution is replete with neural enhancements.

The brain itself is a glucose eating machine and always active. Because sustaining and enhancing neural capability is part of our evolution, even as we sleep the brain continues its computing. It's an expensive organ.

Sustaining the brain uses up a disproportionate amount of our caloric intake, so it's no surprise that we seek out what may enhance its development and sustenance. Enhancing diverse capabilities is something we know how to do quite well; the issue is how to do so fairly and responsibly within the law (Greely 2008) as well as within a broader sense of what's fair (Farah 2005).

Enhancement is pervasive across many life events. One vital aspect is memory. Ampakines, a new class of drug designed to facilitate memory consolidation, is being investigated primarily for the purpose of treating diseases that impair memory. These drugs are also likely to enhance the memory of healthy individuals for whom normal aging is accompanied by a decline in memory ability. Greely (2008), echoing something Holmes might have said, writes that the aim is not to hide or bury enhancement but to welcome it within the bounds of reason. Enhancement has to be legitimated. So expanded use of the term has become an ongoing enterprise, e.g., performance enhancement. Most cognitive enhancement so far is nonspecific (Adderall, Ritalin); but enchanted attention and focus is a real neural enhancement. Of course, real neural enhancement happens under diverse conditions of adversity, such as at the sight of predators.

Giving diverse forms of stimulants is, at first, counter-intuitive except when one considers a typical experimental curve in which stimuli at first enhances response but then, with increased intensity, decreases diverse forms of behavioral responses. Diverse stimulants act in much the same way. The brain is hyperactive and attention and cognitive constraints are limited. When there is no filter, lots of distracting noise can result in poor performance.

Adderall, for example, is a legal therapeutic agent designed to enhance attention and performance. The issue around the evolution of such drug-related enhancement is the fairness for all, along with the health considerations *when the drug is taken* at a maximal level. The issue of fairness is moral, but the legal issue is something else. Once fairness and legality merge, the regulation of the diverse forms of enhancement in the larger culture becomes the new issue. Who has access? In theory, everyone should. In reality, it's only the well-off and well-informed who are able to gain access.

Of course, neuroscience pervades all aspects of the human condition, including the legal system. Regulation is by nature a feature of the law (and the brain) and is accomplished through enforcing control and the monitoring of safety, abuse, coercion—team or otherwise—and consent. This is the philosophical orientation; and one place in which this is considered is when we decide together on review boards for neuroscientific research. Here on these institutional review boards consensus is sought (Moreno 1995). It is brought on by collective deliberation on whether a piece of research is to move forward and then voted on; is the research worth the risk to the person? This is again part of what Jonathan Moreno calls "bioethical naturalism" and co-exists within political liberalism, very Dewey-like in scope and perspective and sagacity. Seeking consensus is a predilection tied to pedagogy; considerations of neuroscientific research is one way, considerations on a jury and many other review boards are others. The scientific review looks for a kind moral consensus, continuous with the scientific aims; real continuity in principle. Ethical committees make manifest concerns or not of the neuroscientific endeavor. It is a natural predilection to seek consensus in a deliberative body, cultivated by pedagogy and habit formation, which is why Dewey (1916) tied democratic sensibility and deliberative into a coherent normative goal.

This idea of "moral" enhancement is attractive in a culture in which responsibility is a core value, a feature holding one accountable. It is a moral virtue in the culture in which considerations overlap without conception of society and the larger consideration of neuroscience, the body politic, and ethics. But our categories will need to be much richer than they are in terms of meaningfully relating neuroscience to issues of law or ethics (Moreno 1999/2003, 2003). Enhancement, natural or unnatural, is here to stay, however. Choosing what we want to enhance, the legality, and the fairness of access to that enhancement is an important feature of discussion.

Tools and Neuroscience: Foraging for Coherence

Tool use is not just a feature of our species. Diverse species make primitive tools and pass on their knowledge to those related in some form or to social groups, probably due to epigenetic gene expression and a reflection of the diverse and rather extensive plasticity built into our brains.

Tool use is also the continued expression of our adaptive capabilities. Of course, tool use is at the heart of science. Engineering, for a pragmatist, is not something less than theory. After all, theory runs through tools, which is part of materializing concepts in engineering, such as understanding how something works, perhaps how to build something. Culture and science are continuous with what we are capable of and of the tools we generate. Tool generation and use are key to our cultural evolution, but it always goes both ways, devolving and advancing.

An instrumentalist view of neuroscience highlights the tools of use, but not at the expense of theory. In fact, theory is often presupposed by tools. This is part of the message as "seeing as" (Hanson 1958/1972). Tool use blends with understanding. In neuroscience, diverse tools that might be used are ones like lie detectors, EEG, MRI or fMRI. In *A Primer on Criminal Law and Neuroscience*, edited by Morse and Roskies, the first part of the book is just that, a primer or introduction to some basic concepts in neuroscience. The second part is what it might mean in the context of linking law and neuroscience. The book is sanguine for the methods that are not oversold; there is no grand delivery at this point with what we have with regard to what might be useful in the courtroom.

Even lie detection, much heralded by law enforcement, is fraught with limitations and ambiguities that are not particularly useful in the context of an arbiter of truth, particularly with regard to brain imaging studies. Greely (2013) notes something that Holmes would have tended towards, namely "statistical methods are crucial,"—that is, crucial in making sense of reading other minds, for which neuroscience and the law are interested in, the former for understanding, the latter in adjudicating.

Darwin, in his book on the emotions, described facial and bodily responses that reveal diverse forms of emotions. Emotions, for a biologist are adaptive and certainly are not construed as part of pathology, as is often suggested by rationalists of diverse persuasion (Spinoza or Freud, or an existentialist, Sartre).

The study of facial responses became a major focus in biology many years after Darwin wrote his book. Many scientists felt that understanding and outlining facial topography would reveal the syntax of emotion and could be used in diverse contexts including tools like the lie detector, something that might be useful in the context of determining truth. Facial expressions, of course, are modulated in a social context, so they are not a fingerprint of emotion or even deception. fMRI or other brain analysis tools may help someday as some evidence suggests (Kozel et al. 2005; Kolber 2014). But for now, it is not overwhelming (Greely 2011; Pardo and Patterson 2013), at least broadly understood.

While facial expressions are loaded with meaning as Darwin well understood, their role in lie detection or other forms of emotional reading in neuroscience is still underwhelming, as Greely and Illes outlined in a systematic view (Greely and Illes 2007).

Indeed, a great deal of neural tissue is tied to facial recognition and if one looks at cortical tissue a great deal is tied to visual expansion. Bodily and facial responses are important information or signaling systems. But they are no necessary barometer for the emotions, though they contribute to detection. In a situation in which deception interacts with such events, the issue is one of war and intelligence—of detection and non-detection.

Neuroscience aims to capture features of the brain that reveal what underlies how we forage and compute the world we are adapting or trying to adapt to. Ultimately, mind reading is what we want to know. That is, we want to be able to read minds by looking at neural signatures. In the law, the prospects would be quite helpful. We are not there yet by any means though (Greely 2011; Davis 2017). What are the reliable techniques, subject to rigor and peer review? Most of the techniques now in use do not yet meet such standards. Even the various issues that surround considerations of the mind, namely pain, consciousness, bias, pathologies like addiction or even criminality, have varying degrees of acceptability. An important question is *what are the states of the brain?* As they all have relevance to potential considerations in the law (see Jones et al. 2014b).

There have only been a handful of cases that show evidence of using neuroscience as a tool, as captured in a recent substantive book, *Law and Neuroscience* (Jones et al. 2014b) (Table 10.2).

The cases range across a wide area of brain dysfunction, or simply brain development (*Roper v. Simons*), to a range of neuroscientific considerations (memory, pain, honest and dishonest, punishment and reward, juries and reasoning and persuasion, racial discrimination, enhancement). As Greely (2013) rightfully notes, "judges will also need to be wary of arguments that fMRI or EEG or some other neuroscience technology is generally well accepted." This caution comes from someone who has embraced technology in biology and the neurosciences. And it is the technological innovations in neuroscience that are going to matter to the law (Kolber 2014). And the pace of the technology in this age is outstanding (Schulkin 2015).

Of course, issues about brain development have already impacted considerations in adjudication of events and accountability (e.g., *Melender, Roper v. Simmons*). Frontal cortical inhibition, as parents know of adolescents, is something of a work in progress. Neuroscience just builds on that. But the law takes account, as Holmes did with regard to the larger culture, of what we believe, of what we value, and of what we think we know.

Issues of capability or competence are issues of law and adjudication in which neuroscience, genetics, social contexts, etc. play a role. Roskies and Morse hit a judicious tone with "neuroscience is poised to have a significant effect on Law" (2013, p. 240). As they note, a combination of the cultural evolution of new tools or techniques with possible medical therapeutics yields an expanding sense of what interested Holmes, namely a further understanding of our nature. Indeed, as neuroscience progresses, some factors on the voluntary/non-voluntary continuum will be diluted. The contrast is not one of absolute determinism vs libertarianism or radical freedom, as it has too often been couched (see Greene and Cohen 2004).

Table 10.2   Use of neuroscientific evidence in court

Case type	Name of case
Using neuroscientific evidence in court (criminal and non-criminal)	Van Middlesworth v. Century Bank & Trust Co.; Florida v. Nelson; Ross v. Schrantz; LaMasa v. Bachman
Murdering brain	People v. Weinstein; People v. Spyder Cystkopf
Adolescent brain	Roper v. Simmons; Graham v. Florida; State v. Andrews; Miler v. Alabama; United States v. Juvenile Male #2
Assessing scientific evidence—"General acceptance" test	Frye v. United States
Assessing scientific evidence—"Gatekeeper" approach	Daubert v. Merrell Dow Pharma., Inc.; Gen. Elec. Co. v. Joiner; Kumho Tire Co., Ltd. v. Carmichael
Brain death in criminal context	People v. Eulo; Hawkins v. DeKalb Med. Ctr., Inc.
Brain injury in amateur sports	Parker v. S. Broadway Athletic Club
Proving causation in brain injury	Hendrix v. Evenflo
Pain	Luchansky v. J.V. Parish, Inc; Wiltz v. Barnhart
Distinguishing "bodily" and "mental" injuries	Garrison v. Bickford; Allen v. Bloomfield Hills Sch. Dist.
Eyewitness memory in court	State v. Henderson; Perry v. New Hampshire
False memories	State v. Michaels
Cross-racial identification	Smith v. State
Emotional defendant	State v. Thornton; State v. Quick
Emotional juror	Booth v. Maryland; Payne v. Tennessee
Lie detection and polygraphs	United States v. Semrau; Selvi v. Karnataka; State v. Lyon; United States v. Scheffer; Maharashtra v. Sharma and Khandelwal
Addiction	Robinson v. California; Powell v. Texas; Traynor v. Turnage; State v. Little

Source Adapted from Jones et al. (2014b)

Integrating Neuroscience

Issues about memory in the court have a mixed history (Nadel and Sinnott-Armstrong). Clear cases fog and fade as memory turns one more difficult avenue of evidence. Of course, it is just one issue amongst others when making a case, driving an argument. Indeed, the continuity in law and training is the consideration of the diverse forms of memory, where they might be reliable and where less so (see Guidelines on Memory and the Law, 2008). Of course, there is the further consideration of the utter continuity of memory and age, or psychopathology (Kiehl and Sinnott-Armstrong 2013). The continuity of law and the behavioral sciences is something Holmes appreciated, as well as what is reliable and what is not.

Indeed, there are many kinds of memory: short and long term, autobiographical, priming, etc., most of which are not conscious, at least the mechanisms are not (Schacter 1996). James understood this and set the stage for the discussion of modern memory in his *Principles of Psychology.*

Neuroscience is just part of the knowledge that has become part of the law, but not replacing features of the law or ethics (cf. Smith 2012; Scott 2012). Moreover, neuroscience will not solve the free will issue. Yet knowing the range of choice and the role of dopamine as one neurotransmitter tied too choice and action is relevant; though there are philosophical conundrums (Greene and Cohen 2004). Running intellectual threads together, capturing human experiences with one another, and certainly not reducing the language of choice to neuroscience, but just linking them, is a theme of this book.

We value and think it virtuous to be trustworthy. Trustworthy or not is a feature in legal reasoning. Neuroscience contributes to the law in understanding something about how trustworthiness (O'Hara 2004) might interact with neural judgment and, for instance, the brain regions or the chemical milieu that underlie these events. Staying the course, a feature of the will or effort (Schulkin 2007) can be linked to trustworthiness. For instance, those individuals that might be likely to tell the truth or be honest: do they do so because they have strong will or because they avoid conflict or vulnerability? It turns out that it is both.

Trustworthy truth tellers experience less conflict and more automatism in telling the truth; less conflict results in less neural innervation. By contrast, greater neural activity, as measured in the brain by functional MRI, is typical when lying; it takes more effort (Greene et al. 2004). None of this is courtroom ready, not remotely. But it is part of the growing epistemic continuity of the neural sciences and understanding ourselves.

Consider what Greene and Paxton (2009) have called "the grace hypothesis." It is the idea that measuring honest moral decisions requires less cognitive control in brain regions (prefrontal cortex) to insure honest moral judgments. That is, gaining money dishonestly, the paradigm used in this context, requires more cortical activity. Individuals, in other words, who are willing to lie for gain require more cortical activity. That should not be surprising perhaps; extra steps as opposed to ease of decision making should reflect greater cortical expression and activation. Unless, perhaps, you are a sociopath, and then there is no conflict. Frankly, neural science issues that impact law leave many questions to engage and many unanswered. Anticipated rewards can be construed independently from honesty (Abe and Greene 2014) or the brain adapts to dishonesty; it gets easier because specific neural signals maybe reduced which generate alarm. (Garrett et al. 2016) by brain activation and in regions of the brain tied to dopamine expression and anticipated rewards.

Of course, modulating factors in social perception, such as whether we perceive someone to have moral character, figures in neural reward activation (Delgado et al. 2003). But it ties the continuous assessment of ourselves to our sense of moral worth, of decision making. Neuroscience considerations will not replace our notion of what moral character is. It will, however, provide a material understanding how our brain underlies moral character in suitable social context and cultural evolution. The fluidity of our understanding of ourselves as agents who are held responsible, and can sustain responsibility is the continuous historical integration or our epistemic and moral ends. This is something understand by diverse pragmatists (Moreno and Schulkin 2020).

Brain: Being There and with Others

We imagine that there are responsible agents; but some temperaments are better suited for moral appraisal and moral action than others. Imagining is one thing, taking action is quite another. Like other forms of courage, moral courage or duty, need discipline for expression. We mirror it in each other; diverse cortical and subcortical regions underlie the cohesiveness with others.

In fact, one core way in which we learn about others and solidify our views is through the imitation of others. Imitation allows us to gain entry to a world of action through the deepening of experience that comes from connecting with others. We now know that when we watch one another, diverse regions of the brain are active (Jeannerod 1999). The organization of action is tied to a social milieu—the larger body politic. It comes down to an ability to interpret those around us, and to do so quickly. The deepening of human experience is with the solitary self (Emerson 1855; Whitehead 1933) and through social contact with others (Jaspers 1913/1997).

A view about the evolution of our social capability is now tied to our capacity to predict the desires and needs of other people in the context of predicting their behaviors. The neural mechanisms are neutral; they can either broaden or narrow toward others. The range of participants can be all of one kind but what Dewey particularly valued was the broadening and deepening of experience, which is inclusive and acknowledged with our biological past and our cultural evolution (see his *Experience and Nature*).

Recognizing other people's perspectives is fundamental to our neural design. The machinery is automatic. The issue is the *narrowness* or broadening of this mechanism for social cohesiveness and social contact. Diverse neural sites fire in rhythmic response to my movement and to yours—the trend is fast synchronicity in neural systems. Social binding and recruiting are features of a prepared brain, which is designed for action and for being with others.

Not surprisingly, regions of the temporal lobe are tied to the perception of others and to the meaning of actions (Rolls 1999). The brain is neutral; our experiences are not. Experiences can be broad or not.

One *feature* goal—towards the end of tying the tools of the law to the enhancement of the human condition in a vision of participatory democracy, frail as it might be and as inept as it is in expression—is the cultural inclusion of others, the broad array of individuals, into some coherence. But that is hard, but so is winning the butterfly in the Olympics.

One interesting finding is the over-exaggeration of a predilection. We are prepared to attribute mind to others. Mind represents, among other things, the beliefs, desires and habits of others. Perhaps this predilection for maintaining social contact is easily overextended (Atran 2002). A metaphor for this attribution is the super-sizing of the mind to mythic proportions and in our age, this doesn't seem so farfetched (Bostrom 2014). "Supermind" is a cognitive metaphor in the modern age of machines and machine capabilities, enhanced, perhaps, by the fusion by insights by the neural sciences. These are modern extensions.

Meaning and Well-Being: Brain Craving

A core pragmatist perspective (Dewey 1896) with regard to the brain and action is that motor systems tied into adapting to the social milieu are pregnant with cognitive expectations (Engel et al. 2015). These are expectations about what is likely to follow in the organization of movement and the interactions and transactions with others. The organization of action requires forms of syntactical organization that reflect cognitive design (Berridge 2004) and descriptions of human action and human cognitive operations are systemic and pervasive (Lakoff and Johnson 1999).

Note that motor systems in the brain are tied to perception in the organization of action (Schulkin 2015; Swanson 2000a, b). In some cases, it is duty or empathy for others or a respect for laws that is binding—self-binding, legally binding, morally binding. There is no conservative or liberal brain, no moral brain. There are regions tied to reason and action as well as clear indicators of activation or devolution of function that underlie moral, legal or any other obvious kind of decision making.

The brain is a very active and highly anticipatory cognitive system. Put otherwise, and more broadly, information systems are endemic across brain function at all levels, with predictive coherence

(Buzsaki et al. 2014). This, of course, goes for the social context as well. Neurons that depict our social coordination, sometimes called "mirror neurons," highlight the social dimension of our actions (Dewey 1925)—an absolute consideration in integrating considerations of the law, the social milieu and neuroscience (Winter 2001).

We crave meaning and well-being in the same way we crave oxygen and tranquility. Of course, the world we are living in is in flux and uncertainty exists among the pockets of clarity and reliable order. Expectation exists. While the brain remains a very active organ that consumes metabolic fuels at great neural rates, we search for human meaning (Jaspers 1913/1997). One basic feature is social contact and our survival is tied to it (Silk 2007). With massive amounts of the brain devoted to social capability, it shouldn't surprise us that searches for meaning and well-being play key roles in our sense of ourselves. Some individuals are better at it than others.

Human well-being and what we owe each other impact the brain, but we know that well-being isn't tied to material wealth. Rather, it's a feature of the human experience that encompasses the ways in which we see value in ourselves and others. It is also about finding value in our bonds with others and in what is around us as well as being considerate of future generations.

We are a social species. Cooperative behaviors are fundamentally linked to cortical expansion (Dunbar 1992, 1996, 2016), and cortical expansion is strongly linked to diverse forms of social life. The consideration of others is basic to human survival and survival is tied to our social link to others, particularly under extreme, even war-like conditions.

Primate Evolution: We Are in It Together

Dewey understood our social evolution reveals we are all in it together. Aristotle noted that we are by nature social. Primate evolution is tied to social capability. Counter to what Darwin believed, we now know that our primate evolutionary history is not a continuous linear thread. Thirty-thousand years ago, there were at least three hominids that co-existed and competed with each other within the same time frame

(Mellars 2006; Dunbar 2016). Some only survived for a short time, but we continued. Social contact and cooperation, we have learned, are essential to evolutionary development. Our evolutionary ascent is tied to our ability to form alliances, maintain contact and to groom and please each other (Cheney and Seyfarth 2007; Dunbar 1992).

A feature of our cultural evolution is the ideal of greater participation, and most likely, it isn't something in which Holmes would have had much faith. Diverse social avenues that pervade are what marks our evolution and biological natures. We are programmed to be socially interactive and to acknowledge the people we need and who also need us. Holmes understood some of this through his experience of war. Partly within this context there is an evolution in a participatory democracy where law takes precedence, in order to reduce the incidence of violence by expanding participation (Pinker 2011). It is participation within an ideal, for Holmes and for many of us, in the context of democracy (as imperfect as it is), that other choices are less sanguine, except under conditions of necessity, which Holmes understood.

As pragmatists, both Holmes and James understood the active sense of the mind; not the passivity associated with associations of the eighteenth century or the rationalists. In a changing world, minds are in competition, and law stabilizes and insures valued forms of behavior. But nothing is a given with the exception of social bounds, which are essential for survival, among other things. Like fellow pragmatists Peirce, Dewey, and Mead, Holmes understood that the larger role of the community and the law evolved in the context of codes. Our laws are just an extension of group formation that's grounded in our biology. We are prepared to exact consequences when rules that matter and ultimately impact group cohesion and personal viability are broken. Holmes was keen on this piece of naturalization of the human condition. He may not have known about variation but appreciated the continuity of nature and culture; naturalism is not the same as excessive reductionism.

As knowers foraging for understanding, we enhance our capabilities where we can and especially where it is safe, legal, and, ideally, fair. Our evolution requires that we do what can to enhance our capabilities even though most of us are not optimizing machines of perfection.

What Holmes felt most comfortable with, in the naturalization of the knowing process, was the sense that capabilities could be adapted and expanded. Limited as it might be, law is an expression. This process took us from the epistemic grains we cultivated to the herbs we garnered and finally, to the methods developed in order to preserve and extend the necessary resources. All parts of the naturalization of the knowing subject are survival, capability and viability. We search for what might help capability. Custom and law come in to make it fair and safe. It is fairness that binds Holmes to the better part of jurisprudence.

At the heart of Peirce's (1898/1992) and Dewey's (1925) versions of pragmatism was the the belief in cooperative endeavor among communities of individuals. Dewey placed an emphasis on what we can and need to do for each other and why it is in our own self-interest to be cooperative. In fact, our large-scale cooperative tendencies have evolved, and our brains are massively devoted to social behaviors (Dunbar 1996; Tomasello 2014). Cooperative inquiry, and settled, hard-fought-for beliefs dominate the view from Peirce to Holmes to Dewey.

The social features of our species—the collective sense of others and oneself—is found particularly in G. H. Mead (1934; see also Vygotsky 1926). Holmes acknowledged the larger body politic in that the law is an expression, but puts less emphasis on "deciding together" (Moreno 1999/2003, 2003), which is the collective cognitive capability in communicative and rational adjudication in individuals and a share public space (Tomasello 2014).

The evolutionary trend in our species, in addition to individual wants and needs, is toward shared effort and shared intention. Our reasoning and communicative capabilities are pervaded by social inferences that are fast and fluid. We are also anchored to collective standards of diverse sorts and giving reason to thought is anchored to these external trappings that permeate the brain.

In theories of knowledge, experience and reality (Shook 2000) are not pitted against each other but integrated into meaningful pockets of understanding. Nature is not on one side and culture on the other; "social hope," a phrase of Dewey's, is tied to the broad arena of social solidarity and the requisite space for the celebration of differences. Needless to say, this is not Holmes's world but it was Dewey's. Holmes had to come appreciate him. Like all of us, Dewey needed others to help him fulfill his vision of a cooperative ambiance as a normative goal.

Jane Addams started Hull House and Dewey, The Laboratory Schools at the University of Chicago, which were both linked to and excited by these concepts (Seigfried 1996; Green 2008). Their ideal is to aim toward cooperative behaviors in the battleground of ideas, a metaphor that is ripe with contradictions. Dewey celebrated an expansion of human participation as the normative goal—a goal centered in inclusiveness (Seigfried 1996).

Holmes understood the permeable relationship between individual reasoning and the collective world in which we breathe the larger body politic. But, eager to diminish intentionality, he missed the inherent "shared intentionality" of our species (Tomasello 2014). In cooperative contexts, the emphasis is on other minds. We are prepared to recognize others and put that knowledge to use in the context of surviving with them (Baron-Cohen et al. 2000). This capability has to do with our cultural evolution.

Evolving Together and What We Owe Each Other: Human Social Evolution

Cooperation is inherent in our species (Dunbar and Shultz 2007). Our evolutionary success is tied to our social capability. As Dewey noted in *A Common Faith*, "there is only one sure road to truth, and that is 'observation, experiment, record and controlled reflection'" (p. 32)—that is, the road of cooperative inquiry.

Holmes understood what he owed other soldiers, what he remembered about that war, what they owed each other, the memory of trauma and social solidarity. This is one feature of the consideration of others. When it came to the plights of the less fortunate, unlike his mentor Wright, he appeared more Spencerian.

Indeed, "considering others" is its own reward, a value we have worked hard to instantiate in a culture where so much is frail and vulnerable (Kitcher 2014; Singer 2015). Life matters as it also emboldens others. Human meaning dwells with others and their sense of being and dwelling with others (Jaspers 1913/1997). So much of human meaning is embracing both a moral insight and an important contract between ourselves and "what we owe each other" (Scanlon 2014).

Contractualism, or at least some parts of it, is the consideration of others and a defense of reasons for action. We owe each other. We are a social species bound to each other. Contractualism is not blind (Rawls 1971; Kant 1788); others are in full sight, others are rich in the meaning of life, hence what we owe each other is what is good for each other.

For Holmes, what we owe each other, or the elevation of one another, cannot be mandated; it has to be voluntary. Nonetheless, we are forced to justify our actions, in part, by the consideration of others. This is true in law, and it is in our self-interest. Our biology is rooted with others. After all, the evolution of the cortex is wholly in a social contact. The expanded cortex is knotted to the vast array of social meaning (Dunbar 1996).

Adaptive Capabilities

Reasoning always requires that our beliefs are subject to criticism, revision, and even ultimately elimination through the development of its own implications by intelligently directed actions, as Dewey (1910) and Holmes well understood.

Adaptative capability is about living things struggling to persevere, and "importance depends upon endurance." Experience, as Whitehead (1925) poetically put it, "is the retention through time of an achievement of value." Law, in particular, reflects this idea, which in turn reflects the fact that "endurance requires favorable environments" (*Science and the Modern World*, p. 194; see also Hall 1973).

Experimentalism and science dominate Holmes's intellectual landscape, along with the idea of lack of intrusion of government on human effort. States develop a wide variety of practices; judges follow the law, while the law can be unjust, even unfair. We retain a notion of what is, as Holmes understood, beyond the law. The law is a hodge-podge of tools, statutes, case precedent, judgments embedded in frames of reference that are fundamental in guidance and structure (e.g., constitution). Holmes was mostly a minimalist about social change and practice.

The pragmatism around Holmes, through the likes of Peirce and James and ultimately Dewey, was nascent, present and inconsistent (Grey 1992). Holmes breathed the intellectual currency of the nineteenth century of utilitarianism and positivism (Kelley 1983, 1990). Holmes would eventually identify with the sense of nature that Dewey depicted in *Experience and Nature*.

Classical pragmatists make frequent reference to the deepening of human experience (James 1907; Dewey 1925) and of human meaning (McDermott 2007; Smith 1970). It is human and social contact that ties our cultural and social evolution (Jaspers 1913/1997). Human social contact and social meaning, the consideration of others, the help of others, is bound to human well-being. Safety from duress and predictability in the social milieu are key features in the purpose of the law.

The deepening of human experience is what Holmes thought he most identified with in Dewey's *Experience and Nature*. Continuous social contact through the evolution of tools are a continuous theme. Law as one set of disciplines is critically important towards this goal. Law is critical, without doubt. And, by definition, law is social; it involves others.

Shared cooperative behaviors underlie our competitive advance in our evolutionary history. We group together, but innovation often takes place in solitude and creativity is generally a lone pursuit (Tomasello 2014). Those innovations matter, and they provide selective advantages (tools, domestication of animals, representations, writing, law).

We still do not know how the brain changes in ways that lead one set of individuals towards the worthy ends of greater participation. A necessary sentiment, as Rousseau and an endless array of thinkers have asserted, is that a "first care is that of self-preservation." Less idealization are individuals in a social world with others, in communities, trains, towns, across boundaries of ethnicity, within ethnicity, porous and permissive relationships; we all navigate a labyrinth of others, mostly familiar, and some not or less so.

It is not surprising that social intelligence is such a premium. A critical feature in the deepening of human experience is the attachments to others, not an overly dependent attachment, just some measure of social relatedness. Law evolved to promote independence, the conservation of wealth, property, religious and social expression, self-preservation, fairness, reduction of fear and violence, and enhancement of self-expression. In one way, law is the balance between freedom to act and the prevention of harm.

Evolution of discovery of inquiry is the larger hodge-podge in both law and neuroscience. Neuroscience is as diverse as the law. What binds us is the sense of discovery and that somehow these events add to the larger goals of civilization. Civilizing through the law or through science without a moral compass of disputation and the deepening of human experience is feeble and short sighted.

Civilizing through law coexists with diverse examples of devolution of civility through law, such as religious fundamentalism or other variants of fundamentalism in our time.

Indeed, human progress and corticalization of function is both a piece of natural and cultural evolution. The emphasis is on neocortex and human progress, which is a familiar and overly used theme in neuroscience. But, of course, this idea is a useful heuristic.

Human progress is a thin veil across a web of moral transgressions. Within this idea of human progress, one can not engage moral issues directly. So one can understand the context of being a judge and, as a result, understand within a wider context of Holmes's legal predilections and his link to the pragmatists. The social or political context and the competition of ideas in a participatory democracy are central to both Holmes and Dewey. A predictive theory of law or science involves the role of theory in human decision making. The normative goal of enlarging human experience and the sense of duty and belonging are essential towards this end (see Wells-Hantzis 1988; Grey 1992).

Group formation requires social stability. At the heart of our cultural evolution is the idea of not staying the same but adapting to changing circumstances. Decision making mechanisms in the brain are oriented to things that change expectations, from what is familiar to the less familiar.

Most features in biology are not about staying the same but about coping with change within finite limits. Our species is perhaps most adept at surviving across domains with great diversity. This is both a cultural achievement and a biological precondition. Moreover, enhancement and recovery in times of war or of peace are not something new; they were part of the war that Holmes fought in (Kamienski 2016).

Anticipatory cognition predominates as a fundamental part of our brain, something that the classical pragmatists understood, because biology figured in their reasoning about problem solving. The experience was not fixed; there is no chain of being. Flexibility and anticipation predominate cephalic expression across diverse neural sites. Competition is one alluring motivation, and cooperation with others is another. Self-preservation is not the only motivation heralded by stoics like Hobbes, Spinoza, Nietzsche or Darwin, yet it is dominant.

The brain is prepared to manage fear. It is an emotion that is all too familiar and the opposite of well-being. Holmes understood fear, particularly the fear that is pervasive in battle. He never forgot the feeling and instinctively found common currency through the adaptation of finding solidarity with his fellow soldiers in group social cohesion.

Conclusion

As noted, Holmes's deep influence was philosopher Chauncey Wright (1877). Within his work is what he called an "experimental philosophy," a deeply sophisticated variant of biology and of Darwinism, a critique of its abuse in Spencer, a biological notion of the evolution of consciousness with a comparative perspective, and a sense of causation. Wright was one of the first philosophers of biology in general, and in the US in particular (Wiener 1949; Schulkin 2012). Neuroscience is a part of biology. Issues about choice, effort, and responsibility are a reflection of the brain.

Biology, issues of the evolution of cognitive capability, and pragmatism run deep in the culture of Holmes. The emphasis is on tools, inventions, patents, discovery. Tools dominate the field of neuroscience, but none of them have reached a level of use for the court room. The treatment of what leads to crime, and the use of neural scientific knowledge to predict crime, is still in its infancy. But issues about mental competence, brain death, addiction, appraisal systems, and reasoning are deeply pregnant within neuroscience.

The fusion of engineering, genetics, and epigenetic information in the modern age fuels our imagination in the age of neuroscience and its embodiment of possibilities across the larger body politic. The range of tools are mainly in the basic sciences and clinical application.

Pragmatism always sees new technology as a lens for seeing more and doing more where possible. New technologies are new tools. The same issues arise in grounding technology as an extension of ourselves and our experiences (Heelan and Schulkin 1998). Part of our adaptive capabilities and continued realism about objects and about us is an anchoring of reference in practice and participation.

With nature and culture, and science and inquiry viewed as a continuous function from adapting to nature, to creating diverse expressions of culture, there is no problem about objectivity, or at least an unmistakable and impenetrable barrier. Inquiry is hard; empiricism is hard. Objectivity lies in contexts of meaning and utility.

If Holmes were to encounter an imaginary neuroscientist of our era, he would caution: this is all interesting and relevant to understanding ourselves, but it is not quite ready. Discussions of consciousness would leave Holmes pale, and the mechanisms surrounding neural circuitry would draw a blank. After all, it was his father who was interested in science. What Holmes was interested in was statistical inference: could we use neuroscience facts to predict behavior? We can. It is just not courtroom ready.

Liberty and freedom are earned and lost; both come with costs. Danger is the omnipresent vapor that raps attention, within are the moments of reprieve and social solidarity. In understanding the human condition, it is necessary to begin with a consideration of anthropology and then a historical grasp of the diverse forms of human social formation (Holmes 1881/1952; Dewey 1893). Understanding an evolution of evidence and regulation and the law set in the context of human conditions is marred by and exalted in human expression. Of course, the everyday habits of stability and viability were and remain central to pragmatist understanding.

Neuroscience, and science more generally, is a mixture of cooperative and competitive behaviors. This mixture is a reflection of our cephalic capabilities. We want to both link with others, and to compete with others. Compatibility of levels between different groups of humans are, of course, variable. In the end, neuroscientific inquiry is a public endeavor, a community effort of participating inquirers. Cephalic basics include shared intentional orientations to events, joint capabilities embedded in hardware systems in the brain that undergo epigenetic changes, that is to say, in neuroscientific terms, plasticity.

References

Abe, N., & Greene, J. D. (2014). Response to Anticipated Reward in the Nucleus Accumbens Predicts in an Independent Test of Honesty. *The Journal of Neuroscience, 34,* 10564–10572.

Alces, P. A. (2018). *The Moral Conflict of Law and Neuroscience.* Chicago: University of Chicago Press.

Atran, S. (2002). *In Gods We Trust: The Evolutionary Landscape of Religion.* Oxford: Oxford University Press.

Baron-Cohen, S., Tager-Flusberg, H., & Cohen, D. J. (1993/2000). *Understanding Other Minds.* Oxford: Oxford University Press.

Bechara, A. (2003). Decision Making, Impulse Control and the Loss of Will Power to Resist Drugs: A Neurocognitive Perspective. *Nature Neuroscience, 8,* 1458–1463.

Berridge, K. C. (2004). Motivation Concepts in Behavioral Neuroscience. *Physiology and Behavior, 81,* 179–209.

Bostrom, N. (2014). *Superintelligence.* Oxford: Clarendon Press.

Boyd, R., & Richerson, P. (2005). *Not by Genes Alone: How Culture Transformed Human Evolution.* Chicago: University of Chicago Press.

Brower, M., & Price, B. (2002). Neuropsychiatry of the Frontal Lobe. Dysfunction in Violent and Criminal Behavior. *Journal of Neurology, Neurosurgery & Psychiatry, 71,* 720–729.

Buzsaki, G., Peyrache, A., & Kubie, J. (2014). Emergence of Cognition from Action. *Cold Springs Harbor Laboratory Press, 23,* 1–15.

Catani, M., & Sandrone, S. (2015). *Brain Renaissance.* Oxford: Oxford University Press.

Cheney, D., & Seyfarth, R. M. (2007). *Baboon Metaphysics.* Chicago: University of Chicago Press.

Clark, A. (2013). Whatever Next? Predictive Brains, Situated Agents and the Future of Cognitive Science. *Behavioral and Brain Sciences, 36*(3), 181–204.

Clark, A. (2017). *Surfing Uncertainty.* Oxford: Oxford University Press.

Davis, K. (2017). *The Brain Defense.* New York: Penguin Press.

De Kogel, C. H., Schrama, W. M., & Smit, M. (2014). Civil Law and Neuroscience. *Psychiatry, Psychology and Law, 21,* 272–285.

Delgado, M. R., Frank, R. H., & Phelps, E. A. (2003). Perceptions of Moral Character Modulate the Neural Systems of Reward During the Trust Game. *Nature Neuroscience, 8,* 1611–1616.

Dewey, J. (1893). Anthropology and the Law. *Islander, 3,* 305–308.

Dewey, J. (1896). The Reflex Arc Concept in Psychology. *Psychological Review, 3*, 357–370.

Dewey, J. (1910/1965). *The Influence of Darwin on Philosophy.* Bloomington: Indiana University Press.

Dewey, J. (1916). *Democracy and Education: An Introduction of the Philosophy of Education.* New York: Macmillan.

Dewey, J. (1925/1989). *Experience and Nature.* LaSalle, IL: Open Court Press.

Dunbar, R. I. M. (1992). Neocortex Size as a Constraint on Group Size in Primates. *Journal of Human Evolution, 22,* 469–493.

Dunbar, R. I. M. (1996). *Grooming, Gossip and the Evolution of Language.* Cambridge, MA: Harvard University Press.

Dunbar, R. I. M. (2016). *Human Evolution.* Oxford: Oxford University Press.

Dunbar, R. I. M., & Shultz, S. (2007). Evolution in the Social Brain. *Science, 317,* 1344–1347.

Emerson, R. W. (1855/1876). *Nature, Addresses and Lectures.* Cambridge: The Riverside Press.

Engel, A. K., Friston, K. J., & Kragic, D. (2015). *The Pragmatic Turn: Toward Action-Oriented Views in Cognitive Science.* Cambridge: MIT Press.

Farah, M. J. (2005). Neuroethics: Trends in Cognitive. *Science, 9,* 334–340.

Gallison, P. (1988). History, Metaphor and the Central Metaphor. *Science in Context, 2,* 197–212.

Garland, B., & Glimcher, P. W. (2006). Cognitive Neuroscience and the Law. *Current Opinions in Neurobiology, 16,* 130–134.

Garrett, N., Lazzaro, S. C., Ariely, D., & Sharot, T. (2016). The Brain Adapts to Dishonesty. *Nature Neuroscience, 19,* 1727–1732.

Gauthier, D. (1990). *Moral Dealing: Contract, Ethics and Reason.* Ithaca: Cornell University Press.

Greely, H. T. (2008). Towards Responsible Use of Cognitive Enhancing Drugs by the Healthy. *Nature, 456,* 702–705.

Greely, H. T. (2009). Law and the Revolution in Neuroscience: An Early Look at the Field. *Akron Law Review, 42,* 687–715.

Greely, H. T. (2011). Reading Minds with Neuroscience—Possibilities for the Law. *Cortex, 47,* 1254–1255.

Greely, H. T. (2013). Mind Reading: Neuroscience and the Law. In S. J. Morse & A. L. Roskies (Eds.), *A Primer on Criminal Law and Neuroscience.* Oxford: Oxford University Press.

Greely, H. T., & Illes, J. (2007). Neuroscience-Based Lie Detection: The Urgent Need for Regulation. *American Journal of Law and Medicine, 33,* 377–431.

Green, J. M. (2008). *Pragmatism and Social Hope.* New York: Columbia University Press.

Greene, J. D., & Cohen, J. (2004). For the Law, Neuroscience Changes Nothing and Everything. *Philosophical Transactions of the Royal Society of London. Series B, 359,* 1775–1785.

Greene, J. D., Nystrom, L. E., Engell, A. D., Darley, J. M., & Cohen, J. D. (2004). The Neural Bases of Cognitive Conflict and Control in Moral Judgment. *Neuron, 44,* 389–400.

Greene, J. D., & Paxton, J. M. (2009). Patterns of Neural Activity Associated with Honest and Dishonest Moral Decision. *Proceedings of the National Academy of Sciences of the United States of America, 106,* 12506–12511.

Grey, T. C. (1992). Holmes, Pragmatism and Democracy. *Oregon Law, 71,* 521–542.

Gross, C. G. (1998). *Brain, Vision, Memory: Tales in the History of Neuroscience.* Cambridge: MIT Press.

Hall, D. L. (1973). *The Civilization of Experience.* New York: Fordham University Press.

Hanson, N. R. (1958/1972). *Patterns of Discovery.* Cambridge, MA: Cambridge University Press.

Heelan, P. A., & Schulkin, J. (1998). Hermeneutical Philosophy and Pragmatism: A Philosophy of Science. *Synthese, 115,* 269–302.

Holmes, O. W., Jr. (1881/1952). *The Common Law.* New York: Dover.

Holmes, O. W., Jr. (1891). Agency. *Harvard Law Review, 5.*

Illes, J. (2006). *Neuroethics: Defining the Issues in Theory, Practice and Policy.* Oxford: Oxford University Press.

James, W. (1890/1952). *The Principles of Psychology.* New York: Dover Press.

James, W. (1907/1955). *Pragmatism.* New York: Meridian.

Jaspers, K. (1913/1997). *General Psychopathology* (Vols. I & II, J. Hoenig & M. W. Hamilton, Trans.) Baltimore: The Johns Hopkins University Press.

Jeannerod, M. (1999). To Act or Not to Act: Perspectives on the Representation of Action. *Quarterly Journal of Experimental Psychology, 52,* 1–29.

Johnson, M. (2007). Mind, Metaphor, Law. *Mercer Law Review, 58,* 845–868.

Jones, O. D., & Goldsmith, T. H. (2005). Law and Behavioral Biology. *Columbia Law Review, 105,* 405–502.

Jones, O. D., Marois, R., Farah, M. J., & Greely, T. H. (2013). Law and Neuroscience. *The International Journal of Neuroscience, 33,* 17624–17630.

Jones, O. D., Bonnie, R. J., Casey, B. J., Davis, A., Faigman, D. L., Hoffman, M., et al. (2014a). Law and Neuroscience: Recommendations Submitted to the President's Bioethics Commission. *Journal of Law and the Biosciences, 1,* 224–236.

Jones, O. D., Schall, J. D., & Shen, F. S. (2014b). *Law and Neuroscience.* Frederick, MD: Kluwer.

Kagan, J. (2002). *Surprise, Uncertainty and Mental Structure.* Cambridge: Harvard University Press.

Kamienski, L. (2016). *Shooting Up: A Short History of Drugs and War.* Oxford: Oxford University Press.

Kamm, F. M. (1993). *Morality, Morality.* Oxford: Oxford University Press.

Kant, I. (1788/1956). *Critique of Reason.* Indianapolis: Bobbs-Merrill.

Kelley, P. J. (1983). A Critical Analysis of Holmes Theory of Torts. *Washington University Law, 61,* 681–744.

Kelley, P. J. (1989–1990). Was Holmes a Pragmatist? Reflections on a New Twist to an Old Argument. *Southern Illinois University Law Review, 14,* 427–467.

Kellogg, F. R. (2004). Holistic Pragmatism and Law: Morton White on Justice Oliver Wendell Holmes. *Transactions of the Charles S. Peirce Society, 40,* 559–567.

Kellogg, F. R. (2007). *Oliver Wendell Holmes: The Legal Theory as Judicial Restraint.* Cambridge: Cambridge University Press.

Kellogg, F. R. (2018). *Oliver Wendell Holmes Jr. and Legal Logic.* Chicago: University of Chicago Press.

Kiehl, K. A., & Sinnott-Armstrong, W. P. (2013). *Handbook on Psychopathology and Law.* Oxford: Oxford University Press.

Kitcher, P. (2014). *Life After Faith.* New Haven: Yale University Press.

Kolber, A. J. (2014). Will There Be a Neurolaw Revolution. *Indiana law Journal, 89,* 808–845.

Kozel, F. A., Johnson, K. A., Mu, Q., Grenesko, E. L., Laken, S. J., & George, M. S. (2005). Detecting Deception Using fMRI. *Biological Psychiatry, 58,* 605–613.

Lakoff, G., & Johnson, M. (1980/2003). *Metaphors We Live By.* Chicago: University of Chicago Press.

Lakoff, G., & Johnson, M. (1999). *Philosophy in the Flesh.* New York: Basic Books.

Lerner, M. (1943). *The Mind and Faith of Justice Holmes.* New York: Random House.

Margolis, J. (2002). *Reinventing Pragmatism.* Ithaca: Cornell University Press.

Markowitsch, H. J., & Staniloiu, A. (2011). Neuroscience, Neuroimaging and the Law. *Corte, 47,* 1248–1251.

McDermott, J. J. (2007). *The Drama of Possibility: Experience as Philosophy of Culture*. Bronx: Fordham University Press.

Mead, G. H. (1934). *Mind, Self, and Society*. Chicago: University of Chicago Press.

Mellars, P. (2006). Why Did Modern Human Populations Disperse from Africa 60,000 Years Ago? *Proceedings of the National Academy of Sciences of the United States of America, 103*, 9381–9386.

Mithen, S. (2006). *The Signing Neanderthal*. Cambridge: Harvard University Press.

Moreno, J. D. (1995). *Deciding Together*. Oxford: Oxford University Press.

Moreno, J. D. (1999/2003). Bioethics Is a Naturalism. In G. McGee (Ed.), *Pragmatic Bioethics*. Cambridge: MIT Press.

Moreno, J. D. (2003). Neuroethics: An Agenda for Neuroscience and Society. *Nature Reviews, 4*, 149–153.

Moreno, J. D., & Farah, M. J. (2012). Neuroethics. *Science Progress, 3*, 13–18.

Moreno, J. D., & Schulkin, J. (2020, in press). *The Brain in Context: A Pragmatic Guide to Neuroscience*. New York: Columbia University Press.

Morse, S. J. (2008). Determinism and the Death of Folk Psychology: Two Challenges to Responsibility from Neuroscience. *Minnesota Journal of Law, Science & Technology, 9*(1), 1–36.

Morse, S. J. (2011). *Avoiding Irrational NeuroLaw Exuberance: A Plea of Neuromodesty*. Neuroethics Publications.

Morse, S. J. (2013a) Common Criminal Law Compatibilism. In N. A. Vincent (Ed.), *Neuroscience and Legal Responsibility*. Oxford: Oxford University Press.

Morse, S. J. (2013b). Brain Overclaim Redux. In *Law and Equality*. Minneapolis: University of Minnesota Law School.

Neville, R. C. (1974). *The Cosmology of Freedom*. New Haven: Yale University Press.

O'Hara, E. A. (2004). How Neuroscience Might Advance the Law. *Philosophical Transactions of the Royal Society of London. Series B, 359*, 1677–1684.

Pardo, M. S., & Patterson, D. (2013). *Minds, Brain and Law: The Conceptual Foundations of Law and Neuroscience*. Oxford: Oxford University Press.

Peirce, C. S. (1898/1992). *Reasoning and the Logic of Things: The Cambridge Conferences Lectures of 1898 (Harvard Historical Studies)* (K. L. Ketner & H. Putnam, Eds.). Cambridge, MA: Harvard University Press.

Pinker, S. J. (2011). *The Better Angels of Our Nature*. New York: Viking Press.

Posner, R. A. (1992). *The Essential Holmes*. Chicago: University of Chicago Press.

Rakic, P. (2000). Setting the Stage for Cognition: Genesis of the Primate Cerebral Cortex. In M. S. Gazzaniga (Ed.), *The New Cognitive Neurosciences*. Cambridge: MIT Press.

Rawls, J. (1971). *A Theory of Justice*. Cambridge: Harvard University Press.

Rolls, E. T. (1999). *The Brain and Emotion*. Oxford: Oxford University Press.

Roskies, A. L., & Morse, S. J. (2013). Neuroscience and the Law: Looking Forward. In S. J. Morse & A. L. Roskies (Eds.), *A Primer on Criminal Law and Neuroscience*. Oxford: Oxford University Press.

Salvato, G., Dings, R., & Reuter, L. (2014). Culture, Neuroscience and Law. *Frontiers in Psychology, 5*, 1–3.

Sapolsky, R. M. (2004). The Frontal Cortex and the Criminal Justice System. *Philosophical Transactions of the Royal Society of London. Series B, 359*, 1787–1796.

Scanlon, T. M. (2014). *Being Realistic About Reasons*. Oxford: Oxford University Press.

Schacter, D. L. (1996). *Searching for Memory*. New York: Basic Books.

Schulkin, J. (2007). *Effort: A Behavioral Neuroscience Perspective on the Will*. Mahway: Erlbaum Press.

Schulkin, J. (2012). *Naturalism and Pragmatism*. London: Palgrave Macmillan.

Schulkin, J. (2015). *Pragmatism and the Search for Coherence in Neuroscience*. London: Palgrave Macmillan.

Schultz, A., & Luckman, T. (1973). *The Structures of the Life World*. Evanston: Northwestern University Press.

Scott, T. R. (2012). Neuroscience May Supersede Ethics and Law. *Science and Engineering Ethics, 18*, 433–437.

Seigfried, C. H. (1996). *Pragmatism and Feminism*. Chicago: University of Chicago Press.

Shappin, S., & Schaffer, S. (1985). *Leviathan and the Air Pump*. Princeton: Princeton University Press.

Shook, J. R. (2000). *Dewey's Empirical Theory of Knowledge and Reality*. Nashville: Vanderbilt University.

Silk, J. B. (2007). The Adaptive Value of Sociality in Mammalian Groups. *Philosophical Transactions of the Royal Society, 362*, 539–559.

Singer, P. (2015). *The Most Good You Can Do*. New Haven: Yale University Press.

Smith, J. E. (1970). *Themes in American Philosophy: Purpose, Experience and Community*. New York: Harper & Row.

Smith, S. R. (2012). Neuroscience, Ethics and Legal Responsibility. *Science and Engineering Ethics, 18,* 475–481.

Sterling, P., & Laughlin, S. (2015). *Principles of Neural Design.* Cambridge: MIT Press.

Swanson, L. W. (2000a). What Is the Brain? *Trends Neuroscience, 23,* 519–527.

Swanson, L. W. (2000b). Cerebral Hemisphere Regulation of Motivated Behavior. *Brain Research, 886,* 113–164.

Taylor, J. S., Harp, J. A., & Elliott, T. (1991). Natural Psychologists and Neurolawyers. *Neuropsychology, 5,* 293–305.

Tinbergen, N. (1951/1969). *The Study of Instinct.* New York: Oxford University Press.

Tomasello, T. (2014). *A Natural History of Human Thinking.* Cambridge: Harvard University Press.

Vygotsky, L. (1926/1962). *Thought and Language.* Cambridge: MIT Press.

Weissman, D. (1993). *Truth's Debt to Value.* New Haven: Yale University Press.

Wells-Hantzis, C. (1988). Legal Innovation Within the Wider Intellectual Tradition: The Pragmatism of Oliver Wendell Holmes Jr. *Northwestern Law Review, 82,* 541–595.

Whitehead, A. N. (1925/1997). *Science and the Modern World.* New York: Free Press.

Whitehead, A. N. (1933). *Adventures of Ideas.* New York: Free Press.

Wiener, P. P. (1949). *Evolution and the Foundations of Pragmatism.* Cambridge: Harvard University Press.

Winter, S. L. (2001). *A Clearing in the Forest: Law, Life and Mind.* Chicago: University of Chicago Press.

Wright, C. (1877/1971). *Philosophical Discussions.* New York: Burt Franklym.

11

Conclusion: Pragmatism and the Law in the Age of Neuroscience

This book has had a two-fold goal. One is to link a consideration of Oliver Wendell Holmes Jr. to the larger culture of pragmatism, inquiry, and the body politic. This has been the larger part of the book. The second is a consideration of neuroscience and the continuity of the law and science which Holmes represented. John Dewey is the companion throughout; essential pragmatist philosopher, who Holmes embraced towards the end of his life and who was well prepared to integrate neuroscience into the larger culture of inquiry, including law and ethics.

In the conclusion, I return to neuroscience and, of course, Holmes's ties to pragmatism, a perspective on neuroscience and the law, as well as the fluidity of our sense of nature and culture.

In order to understand Holmes, I suggested one speculative way is in light of the way he was shaped by his experiences in the war: the glory and the defeat, the senseless loss of life, the agony of cries heard and unheard, all endlessly broadcast against a universe of difference and indifference. There was so much pain and suffering for a war that many did not understand but who fought anyway. The peril of what is right amidst the duty of following underlies the prose of Holmes—crossing a field and shot and mortared and mortified by the screams and the

© The Author(s) 2019
J. Schulkin, *Oliver Wendell Holmes Jr., Pragmatism and Neuroscience*,
https://doi.org/10.1007/978-3-030-23100-2_11

horror. War shaped his brain and his judgment. Perhaps the detachment and the adaptation of clinging to closeness in letters were expressions of adaptation. His intellectual voracity only grew with age. It was also a convenient to begin a book partly about the brain. The event of that war no doubt impacted his brain, and we know it did by what he said, profoundly.

One epistemic theme for Holmes, like other pragmatist, is the anchoring to external events, external predictive events. The brain, an ideal external standard of the sort that Holmes was looking for, would, and perhaps will, be an anchor one day in the law, but not yet—at least not courtroom ready for many things we care about. We know a lot about PTSD and trauma, the breakdown of neural tissue, resilience and the treatment of war veterans, something hidden in the past (Horn et al. 2016). But those that break laws, wreak damage on others, and abuse others as distinct from those who do not commit crimes, are not clear for examining the brain: not yet, but someday.

Law and biology, of which neuroscience is a part, are indeed a hodge-podge of events. Science and the law are continuous with one another, something Holmes very much appreciated. Our advances in neuroscience, while not necessarily applicable to the courtroom, are prevalent and growing in the body politic of settled knowledge (Aharoni et al. 2008; Schulkin 2015). Pragmatist orientation grounds the neuroscience in the larger predictive and epistemic considerations for adjudication within the law, and human understanding and ideals. This book like other books I have written binds considerations of neuroscientific inquiry within the pragmatist sense of inquiry.

For neuroscience and the law right now, these are the relevant or mostly relevant considerations (e.g., Greely 2011; Morse 2010; Markowitsch and Staniloiu 2011; Pardo and Patterson 2013). As I have indicated throughout the book, the tools in neuroscience are not quite ready for the courtroom (e.g., fMRI, neuroscience-based lie detection, etc.; cf. Greely 2010). As I have immediately thought, the tools are meager, although the prospects are high. The continual fusion of science and the law remains the trajectory, and Holmes was part of this.

We are still left thinking about mental capability, the sense of agency, and devolution of neural function (Hirstein et al. 2018; Alces 2018). In the courtroom, insights into brain function enhance our understanding of these events, diverse as they are; we are left in the law and elsewhere with variants of compatibilism (Morse 2013) and, more importantly, continuity of science and the law. Ethics or the law has not been reduced to the brain. No one site in the brain is devoted to ethics, or music, or the law. Diverse regions of the brain underlie problem-solving capabilities and the utter fluidity of culture and nature.

While there is a neuroscience of guilty minds (e.g., Shen et al. 2011), and of compromised memory (Nadel and Sinnott-Armstrong 2012), it is not courtroom ready. An experienced lawyer or investigator could poke endless holes in the data. Neuroscience will only ever be part of how we think about ethics or the law, but it will also come to be part of how we understand the wars we fight and the weapons we use (White 2008; Tennison and Moreno 2012).

Neuroscience will not replace ethics or the law, though it might deepen our sense of the continuity of nature with culture and our common bounds, with common and yet individuated brains (Tsomo 2012; Scott 2012; Damasio 2007; Freeman 2006; Petoff 2015). We are still left with all the hard problems: responsibility, guilt, context, allowance, deterrence, etc. Holmes understood this. We are looking for reasonable ways in which people can be held accountable and responsible and be viable members of the communities in which we participate. It is in this larger social epistemic space in which neuroscience and the law will find common context and contribute towards an understanding of us as human beings (e.g., Pardo and Patterson 2013; Sinnott-Armstrong 2008). Indeed, over time neuroscience will get more integrated with our considerations of the law, of how we treat others, and hold them accountable; indeed, we are within a period in which these events are occurring (Hirstein et al. 2018). And this will only grow into the episemic fabric of our considerations of others. This is the philosophical orientation of pragmatist sensibilities of the law and neuroscience.

The issue about law and neuroscience is in regulation: the regulation (and fairness) of cognitive enhancement, and thus in understanding the actions in the brain (Greely 2010). The question is keeping track of the advances and considering their implications for our understanding.

Anti-social Vulnerabilities: Some Interesting Scientific Findings Eventually for Adjudication in the Context of the Law

Studies suggest that brain regions—rather white matter capability in the prefrontal cortex—might be linked to pathological lying, which is quite interesting (Yang and Raine 2009). Indeed, anti-social behavior, which is how lying is categorized, may be linked to the right orbito-frontal cortex, anterior cingulate and left dorsolateral prefrontal cortex (Yang and Raine 2009); but then so are other behavioral dispositions, such as shyness.

Decreases in autonomic function are a feature of sociopaths or anti-social individuals (Damasio 1996); low autonomic features can occur under other conditions (enhanced attention for instance). These are massive brain regions linked to many behavioral functions.

There are also other brain abnormalities or potential abnormalities in antisocial individuals, and they include enhanced colossal interactions between the two hemispheres. The corpus callosum ties the great hemispheres of the neocortex together; commissures are connective between the two hemispheres. There is a small one sometimes called the anterior commissure. There is a massive one called the corpus callosum. In this case, with regard to anti-social individuals, there is a reported increase in connectivity. There appears to be enhanced prefrontal white matter in pathological liars (Yang and Raine 2009).

This is interesting because, using fMRI, the investigators are suggesting that more connectivity between the hemispheres sets a condition of vulnerability for psychopathic individuals (Raine et al. 2003). Anti-social behaviors are typically linked to less cortical inhibition of the amygdala (Damasio 1996), since damage to this region or less cortical inhibition, the theory and some results suggests, results in more expression, and less inhibition. Increased aggression or inappropriate aggression is due in part to frontal cortical damage, less autonomic input and enhanced amygdala expression. All of this may be somewhat true, but for now it is of little help to the courts in terms of neuroscientific evidence in adjudication.

Of course, deception is also an evolutionary adaptation, predicting behavior by attributing beliefs and desires to others, as it figures, perhaps, in the evolution of our cognitive capability. Deception is a feature in human development (Kagan 2002), at least in the children and the cultures that have been studied thus far. Lying to or misleading others is a feature of the human condition. Indeed, it takes more cortical space to deceive in some experimental paradigms (Greene 2014).

We come prepared to deceive just as we come prepared to be honest. The question is context, capability, training, and what the pragmatist emphasized, following Aristotle, as habit. However, it is interesting that, in a number of experimental contexts, one can predict who is likely to deceive or lie, which is reflected in a pattern of neural configurations (Kozel et al. 2005).

Fuller understanding of the human condition and the law takes into account an understanding of these neural events. Neuroscience is just part of the conversation about the human condition, when are we responsible or not, in pain or not, lying or not. A pragmatist conception looks for the greatest breath of meaning as well as the local context of viability.

Holmes's Myopia: Insensitive to the Less Able

Genetics and the over-determination of behaviors would be something Holmes would welcome, but the neural signature, if it were not substantial, would not necessarily count in his courtroom as good evidence. Common behavioral assessment, such as whether or not a person is mentally ill, would be used, as it was in the 1927 *Buck v. Bell* case, a Virginia case about forced sterilization of a minor, a not uncommon form of birth control still active today.

When Holmes stated that "three generations is enough" of imbecility in the Buck case, he was embedded in the progressive era sentiment of the possible correction for such continuity. As previously mentioned, eugenics was very much in vogue; Margaret Sanger, WEB Dubois, Alexander Graham Bell, to name just three notables, all supported eugenics. It is what Jonathan Moreno has called the "dark side" of progressivism. The larger culture embraced varying degrees of social Darwinism; after all, life was indeed, as Hobbes noted, "nasty, brutish and short."

Of course, for Holmes and others, eugenics was bringing consideration into the larger culture of science. For them, a culture with imbeciles witnessed a devolution of function, a term that Herbert Spencer coined for referring to one sort of change in brain structure over time. The devolution of function is the opposite of evolution of function, or corticalization of function, something that Victorian sensibilities prized, namely inhibition of the proximate responses orchestrated by more ancient regions of the brain.

This view has prevailed and is not totally false. Inhibition is an important adaptation. Nathan Comfort has made clear eugenics was a piece of genetics, which is tied to the progressive sensibility of public health such as the improvement of the human condition now through genetics and genetic amelioration of human suffering. Of course, that is the good side. The ugly side is the Nazi destruction of what they took as the non-desirable, the purification of the race. Holmes was cited in the infamous trials at Nuremburg.

Holmes (1881) himself, in a famous quote, acknowledges that "the history of what the law has been is necessary to the knowledge of what the law is." Law, like science and most human activity, needs to be understood in the context of the historical times: context, venue, understanding, economics, etc.

Eugenics, an old view formalized into a public intellectual commodity, was part of thinking about the economics of the human condition in terms of removing the burden of the "feeble minded" from society. Holmes chose to uphold in the Supreme Court case of *Buck v. Bell* in 1927. And Holmes went further. There was no ambivalence in his judgment and he enjoyed the decision (Kaplan 2016).

Buck v. Bell was a case of sterilization of a minor with mental difficulties, who had been raped by her adopted mother's nephew. Holmes would write the majority opinion in this case, a vote 8-1 (including Brandeis, upholding the sterilization in the Virginia case) that "three generations of imbeciles are enough" and made the case to protect society from "feebler minded offspring." For Holmes, the cost to the public good and the public purse was a real consideration in his adjudication. Such a eugenic sensibility that permeated this age infected the intellectual space of many thinkers in his day.

For Holmes, this decision was a practical thing. He argued that we vaccinate people for the common good. He wrote in his majority opinion, "the principle that sustains compulsory vaccinations is broad enough to cover cutting the fallopian tube" (*Buck v. Bell* 1927). Even if Carrie Buck, as many have since thought (Gould 1984), was of perfectly normal intelligence, the point is that Holmes's view was common, a product of history as is the law and our situations in the world.

Holmes reached back to his core experience, war, and applied it to his decision making. In his decision, he wrote "we have seen more than once that the public welfare may call upon the best citizens for their lives. It would be strange if it could not call upon those which already sap the strength."

In the hands of Darwin's cousin, Francis Galton, eugenics became part of the science of biology, the biology of individual differences. Then, with Karl Pearson, the statistician and like Irving Fisher, a contributor to statistical inference (Pearson coefficient Fisher test, see Hacking 1965), public health and statistical inferences were linked.

Holmes was unwise in this decision. So was Brandies and 6 others. Sterilization—for which Holmes could assert his views and perhaps even mistaken view on whether Ms. Buck was feeble, for she seemed to have lived a full non-feeble life was simply blinded—illustrates a narrow view of the human condition (Gould 1984; Budiansky 2019). The reflex to sterilize as a means of birth control is still a human predilection in diverse cultures, including our own. Holmes "three generations of imbeciles is enough," and just one of the many decisions that are uncomfortable and conflicted when it comes to making decisions about reproductive rights. There is a shared and disturbing history of genetics, neuroscience and the law (Garland and Frankel 2006).

For Holmes, there is no perfection; Holmes, like the rest of us, was vulnerable and made decisions that were regrettable (Haack 2011). And, as I have indicated, one did not have to be a raging eugenic extremist to have supported the majority decision in *Buck v. Bell*. For instance, Clarence Darrow (1926), the famed lawyer of the Scopes trial, would talk about the "eugenics cult" in his time.

Moral fallibilism is a feature applicable to Holmes himself (Haack 2011). There is no mythology about him or anyone else for that matter. It is part of being a grown up, at least trying to be. Holmes might understand this. Holmes, an imperfect human being, climbed to the highest court in the United States. He was a product of his times, conveying many views in private letters and public statements (White 1993). One consistent view of Holmes is a form of elitism. He says in 1873, "why should the greatest number be preferred? Why not the greatest good of the most intelligent and most highly developed?" He is not charitable to the less able. And this is a generous reading about Holmes.

He also took a narrow perspective, on one of the "bad men," who cares for very little about what is right. In Holmes's (1897a, b) view, the concern is what to avoid and to predict; material consequences of choice. This indeed might be a useful heuristic and is often cited: *what would the "bad man" do?* Of course, there are many degrees and context in which to consider such a phrase. Perhaps this is just a piece of the considerations; indeed, it often is. Of course, we know little about the "bad man's brain" or what about the consideration of the brain would matter in considering the bad man.

Pragmatistic sensibilities without worthy ideals is dangerous; A conception of what is worth striving for with your epistemic quests with the science of the mind, of the brain is what one wants to mature, to materialize as we engage issues of neuroscience and the law. The fall beneath high ideals is always available; the fall downward are the myriad of ways across human history we have diluted the worth of human beings, disabled or not.

Some Ways Forward

Fortunately, law, like science or neuroscience, evolves. Yet scientism, namely the exaggerated emphasis on science in the context of its use remains pernicious, whether in the form of the natural or social science (Feldman 2009). There is this tendency to think that a better explanation is found elsewhere; so too with the law and an exaggerated sense that social science will solve many of the law's issues.

The abuse litters the human fabric of expression, including that of law (Feldman 2009). There is no mythologizing about science: "we are human all too human" to quote Nietzsche, and therefore subject to the highs and lows of human expression. Indeed, the richer the sense of science and its many faces and failures and over-reaching is, the better the epistemic positioning.

Judge Rakoff (2011) characterized the relationship between science and the law "as uncomfortable bed fellows" and "we may expect therefore that jumbled together, they will toss and turn for a long time to come." But then many things are uncomfortable. Comforts are moments to be appreciated in epistemic advances but not to be expected.

Law, for some centuries, has always had a fascination with science, and it should continue to do so as long as it is not naïve about it, abusive about it, or blind about it. Scientism is an abusive, narrow view of knowledge acquisition. Self-corrective inquiry is a larger category than a narrow notion of science in characterizing what is essential for human investigation. With enough theory tied to practice, predictive outcomes come to be a part of a larger community of investigators. This is consistent with a view of pragmatism that places Holmes within a larger community of his pragmatist contemporaries, like Peirce, James, and Dewey.

What the body politic values is the intellectual landscape. Here neuroscience and the law are bed fellows in this larger public sphere. How we understand ourselves and what we value have to be seen from this context. This Holmes understood. And here Dewey developed a pragmatism grounded in wisdom, with great value placed in understanding the brain and linking it to education, the law and the larger cultural concerns.

Dewey in Liberalism and Social Action used metaphor and neural language when he poetically suggested that "the economic-material phases of life which belong to the basal ganglia of society has usurped for more than century the cortex of the social body". He was arguing for more cortical and sanguine conception within the working out carried out by the basal ganglia and the materialization of ideas. Ideas that enhance the human condition.

The integration of biology and law was and is a normative goal. Holmes, the naturalist, had a sense for that. After all, the individuals he most admired, such as Chauncey Wright, had embraced Darwin early on and Gregor Mendel added a sense of genetics, something long noted but not understood. Holmes lived in an era of the exciting beginning of understanding genes.

Thought disorders, repetitions of behavior, fixation, and tyrannical action can be found in the range of human behavior. Here it meets the barbaric: the slaughter of the innocent, the unarmed, the defenseless. Even if someone knew what he was doing, when he did it he was in some sense insane, and indeed he looked the part. People can learn to blow other people apart rather easily under the guise of a system that legitimates their actions, such as extreme variants of Islam, the emperor worship of the kami kazi pilots, and other radical belief systems. It always comes down to what beliefs, what ends, what goals.

If we had a signature of insanity of the sort in the brain that we could show off, that would be fantastic; but we do not as of now. As Judge Rakoff (2015) at a Neuroscience Society meeting in Chicago in 2015, where nearly 30,000 people attended from across the world, said, the courts should not be burned again with expectations about neuroscience that would impact the court. As the judge writes in his article in *The New York Review of Books*, "not so fast."

The philosophical issues about free will and responsibility are not, as I have indicated, dissolved by neuroscience; I use the term "effort" and "choice" instead of free will (Schulkin 2007) (cf. Wegner 2002; Libet 2004), and certainly consistent with Holmes and Dewey. Sounding something, one could easily imagine Holmes saying, as Professor of Law Greely put it well, "I am not sure we have the free will to truly believe and act as if we do not have free will" (see also Roskies 2007;

Roskies and Morse 2013). But knowing something about neural degeneration is a consideration in understanding one another, legally and otherwise. Key issues about understanding others are enhanced by knowing how the brain works, but more importantly knowing the diverse social milieu in which they act. Understanding the brain is knowing the social milieu, and for those of us in the field it is knowing the social context in which science, in this case, neuroscience, takes place.

Its place in the court is still promissory. There is little mind reading by looking at neural structure under the best of conditions at the neuroscientific epistemic moment (Morse and Roskies 2013; Moreno and Schulkin 2020). But that does not mean that one day, and indeed, I would thank such neuroscientific facts will emerge in legal/ neuroscientific edifice of prediction and understanding.

But the avalanche of research is volumes full. Although the theory is lacking, many exciting methodologies have emerged, including diverse ways of imaging the brain, capturing the neural networks, and representing the diverse richness of information molecules and their expression in the brain and their regulation.

Holmes (1881) is often quoted for his statement that dogs know the difference between an intentional kick and being accidentally stumbled over—"even a dog distinguishes between being stumbled over and being kicked"—a distinction whether one may go to jail or be set free (Rakoff 2016).

Similarly, most of where neuroscience is in our common sensibility is fueled by diverse sciences of the mind. Without the brain, there is no behavior, but we still know more about behavior and its relevance than we do about the brain. The science is still relatively new and the larger body politic of knowledge will continue to find ways of relevance for it, in the schools, in the courtroom and over the life cycle.

Science shows that males more than females are likely to be more physically aggressive and violent. Yet both are punished, and supposedly equally under the law. Law is necessary to combat our vulnerabilities towards our neighbors. Law is required, but always, as Holmes understood, reflects the body politic, the political milieu. Science, while within the social contexts, looks to go beyond the body politic.

What Holmes perhaps did not appreciate or understand was the impact of a changing brain, not a frozen inevitability (McEwen et al. 2014). Chauncey Wright or John Dewey knew better than Holmes on this. Biology for our species is much more labile. Epigenetic changes appear to be a cardinal feature of basic neural function; plasticity is inherent as we all know in human expression as well as common core features. We still do not know, for the most part, how to measure brain function meaningfully in the law, in the court (Pardo and Patterson 2013).

But the adventure of trying to do so is part of the vernacular and pulse of a knowing process. Neuroscience is not about to overtake our sense of responsibility or the will. That there are endlessly antimonies that underlie our understanding of the brain and choice, of voluntary actions is something that is co-evolving with our understanding of ourselves (Schulkin 2007). An instrumentalist/realist and predictive point of view (Schulkin 2015; Solymosi and Shook 2014) underlies a pragmatist view of neuroscience, for the legitimacy of terms, like free will or good enough and earned freedom of choice in our lexicon of understanding.

We come prepared to represent objects perhaps by their sensorimotor features and functional characteristics in the brain (Martin 2009). Holmes was looking to externalize events, anchored to objects not to internal thoughts. Of course, the fluidity of thought and action, imagination and actual objects recruits many of the same brain regions (Kosslyn 1986).

We come prepared to infer events about others; their minds included. It is a key biological adaptation. A range of computational devices in the neural hardware spread across diverse brain regions renders this possible in multiple contexts. Reading minds is a possibility and will impact the law (Murphy and Greely 2011), but not soon. Exciting fusions that cut across the simulation of brains in machines and machine capability for brains nudges a little closer for discerning what a machine and a brain might look like when looking at a smiling face, or a dejected one.

Continued evolution is the wellspring of our cultural milieu. Law is both stable and evolving and tied to context and interests (Pound 1908, 1921a, b; Di Filippo 1988). One thing evolving is a legal culture that is highly interpretative (Hutchinson 2005). There is no inherent antipathy in acknowledging the interpretative turn in science or in the practice of law (Heelan and Schulkin 1998; Pardo and Patterson 2013).

Holmes was no saint, no sage. He was a thinker, a pragmatist, and a naturalist. He would have considered and appreciated the neurosciences but would have been sanguine and judiciously skeptical in their use. No eternal principles, just good enough practical ones. This is a Holmesian perspective (see his Law in Science). Principles of value and morals are built into the fabric of the body politic. That said, that does not mean they are of equal value and that we do not come into the world with pro-social propensities that ground human action and meaning.

Cultural Evolution: Law and the Larger Body Politic

One does not have to believe in original sin to see the panoply of human expressions in social cooperative behaviors; Holmes was no fool. In his small corner of the universe, under relatively idealist conditions, the universe was not washed clean of diverse social sins, one of which was the original sin of slavery, and the masquerading language of people born free. The Supreme Court had voted that Dred Scott was not really human enough to have rights. After all, in this same social context, some years later, the Bronx Zoo exhibited an African in a cage.

Judge Taney, the Supreme Court chief justice in the Dred Scott case (1856), represents a narrow view on the human condition, a legal one, where the court's role is to uphold the law. Holmes understood that unjust events occur because the body politic has not captured the worth of what matters. He was not brash. Justice Harlan, whom Holmes did not think much of in general, had the high moment or the moral ground in his dissent against separate and equal as inherently unequal and unfair in the decision of 1896 on the Supreme Court, nearly a decade before Holmes arrived. Harlan, who was also a veteran of the Civil War from the North, from Kentucky, was generally much more sympathetic to African Americans than Holmes (Rosen 2007; Grant 2016).

Holmes may have fought for a union, and may have been appalled by slavery as was Lincoln, but he wore few sentiments of social solidarity on his sleeves. He did not worry about the inequality of humans as Rousseau in poetic verse and social solidarity did. But he would agree with Rousseau or Hobbes or Spinoza and certainly Darwin about self-preservation.

Locating Holmes is finding him where he kept his blood-soaked uniform:

> I hope and I believe that long after our tears for the dead have been forgotten, this monument to their memory will still give such help to generations to whom it is only a symbol – a symbol of man's destiny and power for duty, but a symbol also of that something more by which duty is swallowed up in generosity, that something more which let men like Shaw to toss life and hope like a flower before the feet of their country and their cause. (1884, Harvard Lecture)

Our cultural evolution is bound to our laws—in commerce, in precedent, in rights, in arguments, in revision, in the battle for human freedom. It is bound in two senses: freedom to express and freedom to participate. Lincoln was bold. He broke the law to fight for freedom, once the fight had begun.

Unwise and provocative were also part of Holmes's legacy of literature. Holmes once asserted, "I am a philosopher and if people cannot stand the truth it is no concern of mine." He could be rather brash, bold, and obnoxious. He was human; most of us say a load we wished we had not. But Holmes was philosophical, engaged in the adventure of ideas. Holmes, as opposed to his friend, Brandeis, was not political (Urofsky 2015). Holmes was fatalistic in his correspondence, aiming at impartial judgment. His range of pragmatism goes far and wide across the range of pragmatists' positions: naturalism, predictive and anticipatory reasoning, external pillars, science and law continuous and not reductive. The emphasis is on "good enough theory" (Grey 1989) and more than just attention to consequences (Mendenhall 2015), a tendency to find balance, a tendency towards grounded narrow considerations within law.

From an adumbration about human destiny, the political context of Paine and Madison was materialized in ways that an earlier generation dominated by theology was not (Perry 1935; see Rakove 1996). Original meanings are understood by looking at the debates (Rakove 1996), at compromise and principle—a moment of what my colleague Jonathan Moreno calls "Deciding Together"—shaped amongst others by the likes of James Madison. Madison appears to seek some balance amongst extremes. He was an interesting and original thinker (Rakove 1990).

Lincoln, self-educated trial lawyer and literary entrepreneur, was also shrewd. He used the first amendment to target the press, to get his message out; he also knew how to shut down and control it under conditions of adversity (Holzer 2014). Free speech, from the traditions of, for instance, Locke or Montesquieu and the toleration and importance of free expression, nevertheless as Holmes noted, had its limits; context matters (Peters 2005; Stone 2004).

As opposed to a narrow positivism in the law, pragmatists recognize that moral notions pervade the law. Holmes understood that. The diverse dualism, including morals from law, are undercut. The point Holmes was making was not to superimpose endless moral disputations in adjudication in the law. The task at hand is the regulation of individuals and accountability in the culture we have. Perhaps—though, this is not Holmes—perhaps we should have.

Holmes's earlier flirtation with epistemic positivism gave way to pragmatism, where such dualism is eschewed (Howe 1957/1963; White 1986; Kellogg 1984). But, indeed getting clear about rules, not superimposing obfuscations like getting clear in science, being clear about methods, is essential to inquiry. Methods are one thing, and positivism, declawed from ontological purity, is one method amongst others, as instrumentalists with possible use of diverse methods. This holds in neuroscience.

Holmes understood the importance of history and that history is not uniform in its direction. There is no one sense of progress, despite the diverse ways in which we might understand its possibilities (e.g., Burry 1932/1960). It is an endless zig zag of foraging ahead in a competition of diverse ideas in a marketplace of possibilities. And, for Holmes, there are fatalistic expressions of fatigue, barbarity, and vulnerability amid the ideals of enriched experience of human possibilities and discovery.

Law for Holmes is an outgrowth of adaptation and human informa-tion processing—a form of pragmatism in an established part of the social fabric, as it blends into epistemic orientations and understand-ing. And law, like biology or neuroscience, is "a hodge-podge of stuff;" (Hank Greely, personal conversation 2014) life and death, marriage, property, social welfare, trade freedom, punishment, inheritance, con-tract, negligence, etc. There is no unification nor is there a pre-estab-lished harmony.

Of course, Holmes is remembered for his elegant use of language, his keen rhetoric, though duplicitous at times, direct and others not so (Grey 1992; Posner 1992; Luban 1992). Aesthetics runs through legal reasoning as it does most forms of human activity. That is one of the major messages of Dewey's *Art as Experience*. Indeed, aesthetics runs through jurisprudence, through the writing and reasoning, and the encapsulation of what is important to convey (Butler 2003). Holmes is literary. Interesting phrases are part of his judicial vernacular and strength (White 1993).

Literary beauty runs through the pen of Jefferson and Madison, and, some two hundred years later, in the voice of King (West 1985a, b). Romantics and skeptics run through the literature of the law. The law is inherently language. The many senses of language serve as the codifi-cation of human regulation in law-like legal expressions. Language is at the heart of law. Holmes understood that. Of course, he was often sug-gestive in his prose. In discussing whether a treaty is legitimate or not, he says, "acts of congress are the supreme law of the land only when made in pursuance of the constitution" (*Missouri v. Holland*, White 2016). Of course, this is open to endless situational and underdeter-mined meaning.

Holmes's life would be in one way or another a reference back to his war experience. One feature that emerges in some survivors is a form of detachment from others (Rogat 1964; Grey 1989). Survival has its costs, and there are many forms of long-term adaptations and vulner-abilities from trauma. For Holmes, one of the most meaningful experi-ences in his life was that of the American Civil War. He was influenced by those experiences for another 70 years.

Why should we care about Holmes? Well, he represents the diverse ways in which to consider the law, as a literate person who sought to naturalize and give a realistic account of the human condition. He understood the law in terms of context, history, fallibility, and inquiry. His sense of the law is much like what goes on in science, in this case, neuroscience.

He understood war. It impacted his brain. He would have wanted to know something about the brain. But, like Judge Rakoff (2016), he might have said, "not yet." Wonder is one thing and the prospect of integrating neuroscience and the law is quite exciting, like other epistemic endeavors. But skepticism would limit sheer excitement. This is the Holmesian moment. Holmes, the naturalistic realist, would, perhaps also like Judge Rakoff, see neuroscience as "mind boggling" and that neuroscience may be "useful"—will be useful but within the bounds of reason.

The rise and fall of Holmes's reputation (White 1971; Alschuler 2000; Grant 2016), over the last 100 years or so, is instructive of the historical context of understanding the individual. As I have indicated, he was no saint. His views are in favor of independence and maximal freedom. He was not ideological either. Freedom for him had to be placed in a context where choice might be limited (e.g., war). He can be vilified for what he did in his opinion on the *Buck* case (Kaplan 2016) or elevated as he often still is with regard to free speech, including his change in position with regard to some of the limits of speech (cf. Schenck, Abrams, Doolinig, *Wall Street Journal*, October 13, 2017).

With that came the accusations of his amorality. Science is about facts (Weber 1903), but if science is laden with value—as Holmes might have evolved into holding and which other pragmatists like Dewey might have asserted (Grey 2014)—then our legal system is itself a piece of cultural evolution of attempted problem solving (Elliott 1984). Of course, that does not mean that everything is up for grabs; some things are more formidably in place in our constitution and at its root meaning than others. Getting clear about meaning is a good heuristic. Holmes would agree with that, along with the political contributions coming from the other two branches of government.

Like his colleague, Brandeis, he stood for the integration of science including social science (White 1995) and the law (Pound 1921a, b). Pound, the Dean of the Harvard Law School, idealizing Holmes, interestingly received a PhD in the biological sciences (see LaPiana 1994). But Holmes was no progressive as Brandeis was (Urofsky 2009). While Holmes went to great lengths to separate himself from his colleagues (at Harvard, Dean Langdell), particularly in the emphasis on formalism, they would nevertheless agree about a number of cases (e.g., Northern Securities, Brown and Kimball 2001) that mattered. What does matter is legal pragmatism with a moral theory inherent in some of the laws themselves (Farber 1995).

Holmes was right to admire Dewey. Dewey labored to bridge and promote continuity of law and the larger body of knowledge—the continuity of science and the law, the utter link between the theoretical and the practical, nature and culture, ideals and material adaptation, thought and action. Dewey made knowledge acquisition the fabric of everyday life, ripe with aesthetics. The aim is problem solving and adjudication, the stuff of the law. This was the stuff that impressed Holmes, despite his dislike of Dewey's prose.

Dewey, like Holmes, was part of the transition from premodern to modern thinking. Holmes was an elitist, an uncommon person, writing a "common law;" Dewey was a common person writing a "common faith" about the stuff that brings us together, what we owe each other or should owe each other. Both are wedded to a sense of nature and culture as continuous with science, inquiry, and the larger body public. And Dewey in his treatise Experience and Nature would refer to Holmes as "one of our greatest American philosophers" and would further note that "Justice Holmes touches upon the relation of philosophy (thus conceived) to our scientific and metaphysical insight into the kind of a world in which we live" (p. 338).

While there is no extra of specific part of the brain devoted to ethics, that is not say that the brain does not underlie ethical judgment. We come prepared to make social judgments, to punish and reward behaviors for group solidarity, and for basic adaption and viability. Brain activity is obviously at the heart of all that, but then so is the culture we are living in. Law represents the moral trajectories that runs through our practices, which Holmes understood.

Associated with positivism or utilitarianism, Holmes was looking to be empirical, though there was a period when he seemed more a positivist than not (e.g., the absolute separation of facts from values). As I have reiterated throughout the book, like other pragmatists, he realized that the fluidity of facts and values converge in decision making. Holmes, the pragmatist, takes precedence. But the many faces of pragmatism dilute any substance of meaning.

Indeed, there is no extra region of the brain devoted exclusively to valuation. Valuation is at the heart of decision making. We give value to events. This is done in statistical inferences, as when Quine, perhaps the pragmatist closest to Holmes, uttered, "to be is to be the value of a bound variable." But Holmes, like Dewey, was probably quick to see the large role that valuation plays in human reasoning, and not just in quantification. Valuation inheres in the judgment of whether something is safe, worthy, aesthetic, or repulsive.

Law is couched in the common law that we inherited along with notions of freedom, and freedom from laws that are unjust and unfair: "no taxation without representation." Like everything, history matters; and it certainly matters in understanding the law. Holmes understood that, though he was not that careful about it (Rabban 2013).

Holmes, the conservative, is in favor of judicial restraint. This is a tradition that is also found in Marshall (Kellogg 2003), a primary force in the development and instantiation of power in the judicial branch, the third wing of power in the US constitution. Holmes always believed that "the best truth is the power of the thought be itself accepted in the competition of the market" (*Abrams v. United States*).

The body politic in a thriving democracy allows for the competition of ideas (Blasi 2005). It is not housed in a narrow Spencerian sensibility, but a hodgepodge of conflicting views (Schwarz 1985). Of course, as Holmes (1873) noted early, struggle and strife are a common currency in class conflict, in workers' expectations. "The struggle for life" is a common currency in our transactions. Holmes was always close to this recognition, though not particularly sensitive to the people.

A universe of commerce, competition, and applied intelligence lay behind Marshall as it did even more clearly behind Holmes a century later (Posner 1992). Whether a case involved African Americans or

American Indians, Marshall was always cautious, conservative, and oriented to a union together, stronger. But an ethical sentiment, an opposition to slavery, led to his trying to fight Andrew Jackson with regard to the great trail of tears and what was to become in a few decades a war between the states—a civil war as Marshall adumbrated would occur.

Holmes's mind to the law is what James's was to psychology and what Peirce's was to philosophy: spelling out the logic of inquiry and discovery, all anchored in classical pragmatism and housed in the larger community of inquirers. No doubt both learned from each other during their periods of interactions in the early 1870s at Cambridge, Massachusetts.

Peirce in an essay from 1873 would suggest a pragmatist theme, "living doubt is the life of investigation. When doubt is set at rest inquiry must stop." Well, it does not. But certainly, that is part of it. In an essay on the classification of the sciences, Peirce would suggest that "the social instincts were more sympathetic to reason" when tied to efficient causes that reflect human performance and purpose and cultivated within a community of inquirers.

For Holmes, like Peirce, the proceeds of inquiry, if they became instantiated, became part of the communities that we inhabit (see Wells-Hantzis 1987–1988). But there are important differences; *pragmaticism* was coined to separate Peirce's notion of pragmatism from James's popularization. Holmes, who never spelled out any systematic perspective, except for *The Common Law*, straddled between many forms of pragmatism and philosophical perspectives. What is constant are the emphasis on external anchors, adjudication and competition of ideas, and naturalism within epochal expression of creative energy and creativity, and being carried by the battles of the day. The world was also rich with the irrational. He is not so detached up close and personal in his worldview.

Holmes, like Peirce, perhaps was too fast to eschew anything internal. All measures were external, anchored to something tangible, and placed in a probabilistic bed rock of prediction, coherence, reliability, and viability. Reason is demythologized; and a common law is put in the larger context of culture, inquiry, and our social evolution. Nothing is linear or necessary.

Holmes underestimated what we can learn about internal machinations, from the cognitive architecture, and an understanding of the adaptive fluidity of decision making from emotional to more detached, from immediate to more reflective. The neuroscience is embedded throughout; it is just not as reductive as many would like. In this age of neuroscience, considerations that impact diverse regions of the law, ethics, are symbolic of neuroscience's impact on the larger body of our lives.

Conceptual issues and concerns about theory depth pervade law as they do neuroscience. Both are couched in terms of cultural evolution and continuity from natural selection, given the kinds of brains we have, the kinds of capabilities we have, the degree of plasticity and epigenetic regulation, etc.

It is hard to be "deep epistemically" Again, "is that all there is my friend then let the music..." as the Peggy Lee song goes. Change and living with some viability amidst adaption are themes about law in the regulation of the body politic in which we find ourselves, in which we are trying to blend a "more perfect" place, or set of places.

But with Holmes one wants, as I do, to see the continuity of science within the law, for how we understand people. Our notion of mechanisms impacts our notion of retribution (Shariff et al. 2014), but it does not solve or dissolve issues about free will, except in very extreme cases. In the bulk of human decision making and action, we are left with such issues, despite what we know at the moment in neuroscience.

We also understand, as Holmes knew, the irrationality that pervades human experience, even in the best of us (Vincent 2013), and how the brain expresses diverse forms of irrational decision making or even destructive decision making. While there are some suggestions of using "neuroprotection" for possible future re-arrest (lower levels of neural activation in the anterior cingulate region of the brain, Aharoni et al. 2013) or violent behaviors, and, certainly, for diverse forms of psychopathology (Fulham et al. 2009), we are far from something we can call neurolaw. But it is all new. The science is new. It will evolve as there are signs where one can tell through neural expression and then prediction whether a defendant in a courtroom had knowledge of an event or was simply reckless (Vilares et al. 2017). Indeed, the interface between machine capability and neural expression will be one continuous function in our scientific exploration as well as our judicial. Soon but not yet.

As Judge Rakoff cautioned and Holmes would understand, "Not So Fast." Indeed, I do not see at this time what that means except for the positive sense of further integration of neuroscience into the body politic of knowledge and inquiry and self-understanding (Markowitsch and Staniloiu 2011). But that is no small thing. Judge Holmes would have appreciated that.

Pragmatists, like Holmes, are comfortable with the continuity of science and the rest of life, including the law. Demythologizing the science, neuroscience or otherwise, is being a grown up. Love science, mistrust overzealous scientism. I think Holmes would understand that, as well as the fact that our moral sensibility is embedded in the oxygen of regulation in the law. But it was Dewey that gave us a rich philosophical perspective for a philosophy of law and an orientation to neural sciences that could be captured into the broder family of the human sciences, science or inquiry that serves high ideals towards broad based human well being. We have evolved to come to terms with responsibility, however frail, muted, clouded, and underdetermined. Such is the lot of the human condition.

References

Aharoni, E., Funk, C., Sinnott-Armstrong, W., & Gazzaniga, M. (2008). Can Neurological Evidence Help Courts Assess Criminal Responsibility: Lessons from the Law and Neuroscience. *Annals New York Academy of Sciences, 1124,* 145–160.

Aharoni, E., Vincent, G. M., Harenski, C. L., Calhoun, V. D., Sinnott-Armstrong, W., Gazzaniga, M. S., et al. (2013). Neuroprotection of Future Arrest. *PNAS, 110,* 6223–6228.

Alces, P. A. (2018). *The Moral Conflict of Law and Neuroscience.* Chicago: University of Chicago Press.

Alschuler, A. W. (2000). *Law Without Values: The Life, Work and Legacy of Justice Holmes.* Chicago: University of Chicago Press.

Blasi, V. (2005). Holmes and the Market Place of Ideas. *The Supreme Court Review, 2004*, 1–46.

Brown, R., & Kimball, B. A. (2001). When Holmes Borrowed from Langdell: The "Ultralegel" Formalism and Public Policy of Northern Securities (1904). *The American Journal of Legal History, 45*, 278–321.

Budiansky, S. (2019). *Oliver Wendell Holmes: A Life in War, Law, and Ideas.* New York: Norton.

Burry, J. B. (1932/1960). *The Idea of Progress.* New York: Dover Press.

Butler, B. E. (2003). Aesthetics and American Law. *Legal Studies Forum, 1*, 203–220.

Damasio, A. R. (1996). The Somatic Marker Hypothesis and the Possible Functions of the Prefrontal Cortex. *Philosophical Transactions of the Royal Society of London, 354*, 1413–1420.

Damasio, A. (2007). Neuroscience and Ethics: Intersections. *American Journal of Bioethics, 7*, 3–7.

Darrow, C. (1926). The Eugenics Cult. *The American Mercury, 8*, 5–12.

Di Filippo, T. (1988). Pragmatism, Interest Theory and Legal Philosophy: The Relation of James and Dewey to Roscoe Pound. *Transactions of the C. S. Peirce Society, 24*(Fall), 487–508.

Elliott, E. D. (1984). Holmes and Evolution: Legal Process as Artificial Intelligence. *Faculty Scholarship Series, 5079*(13), 113–146.

Farber, D. A. (1995). Reinventing Brandeis: Legal Pragmatism for the Twenty-First Century. *Berkeley Law Review, 163*, 163–190.

Feldman, R. (2009). *The Role of Science in Law.* Oxford: Oxford University Press.

Freeman, M. (2006). Law and Neuroscience. *International Journal of Law in Context, 2*, 217–219.

Fulham, R. S., McKie, S., & Dolan, M. C. (2009). Psychopathic Traits and Deception: Functional Magnetic Resonance Imaging Study. *British Journal of Psychiatry, 194*, 229–235.

Garland, B., & Frankel, M. S. (2006). Considering Convergence: A Policy Dialogue About Behavioral Genetics. *AAS. Neuroscience and Law* (Winter/Spring), 101–113.

Gould, S. J. (1984). Carrie Buck's Daughter. *Natural History, 93*, 14–18.

Grant, S. M. (2016). *Oliver Wendell Holmes Jr. Civil War Soldier, Supreme Court Justice.* New York and London: Routledge.

Greene, J. D. (2014). Beyond Point and Shoot Morality: Why Cognitive Neuroscience Matters for Ethics. *Ethics, 124*, 695–726.

Greely, H. T. (2010). Enhancing Brains: What Are We Afraid Of? *Cerebrum, 2010, 14*, 8–18.

Greely, H. T. (2011). Reading Minds with Neuroscience—Possibilities for the Law. *Cortex, 47*, 1254–1255.

Grey, T. C. (1989). Holmes and Legal Pragmatism. *Stanford Law Review, 41*, 787–870.

Grey, T. C. (1992). Holmes, Pragmatism and Democracy. *Oregon Law, 71*, 521–542.

Grey, T. C. (2014). *Formalism and Pragmatism in American Law*. Boston: Brill Press.

Haack, S. (2011). Pragmatism, Law, and Morality: The Lessons of *Buck v. Bell. European Journal of Pragmatism and American Philosophy, 2*, 67–87.

Hacking, I. (1965/1979). *Logic of Statistical Inference*. Cambridge: Cambridge University Press.

Heelan, P. A., & Schulkin, J. (1998). Hermeneutical Philosophy and Pragmatism: A Philosophy of Science. *Synthese, 115*, 269–302.

Hirstein, W., Stifford, K. L., & Fagan, T. K. (2018). *Responsible Brains: Neuroscience, Law and Human Culpability*. Cambridge: MIT Press.

Holmes, O. W., Jr. (1873). The Gas-Stokers Strike. *American Law Review, 7*, 582–583.

Holmes, O. W., Jr. (1881/1952). *The Common Law*. New York: Dover Publications.

Holmes, O. W., Jr. (1884). *Sons of Harvard Who Fell in Battle*. Harvard Alumni Dinner Speech.

Holmes, O. W., Jr. (1897a). The Path of the Law. *Harvard Law Review, 457*.

Holmes, O. W., Jr. (1897b). *The Fraternity of Arms*. Remarks at a Meeting of the 20th Regimental Association.

Holmes, O. W., Jr. (1927). *Buck v. Bell*.

Holzer, H. (2014). *Lincoln and the Power of the Press*. New York: Simon & Schuster.

Horn, S. R., Charney, D. S., & Feder, A. (2016). Understanding Resilience: New Approaches for Preventing and Treating PTSD. *Experimental Neurology, 284*, 119–132.

Howe, M. D. (1957/1963). *Justice Oliver Wendell Holmes: Volumes 1 and Volumes 2—The Proving Years*. Cambridge: Harvard University Press.

Hutchinson, A. C. (2005). *Evolution and the Common Law*. Cambridge: Cambridge University Press.

Kagan, J. (2002). *Surprise, Uncertainty and Mental Structure*. Cambridge: Harvard University Press.

Kaplan, A. (2016). *Imbecility: The Supreme Court, American Eugenics and the Sterilization of Carrie Buck*. New York: Penguin.

Kellogg, F. R. (1984). *Formative Essays of Oliver Wendell Holmes Jr*. London: Greenwood Press.

Kellogg, F. R. (2003). Holmes, Common Law Theory, and Judicial Restraint. *The John Marshall Law Review, 36*, 457–502.

Kosslyn, S. K. (1986). *Image and Mind*. Cambridge: Harvard University Press.

Kozel, F. A., et al. (2005). Detecting Deception Using fMRI. *Biological Psychiatry, 58*, 605–613.

LaPiana, W. P. (1994). *Logic and Experience*. Oxford: Oxford University Press.

Libet, B. (2004). *Mind Time*. Cambridge: Harvard University Press.

Luban, D. (1992). Justice Holmes and Judicial Virtue. *Nomos, 34*, 235–264.

Markowitsch, H. J., & Staniloiu, A. (2011). Neuroscience, Neuroimaging and the Law. *Corte, 47*, 1248–1251.

Martin, A. (2009). Circuits in Mind: The Neural Foundations for Object Concepts. In M. S. Gazzaniga (Ed.), *The Cognitive Neurosciences* (4th ed., pp. 1031–1045). Cambridge: MIT Press.

McEwen, B. S., Gray, J. D., & Nasca, C. (2014). Recognizing Resilience: Learning from the Effects of Stress on the Brain. *Neurobiology of Stress, 1*, 1–11.

Mendenhall, A. P. (2015). Pragmatism on the Shoulders of Emerson: Oliver Wendell Holmes Jr.'s Jurisprudence as a Synthesis of Emerson, James and Dewey. *The South Carolina Review, 48*(1), 93–109.

Moreno, J. D., & Schulkin, J. (2020, in press). *The Brain in Context: A Pragmatic Guide to Neuroscience*. New York: Columbia University Press.

Morse, S. J. (2010). Lost in Translation: An Essay on Law and Neuroscience. In *Law and Neuroscience*. Oxford: Oxford University Press.

Morse, S. J. (2013). Common Criminal Law Compatibilism. In N. A. Vincent (Ed.), *Neuroscience and Legal Responsibility*. Oxford: Oxford University Press.

Morse, S. J., & Roskies, A. L. (Eds.). (2013). *A Primer on Criminal Law and Neuroscience*. Oxford: Oxford University Press.

Murphy, E., & Greely, H. T. (2011). What Will Be the Limits of Neuroscience-Based Mind-Reading and the Law. In *Oxford Handbook of Neuroethics*. Oxford: Oxford University Press.

Nadel, L., & Sinnott-Armstrong, W. P. (2012). *Memory and Law*. Oxford: Oxford University Press.

Pardo, M. S., & Patterson, D. (2013). *Minds, Brain and Law: The Conceptual Foundations of Law and Neuroscience*. Oxford: Oxford University Press.

Perry, R. B. (1935). *The Thought and Character of William James* (Vol. 1 & 2). Boston: Little, Brown.

Peters, J. D. (2005). *Courting the Abyss: Free Speech and the Liberal Tradition.* Chicago: University of Chicago Press.

Petoff, A. (2015). Neurolaw: A Brief Introduction. *Iranian Journal of Neurology, 5,* 53–58.

Pound, R. (1908). Mechanical Jurisprudence. *Columbia Law Review, 8,* 609–610.

Pound, R. (1921a). A Theory of Social Interests. *American Sociological Society, 15,* 16–45.

Pound, R. (1921b). Judge Holmes's Contributions to the Science of the Law. *Harvard Law Review, 34,* 449–453.

Posner, R. A. (1992). *The Essential Holmes.* Chicago: University of Chicago Press.

Rabban, D. M. (2013). *Law's History: American Legal Thought and the Transatlantic Turn of History.* Cambridge: Cambridge University Press.

Raine, A., Lencz, T., Taylor, K., Hellige, J. B., Bihrle, S., Lacasse, L., et al. (2003). Corpus Callosum Abnormalities in Psychopathic Antisocial Individuals. *Archives of General Psychiatry, 60,* 1134–1142.

Rakoff, J. S. (2011). Science and the Law: Uncomfortable Bedfellows. *Seton Hall Law Review, 38,* 1379–1393.

Rakoff, J. S. (2016, May). Neuroscience and the Law: Don't Rush In. *The New York Review of Books.*

Rakove, J. N. (1990). *James Madison and the Creation of the American Republic.* New York: HarperCollins.

Rakove, J. N. (1996). *Original Meanings.* New York: Knopf.

Rogat, Y. (1964). The Judge as Spectator. *University of Chicago Law Review, 31,* 213–256.

Rosen, J. (2007). *The Supreme Court.* New York: Henry Holt and Co.

Roskies, A. L. (2007). Are Neuroimages Like Photographs of the Brain? *Philosophy of Science, 74,* 660–672.

Roskies, A. L., & Morse, S. J. (2013). Neuroscience and the Law: Looking Forward. In S. J. Morse & A. L. Roskies (Eds.), *A Primer on Criminal Law and Neuroscience.* Oxford: Oxford University Press.

Schulkin, J. (2007). *Effort: A Behavioral Neuroscience Perspective on the Will.* Mahway: Erlbaum Press.

Schulkin, J. (2015). *Pragmatism and the Search for Coherence in Neuroscience.* London: Palgrave Macmillan.

Schwarz, J. (1985). Oliver Wendell Holmes's "The Path of the Law". *American Journal of Legal History, 29,* 235–250.

Scott, T. R. (2012). Neuroscience May Supersede Ethics and Law. *Science and Engineering Ethics, 18,* 433–437.

Shariff, A. F., Greene, J. D., Karremans, J. C., Luguri, J. B., Clark, C. J., Schooler, J., et al. (2014). Free Will and Punishment: A Mechanistic View of Human Nature Reduces Retribution. *American Psychological Science, 20,* 1–8.

Shen, F. X., Hoffman, M. B., Jones, O. D., Greeme, J. D., & Marois, R. (2011). Sorting Guilty Mind. *New York University Law Review, 86,* 1306.

Sinnott-Armstrong, W. (2008). *Moral Psychology: The Neuroscience of Morality: Emotion, Brain Disorders and Development.* Cambridge: MIT Press.

Solymosi, T., & Shook, J. R. (2014). *Neuroscience, Neurophilosophy and Pragmatism: Brains at Work with the World.* London: Palgrave Macmillan.

Stone, G. R. (2004). *Perilous Times.* New York: Norton.

Tennison, M. N., & Moreno, J. D. (2012). Neuroscience, Ethics, and National Security: The State of the Art. *PLoS Biology,* 10, 1–4.

Tsomo, K. L. (2012). Compassion, Ethics and Neuroscience: Neuroethics Through Buddhist Eyes. *Science and Engineering Ethics, 18,* 529–537.

Urofsky, M. I. (2009). *Louis D. Brandeis.* New York: Pantheon Books.

Urofsky, M. I. (2015). *Dissent and the Supreme Court.* New York: Pantheon Books.

Vilares, I., Wesley, M. J., Ahn, W. Y., Bonnie, R. J, Hoffman, M., Jones, O. D., et al. (2017). Predicting the Knowledge-Reckless Distinction in the Human Brain. *PNAS, 114,* 3222–3327.

Vincent, N. A. (2013). Enhancing Responsibility. In N. A. Vincent (Ed.), *Neuroscience and Legal Responsibility.* Oxford: Oxford University Press.

Weber, M. (1903/1917/1949). *The Methodology of the Social Sciences.* New York: Free Press.

Wegner, D. M. (2002). *The Illusion of Conscious Will.* Cambridge: Harvard University Press.

Wells-Hantzis, C. (1988). Legal Innovation Within the Wider Intellectual Tradition: The Pragmatism of Oliver Wendell Holmes Jr. *Northwestern Law Review, 82,* 541–595.

West, R. L. (1985a). *Jurisprudence as Narrative: An Aesthetic Analysis of Modern Legal Theory, 60,* 145–211.

West, R. L. (1985b). Liberalism Rediscovered: A Pragmatic Definition of the Liberal Vision. *Pittsburgh Law Review, xliv,* 673–738.

White, G. E. (1971). The Rise and Fall of Justice Holmes. *University of Chicago Law Review, 39,* 51–77.

White, G. E. (1986). Looking at Holmes in the Mirror. *Law and History Review, 4,* 439–465.

White, G. E. (1993). *Justice Oliver Wendell Holmes: Law and the Inner Self.* Oxford: Oxford University Press.

White, G. E. (1995). The Canonization of Holmes and Brandeis: Epistemology and Judicial Reputations. *New York Law Review, 70,* 576–621.

White, S. E. (2008). Brave New World: Neurowarfare and the Limits of Humanitarian Law. *Cornell International Law Journal, 41,* 177–210.

Yang, Y., & Raine, A. (2009). Prefrontal Structural and Functional Brain Imaging Findings in Antisocial, Violent, and Psychopathic Individuals. *Psychiatry Review, 174,* 81–88.

References

Abe, N. (2011). How the Brain Shapes Deception. *The Neuroscientist, 17,* 560–574.

Abe, N., & Greene, J. D. (2014). Response to Anticipated Reward in the Nucleus Accumbens Predicts in an Independent Test of Honesty. *The Journal of Neuroscience, 34,* 10564–10572.

Adams, M. C. (2014). *Living: The Dark Side of the Civil War.* Baltimore: Johns Hopkins University Press.

Addams, J. (1911). *Democracy and Social Ethics.* London: Macmillan.

Adolphs, R. (1999). Social cognition and the human brain. *Trends in Cognitive Sciences, 3,* 469–479.

Aharoni, E., Funk, C., Sinnott-Armstrong, W., & Gazzaniga, M. S. (2008). Can Neurological Evidence Help Courts Assess Criminal Responsibility: Lessons from the Law and Neuroscience. *New York Academy of Sciences, 1124,* 145–160.

Aharoni, E., Vincent, G. M., Harenski, C. L., Calhoun, V. D., Sinnott-Armstrong, W., Gazzaniga, M., et al. (2013). Neuroprotection of Future Arrest. *Proceedings of the National Academy of Sciences of the United States of America, 110,* 6223–6228.

Albrecht, J. M. (2012). *Reconstructing Individualism.* New York: Fordham University Press.

© The Editor(s) (if applicable) and The Author(s), under exclusive license to Springer Nature Switzerland AG 2019
J. Schulkin, *Oliver Wendell Holmes Jr., Pragmatism and Neuroscience,*
https://doi.org/10.1007/978-3-030-23100-2

Alces, P. A. (2018). *The Moral Conflict of Law and Neuroscience.* Chicago: University of Chicago Press.

Allen, D. (2014). *Our Declaration.* New York: Norton.

Allison, T. (2001). Neuroscience and Morality. *The Neuroscientist, 1,* 360–364.

Alschuler, A. W. (1997). The Descending Trail: Holmes's Path of the Law One Hundred Years Later. *Florida Law Review, 49,* 353–418.

Alschuler, A. W. (2000). *Law Without Values: The Life, Work and Legacy of Justice Holmes.* Chicago: University of Chicago Press.

Alschuler, A. W. (2009). From Blackstone to Holmes: The Revolt Against Natural Law. *Pepperdine Law Review, 36,* 491–505.

Anderson, B. L., & Schulkin, J. (Eds.). (2014). *Numerical Reasoning in Judgments and Decision Making About Health.* Cambridge: Cambridge University Press.

Ansell, C. (2015). *Pragmatic Democracy: Evolutionary Learning as Public Policy.* Oxford: Oxford University Press.

Arendt, H. (1963/1965). *Eichmann in Jerusalem: A Report of the Banality of Evil.* New York: Penguin Press.

Aristotle. (1953/1974). *Ethics.* New York: Penguin Books.

Aristotle. (1968). *De Anima* (D. Hamylin, Trans.). Oxford: Oxford University Press.

Armstrong, W. S. (2008). *Moral Psychology* (Vols. 1, 2, & 3). Cambridge: MIT Press.

Armstrong, W. S., Roskies, A., Brown, T., & Murphy, E. (2008). Brain Images and Legal Evidence. *Episteme, 208,* 359–373.

Atiyah, P. S. (1983). The Legacy of Holmes Through English Eyes. *Boston University Law Review, 63,* 341–380.

Atran, S. (1990/1996). *Cognitive Foundations of Natural History.* New York, NY: Cambridge University Press.

Atran, S. (2002). *In Gods We Trust: The Evolutionary Landscape of Religion.* Oxford: Oxford University Press.

Atran, S., & Medin, D. (2008). *The Native Mind and the Cultural Construction of Nature.* Cambridge: MIT Press.

Austin, J. (1832). *The Province of Jurisprudence Determined.* London: John Murray.

Austin, J. (1885). *Lectures on Jurisprudence and the Philosophy of Positive Law.* London: John Murray.

Ayer, A. J. (1946/1952). *Language, Truth and Logic.* New York: Dover Press.

Baker, L. (1991). *The Justice from Beacon Hill: The Life and Times of Oliver Wendell Holmes Jr.* New York: Harper & Row.

Balkin, J. M., & Levinson, S. (1998). The Badman—The Good and the Self-Reliant. *Boston University Law Review, 885,* 1–20.

Baron-Cohen, S., Tager-Flusberg, H., & Cohen, D. J. (1993/2000). *Understanding Other Minds.* Oxford: Oxford University Press.

Bartz, J. (2016). Oxytocin and the Pharmacological Dissection of Affiliation. *Current Directions in Psychological Science, 25,* 104–110.

Barzun, C. L. (2018). Three Forms of Legal Pragmatism. *Washington University Law Review, 95,* 1003–1034.

Bayes, T., & Price, R. (1763). An Essay Towards Solving the Problem in the Doctrine of Chance. *Philosophy of the Royal Society 43,* 370–418.

Beard, C. S. (1913/1986). *An Economic Interpretation of the Constitution of the United States of America.* New York: Free Press.

Bechara, A. (2003). Decision Making, Impulse Control and the Loss of Will Power to Resist Drugs: A Neurocognitive Perspective. *Nature Neuroscience, 8,* 1458–1463.

Bechara, A., Damasio, H., & Damasio, A. R. (2003). Role of the Amygdala in Decision Making. *Annals of the New York Academy of Sciences, 985,* 356–369.

Bent, S. (1932). *Justice Oliver Wendell Holmes Jr.: A Biography.* New York: Garden City Publications.

Bentham, J. (1782/1970). *An Introduction to the Principles of Morals and Legislation: Of Laws.* London: University of London.

Berkowitz, R. (2005). *The Gift of Science: Leibniz and the Modern Legal Tradition.* New York: Fordham University Press.

Berlin, I. (1969). *Four Essays on Liberty.* Oxford: Oxford University Press.

Berlin, I. (1976). *Vico and Herder: Two Studies in the History of Ideas.* London: Chatto & Windu.

Berlin, I. (1999). *The Roots of Romanticism.* Princeton: Princeton University Press.

Bernstein, R. J. (2010). *The Pragmatic Turn.* Cambridge: Polity Press.

Berridge, K. C. (2004). Motivation Concepts in Behavioral Neuroscience. *Physiology and Behavior, 81,* 179–209.

Blackman, J., Frye, B. L., & Closkey, M. (2013). Justice John Marshal Harlan: Professor of Law. *The George Washington Law Review, 81,* 1063–1134.

Blasi, V. (1999). Reading Holmes Through the Lens of Schauer: The Abrams Dissent. *Notre Dame Law Review, 72,* 1343–1360.

Blasi, V. (2005). Holmes and the Market Place of Ideas. *The Supreme Court Review, 2004*, 1–46.

Bloom, H. (2002). *Genius*. New York: Warner Books.

Bloom, H. (1983). *Agon*. Oxford: Oxford University Press.

Blumenthal, S. L. (2007). The Default Legal Person. *UCLA Law Review, 54*, 1135–1265.

Blumenthal, S. L. (2016). *Law and the Modern Mind: Consciousness and Responsibility in American Legal Cutlure*. Cambridge: Harvard University Press.

Bonner, T. N. (2002). *Iconoclast: Abraham Flexner and a Life in Learning*. Baltimore, MD: Johns Hopkins University Press.

Bork, R. H. (1980). *The Tempting of America*. New York: Free Press.

Bostrom, N. (2014). *Superintelligence*. Oxford: Clarendon Press.

Bowen, C. D. (1945). *Yankee from Olympus*. Boston: Little Brown.

Boyd, B. (2010). *On the Origins of Stories: Evolution, Cognition and Fiction*. Cambridge: Harvard University Press.

Boyd, R. (1993). Metaphor and Theory Change. In A. Ortony (Ed.), *Metaphor and Thought*. Cambridge: Cambridge University Press.

Boyd, R., & Richerson, P. (2005). *Not by Genes Alone: How Culture Transformed Human Evolution*. Chicago: University of Chicago Press.

Brandeis, L. D. (1914). *Other Peoples' Money and How Bankers Use It*. New York: Stokes.

Brandeis, L. D. (1916). The Living Law. *Illinois Law Review, 10*, 4–22.

Brandom, R. B. (2010). *Reason in Philosophy*. Cambridge: Cambridge University Press.

Brent, J. (1993). *Charles Sanders Peirce*. Bloomington: Indiana University Press.

Brower, M., & Price, B. (2002). Neuropsychiatry of the Frontal Lobe. Dysfunction in Violent and Criminal Behavior. *Journal of Neurology, Neurosurgery & Psychiatry, 71*, 720–729.

Brown, R. B., & Kimball, B. A. (2001). When Holmes Borrowed from Langdell: The "Ultra Legal" Formalism and Public Policy of Northern Securities (1904). *The American Journal of Legal History, 45*, 278–321.

Buber, M. (1937/1979). *I and Thou*. Edinburgh: Clark.

Buckholtz, J. W., & Faigman, D. L. (2014). Promises, Promises for Neuroscience and Law. *Current Biology, 24*, 861–867.

Budiansky, S. (2019). *Oliver Wendell Holmes: A Life in War, Law, and Ideas*. New York: Norton.

Burry, J. B. (1932/1960). *The Idea of Progress*. New York: Dover Press.

Burt, J. (2013). *Lincoln's Tragic Pragmatism*. Cambridge: Harvard University Press.

Burton, D. H. (1979). *Oliver Wendell Holmes Jr.: What Manner Liberal*. New York: Krieger.

Burton, D. H. (1998). *Taft and Holmes and the 1920s Court*. London: Associated Press.

Burton, S. J. (Ed.). (2000). *The Path of the Law and Its Influence: The Legacy of Oliver Wendell Holmes Jr*. Cambridge: Cambridge University Press.

Butler, B. E. (2000). Posner's Problem with Moral Philosophy. *University of Chicago Law School, 7*, 307–326.

Butler, B. E. (2002). Legal Pragmatism: Banal or Beneficial as a Jurisprudential Position. *Essays in Philosophy, 3*, Article 14.

Butler, B. E. (2003). Aesthetics and American Law. *Legal Studies Forum, 1*, 203–220.

Butler, B. E. (2010). Democracy and Law: Situating Law Within John Dewey's Democratic Vision. *Ethics and Politics, 12*(1), 256–280.

Butler, B. E. (2014). Pragmatism, Democratic Experimentalism and the Law. In G. Hubbs & D. Lid (Eds.), *Pragmatism, Law and Language*. London: Routledge.

Butler, B. E. (2017). *The Democratic Constitution*. Chicago: University of Chicago Press.

Butterfield, H. (1981). *The Origins of History*. New York: Basic Books.

Buzsaki, G., Peyrache, A., & Kubie, J. (2014). Emergence of Cognition from Action. *Cold Springs Harbor Laboratory Press, 23*, 1–15.

Byrne, R. W., & Bates, L. A. (2007). Sociality, Evolution and Cognition. *Current Biology, 17*, R714–R723.

Byrne, R. W., & Corp, N. (2004). Neocortex Size Predicts Deception Rate in Primates. *Proceedings of the Royal Society, 271*, 1693–1699.

Cacioppo, J. T., Visser, P. S., & Pickett, C. L. (2006). *Social Neuroscience*. Cambridge: MIT Press.

Carodozo, B. N. (1921). *The Nature of the Judicial Process*. New Haven: Yale University Press.

Casebeer, W. D. (2003/2005). *Natural Ethical Facts: Evolution, Connectionism and Cognition*. Cambridge: MIT Press.

Cassirer, E. (1944/1978). *An Essay on Man*. New Haven: Yale University Press.

Catani, M., & Sandrone, S. (2015). *Brain Renaissance*. Oxford: Oxford University Press.

Cavell, S. (1990). *Conditions Handsome and Unhandsome*. Chicago: University of Chicago Press.

Chan, W.-T. (1963). *A Source Book in Chinese Philosophy*. Princeton: Princeton University Press.

Charney, D. (2013). *Neurobiology of Mental Illness*. Oxford: Oxford University Press.

Charuvastra, A., & Cloitre, M. (2008). Social Bonds and Posttraumatic Stress Disorder. *Annual Psychology, 59*, 301–328.

Cheney, D., & Seyfarth, R. M. (2007). *Baboon Metaphysics*. Chicago: University of Chicago Press.

Cherniak, C. (1986). *Minimal Rationality*. MIT: Cambridge University Press.

Churchland, P. S. (2011). *Brain Trusts*. Princeton: Princeton University Press.

Cicero. (1960). *Selected Works*. New York: Penguin.

Clark, A. (1996). *Being There*. Cambridge: MIT Press.

Clark, A. (2013). Whatever Next? Predictive Brains, Situated Agents and the Future of Cognitive Science. *Behavioral and Brain Sciences, 36*(3), 181–204.

Clark, A. (2017). *Surfing Uncertainty*. Oxford: Oxford University Press.

Cohen, F. S. (1935). Transcendental Nonsense and the Functional Approach. *Columbia Law Review, 35*, 809–849.

Cohen, F. S. (1951). *Judicial Ethics*. Yale Law School Legal Scholarship Repository.

Cohen, M. R. (1931/1959). *Reason and Nature: An Essay on the Meaning of the Scientific Method*. New York: Dover Press.

Cohen, M. R. (1949). *Reason and the Law*. Glencoe, IL: Free Press.

Cohen, M. R., & Nagel, E. (1934). *An Introduction to Logic and Scientific Method*. New York: Harcourt, Brace and Company.

Collingwood, R. G. (1994). *The Idea of History*. Oxford: Oxford University Press.

Collingwood, R. G. (1945/1976). *The Idea of Nature*. Oxford: Oxford University Press.

Comfort, N. (2012). *The Science of Human Perfection*. New Haven: Yale University Press.

Commanger, H. S. (1950). *The American Mind*. New Haven: Yale University Press.

Cornford, F. M. (1912/1981). *From Religion to Philosophy*. Princeton: Princeton University Press.

Craver, C. L. (2007). *Explaining the Brain: Mechanisms and Mosaic Unity of Neuroscience*. Oxford: Clarendon Press.

Crews, D. (2008). Epigenetics and Its Implications for Behavioral Endocrinology. *Frontiers in Neuroendocrinology, 29*, 344–357.

Curley, J. P., & Keverne, E. B. (2005). Genes, Brains and Mammalian Social Bonds. *Trends in Ecology and Evolutio, 20*, 561–567.

Currie, D. P. (1985). *The Constitution in the Supreme Court: The First Hundred Years.* Chicago: University of Chicago Press.

Currie, D. P. (1990). *The Constitution in the Supreme Court: The Second Century.* Chicago: University of Chicago Press.

Dahan-Katz, L. (2013). The Implications of Heuristics and Biases: Research on Moral and Legal Responsibility: A Case Against the Reasonable Person Standard. In N. A. Vincent (Ed.), *Neuroscience and Legal Responsibility.* Oxford: Oxford University Press.

Dailey, A. C. (1998). Holmes and the Romantic Mind. *Duke University School of Law, 48*, 429–510.

Damasio, A. (2007). Neuroscience and Ethics: Intersections. *American Journal of Bioethics, 7*, 3–7.

Damasio, A. R. (1994). *Descartes Error.* New York: Grossett/Putnam.

Damasio, A. R. (1996). The somatic marker hypothesis and the possible functions of the prefrontal cortex. *Philosophical Transactions of the Royal Society of London, 354*, 1413–1420.

Damasio, A. R., Tranel, D., & Damasio, H. (1990). Individuals with Sociopathic Behavior Caused by Frontal Damage Fail to Respond Autonomically to Social Stimuli. *Behavioral Brain Research, 41*, 81–94.

Daniels, N. (1974/1989). *Thomas Reid's Inquiry.* Palo Alto: Stanford University Press.

Danisch, R. (2007). *Pragmatism, Democracy and the Necessity of Rhetoric.* Columbia: University of South Carolina Press.

Danisch, R. (2008). Enthymemes, and Oliver Wendell Holmes Jr. and the First Amendment. *Rhetoric Review, 27*, 219–235.

Danson, N. L. (1980). *Brandeis, Frankfurter and the New Deal.* New York: Anchor Books.

Darrow, C. (1926). The Eugenics Cult: The American. *Mercury, 8*, 5–12.

Darwin, C. (1859/1958). *The Origin of Species.* New York: Mentor Book.

Darwin, C. (1868). *The Variation of Animals and Plants Under Domestication* (2 Vols.). London: John Murray.

Darwin, C. (1871/1874). *The Descent of Man and Selection in Relation to Sex.* Chicago: Rand, McNally & Co.

Darwin, C. (1872/1998). *The Expression of the Emotions in Man and Animals.* Oxford: Oxford University Press.

Davis, K. (2017). *The Brain Defense.* New York: Penguin Press.

De Kogel, C. H., Schrama, W. M., & Smit, M. (2014). Civil Law and Neuroscience. *Psychiatry, Psychology and Law, 21,* 272–285.

Delgado, M. R., Frank, R. H., & Phelps, E. A. (2003). Perceptions of Moral Character Modulate the Neural Systems of Reward During the Trust Game. *Nature Neuroscience, 8,* 1611–1616.

de Montaigne, M. (1958). *The Complete Essays.* Stanford: Stanford University Press.

de Montesque, B. (1748). *The Spirit of the Laws.* Cambridge: Cambridge University Press.

Dennett, D. C. (2003). *Freedom Evolves.* New York: Penguin Books.

Descartes, R. (1628/2000). *Rules for the Direction of the Mind.* Indianapolis: Bobbs-Merrill.

Descartes, R. (1641/1911). *Meditations on First Philosophy.* Cambridge: Cambridge University Press.

Descartes, R. (1989). *The Passions of the Soul* (S. Voss, Trans.). Cambridge: Hackett.

de Tocqueville, A. (1945). *Democracy in America* (Vols. 1 & 2). New York: Knopf.

Dewey, J. (1893). Anthropology and the Law. *Islander, 3,* 305–308.

Dewey, J. (1894). Austin's Theory of Sovereignty. *Politics Science Quarterly, IX,* 31–52. In *Early Essays and the Study of Ethics.* Carbondale: Southern Illinois University Press.

Dewey, J. (1896). The Reflex Arc Concept in Psychology. *Psychological Review, 3,* 357–370.

Dewey, J. (1908/1960). *Theory of Moral Life.* New York: Holt, Rinehart & Winston.

Dewey, J. (1910/1965). *The Influence of Darwin on Philosophy.* Bloomington: Indiana University Press.

Dewey, J. (1916). *Democracy and Education: An Introduction of the Philosophy of Education.* New York: Macmillan.

Dewey, J. (1920/1948). *Reconstruction in Philosophy.* Boston: Beacon Press.

Dewey, J. (1925/1989). *Experience and Nature.* LaSalle, IL: Open Court Press.

Dewey, J. (1929/1960). *The Quest for Certainty.* New York: Capricorn Books.

Dewey, J. (1928). Justice Holmes and the Liberal Mind. *New Republic, 53,* 210. In J. A. Boydston (Ed.), *The Later Works of John Dewey, Volume 3, 1927–1938* (pp. 177–183). Carbondale: Southern Illinois University Press.

Dewey, J. (1931). *Philosophy and Civilization.* New York: Minton, Balach and Co.

Dewey, J. (1934/1970). *A Common Faith.* New Haven: Yale University Press.

Dewey, J. (1935/1963). *Liberalism and Social Action.* New York: Capricorn Books.

Dewey, J. (1939). *A Theory of Valuation.* Chicago: University of Chicago Press Press.

Dewey, J. (1975). *Moral Principles in Education.* Carbondale: Southern Illinois University Press.

Dewey, J. (1999). *Liberalism and Social Action.* New York: Prometheus Books.

Dewey, J. (2012). *Unmodern and Modern Philosophy.* Carbondale: Southern Illinois University Press.

Diamond, A. S. (1935). *Primitive Law.* London: Longmans.

Dicey, A. V. (1882). Review of Holmes's The Common Law. *The Spectator, 3,* 745–747.

Dickstein, M. (1998). *The Revival of Pragmatism.* Durham: Duke University Press.

Di Filippo, T. (1988, Fall). Pragmatism, Interest Theory and Legal Philosophy: The Relation of James and Dewey to Roscoe Pound. *Transactions of the C. S. Peirce Society, 24,* 487–508.

Diggins, J. P. (1994). *The Promise of Pragmatism.* Chicago: University of Chicago Press.

Dolinko, D. (1997). Alschuler's Path. *Florida Law Review, 49,* 421–439.

Domnarski, W. (2016). *Richard Posner.* Oxford: Oxford University Press.

Donald, D. (1947/1984). *Lincoln Reconsidered.* New York: Vintage Books.

Donald, M. (1991). *Origins of the Modern Mind.* Cambridge: Harvard University Press.

Dooling, R. (2017, October 13). Justice Holmes's Free Speech Lesson. *Wall Street Journal.*

DuBois, W. E. B. (1903). *The Souls of Black Folk.* Chicago: A.C. McClurg.

DuBois, W. E. B. (1920). *Darkwater.* New York: Harcourt Brace.

Duggan, M. F. (2007). The Municipal Ideal and the Unknown End: A Resolution of Oliver Wendell Holmes. *North Dakota Law Review, 83,* 463–544.

Dunbar, R. I. M. (1992). Neocortex Size as a Constraint on Group Size in Primates. *Journal of Human Evolution, 22,* 469–493.

Dunbar, R. I. M. (1996). *Grooming, Gossip and the Evolution of Language.* Cambridge, MA: Harvard University Press.

Dunbar, R. I. M. (2016). *Human Evolution.* Oxford: Oxford University Press.

Dunbar, R. I. M., & Shultz, S. (2007). Evolution in the Social Brain. *Science, 317,* 1344–1347.

Dupre, J. (1981). Natural Kinds and Biological Taxa. *The Philosophical Review, 90*, 66–90.

Dupre, J. (1993). *The Disorder of Things*. Cambridge: Harvard University Press.

Durkheim, E. (1974). *Sociology and Philosophy* (D. F. Pocock, Trans.). New York: Free Press.

Duxbury, N. (1995). *Patterns of American Jurisprudence*. Oxford: Oxford University Press.

Duxbury, N. (2004). *Frederick Pollock and the English Juristic Tradition*. Oxford: Oxford University Press.

Dworkin, R. (1977). *Taking Rights Seriously*. Cambridge: Harvard University Press.

Dworkin, R. (1986). *Law's Empire*. Cambridge: Cambridge University Press.

Dworkin, R. (1987). *A Matter of Principle*. Cambridge: Harvard University Press.

Dworkin, R. (1996). *Freedom's Law: The Moral Reading of the Constitution*. Cambridge: Harvard University Press.

Eames, S. M. (1977). *Pragmatic Naturalism*. Carbondale: Southern Illinois University Press.

Elliott, E. D. (1984). Holmes and Evolution: Legal Process as Artificial Intelligence. *Faculty Scholarship Series., 5079*(13), 113–146.

Ellis, R. E. (2007). *Aggressive Nationalism*. Oxford: Oxford University Press.

Elster, J. (1983). *Sour Grapes*. Cambridge: Cambridge University Press.

Elster, J. (1983/1985). *Explaining Technical Change*. Cambridge: Cambridge University Press.

Elster, J. (1989). *Solomonic Judgments*. Cambridge: Cambridge University Press.

Elster, J. (2000). *Ulysses Unbounded*. Cambridge: Cambridge University Press.

Emerson, R. W. (1855/1876). *Nature, Addresses and Lectures*. Cambridge: The Riverside Press.

Emerson, R. W. (1889). *Complete Essays* (Vols. 1–8). Boston: Houghton, Mifflin and Company.

Emerson, R. W. (1893). *Natural History of Intellect*. Boston: Houghton, Mifflin and Company.

Engel, A. K., Friston, K. J., & Kragic, D. (2015). *The Pragmatic Turn: Toward Action-Oriented Views in Cognitive Science*. Cambridge: MIT Press.

Epicurus. (1964). *Letters, Doctrines and Vatican Sayings*. Indianapolis: Bobbs-Merrill.

Epstein, R. A. (1998). Pennsylvania Coal vs Mahon: The Erratic Taking Jurisprudence of Justice Holmes. *Georgetown Law Review, 86*, 875–905.

Epstein, R. A. (2000). *Cases and Material on Torts*. New York: Wolters Kluwer.

Erasmus, D. (1511/1922). *In Praise of Folly*. New York: Perter Eckler Publishing Company.

Faigman, D. L. (2004). *Laboratory of Justice*. New York: Time Books.

Farah, M. J. (2005). Neuroethics: Trends in Cognitive. *Science, 9*, 334–340.

Farber, D. A. (1987–1988). Legal Pragmatism and the Constitution. *Minnesota Law Review, 72*, 1331–1377.

Farber, D. A. (1995). Reinventing Brandeis: Legal Pragmatism for the Twenty-First Century. *Berkeley Law Review, 163*, 163–190.

Farnum, G. F. (1937, May). Holmes—The Solitary Scholar. *Journal of the Law Society of Mass, 1*, 2–9.

Farnum, G. F. (1943). Oliver Wendell Holmes Jr.: Soldier and Philosopher. *American Bar Association, 29*, 17–18.

Faust, D. G. (2008). *The Republic of Suffering*. New York: Knopf.

Feldman, N. (2010). *Scorpions*. New York: Hachette Books.

Feldman, R. (2009). *The Role of Science in Law*. Oxford: Oxford University Press.

Ferguson, R. A. (1988). Holmes and the Judicial Figure. *University of Chicago Law Review, 55*, 506–548.

Finger, S. (1994). *Origins of Neuroscience*. Oxford: Oxford University Press.

Finnis, J. (2002). Natural Law: The Classical Tradition. In J. Coleman & S. Shapiro (Eds.), *Jurisprudence and the Philosophy of Law*. Oxford: Oxford University Press.

Fisch, M. H. (1942/1986). Justice Holmes, the Prediction Theory of Law and Pragmatism. In M. H. Fisch (Ed.) (1986), *Peirce, Semiotic and Pragmatism*. Bloomington: Indiana University Press.

Fisch, M. H. (1954/1986). Alexander Bain and the Genealogy of Pragmatism. In M. H. Fisch (Ed.) (1986), *Peirce, Semiotic and Pragmatism*. Bloomington: Indiana University Press.

Flanagan, O. (1991). *Varieties of Moral Personality: Ethics and Psychological Realism*. Cambridge: Harvard University Press.

Flexner, A. (1910). *Medical Education in the United States and Canada*. New York: Carnegie Foundation.

Flower, E., & Murphy, M. G. (1977). *A History of Philosophy in America* (Vols. 1 & 2). New York: Capricorn Press.

Fodor, J. (1983). *The Modularity of Mind*. Cambridge: MIT Press.

Foner, E. (1976). *Tom Paine and Revolutionary America*. Oxford: Oxford University Press.

Foner, E. (2001). *The Idea of Freedom in American History*. New York: Columbia University.

Foner, E. (2006/2012). *Give Me Liberty*. New York: Seagull.

Foner, E. (2011). *The Fiery Trial: Abraham Lincoln and American Slavery*. New York: Norton.

Frank, J. N. (1930). *Law and the Modern Mind*. New York: Brentano Publishers.

Frank, J. N. (1932). Mr. Justice Holmes and Non-Euclidean Legal Thinking. *Cornell Law Review, 17,* 568–588.

Frank, J. N. (1950). Modern and Ancient Legal Pragmatism. *Yale Law Review, 25,* 207–255.

Frank, J. N. (1954). A Conflict with Oblivion: Some Observations on the Founders of Legal Pragmatism. *Yale Law Review, 9,* 425–463.

Frankfurter, F. C. (1923). Twenty Years of Mr Justice Holmes's Constitutional Opinions. *Harvard Law Review, 36,* 909–939.

Frankfurter, F. C. (1927/1961). *The Case of Sacco and Vanzetti*. New York: W. S. Hein & Company.

Frankfurter, F. C. (1939). *Mr Justice Holmes*. Cambridge: Harvard University Press.

Franklin, B. (1987). *Writings*. New York: Library of America.

Freeman, M. (2006). Law and Neuroscience. *International Journal of Law in Context, 2,* 217–219.

Fried, C. (2004). *Saying What the Law Is: The Constitution in the Supreme Court*. Cambridge: Harvard University Press.

Friedman, L. M. (1973/1985). *A History of American Law*. New York: Simon and Schuster.

Friedman, L. M. (1993). *Crime and Punishment in American History*. New York: Basic Books.

Friedman, M. (1974). Explanation and Scientific Understanding. *Journal of Philosophy, 71*(1), 5–19.

Friedman, M. (1992). *Kant and the Exact Sciences*. Cambridge: Harvard University Press.

Friston, K. (2010). The Free Energy Principle: A Unified Brain Theory? *Nature Reviews, 11,* 127–136.

Frith, C., & Wolpert, D. (2003). *The Neuroscience of Social Interaction*. Oxford: Oxford University Press.

Fromm, E. (1941). *Escape from Freedom*. New York: Farrar & Rinehart.

Fromm, E. (1955). *The Sane Society*. Robbinsdale, MN: Fawcett.

Fulham, R. S., McKie, S., & Dolan, M. C. (2009). Psychopathic Traits and Deception: Functional Magnetic Resonance Imaging Study. *British Journal of Psychiatry, 194,* 229–235.

Fuller, L. L. (1964). *The Morality of Law.* New Haven: Yale University Press.

Fussell, P. (1990). *Wartime.* Oxford: Oxford University Press.

Gallison, P. (1988). History, Metaphor and the Central Metaphor. *Science in Context, 2,* 197–212.

Garland, B., & Frankel, M. S. (2006, Winter/Spring). Considering Convergence: A Policy Dialogue About Behavioral Genetics, Neuroscience and Law. *Law and Contemporary Problems, 69,* 101–113.

Garland, B., & Glimcher, P. W. (2006). Cognitive Neuroscience and the Law. *Current Opinions in Neurobiology, 16,* 130–134.

Garrett, N., Lazzaro, S. C., Ariely, D., & Sharot, T. (2016). The Brain Adapts to Dishonesty. *Nature Neuroscience, 19,* 1727–1732.

Gauthier, D. (1990). *Moral Dealing: Contract, Ethics and Reason.* Ithaca: Cornell University Press.

Gazzaniga, M. S. (1985). *The Social Brain.* New York: Basic Books.

Gazzaniga, M. S. (2005). *The Ethical Brain.* Washington, DC: Dana Press.

Gazzaniga, M. S. (2008). The Law and Neuroscience. *Neuron, 60,* 412–415.

Geary, A. (2001). *Law and Aesthetics.* Oxford: Hart Publishing.

George, R. P. (2000). *Great Cases in Constitutional Law.* Princeton: Princeton University Press.

George, R. P. (2004). Holmes on Natural Law. In J. De Groot (Ed.), *Nature in American Philosophy.* Washington, DC: Catholic University of America Press.

Gert, B. (2012, May/June). Neuroscience and Morality. *Hastings Center Report, 42*(3), 22–28.

Gibian, P. (2001). *Oliver Wendell Holmes and the Culture of Conversation.* Cambridge: Cambridge University Press.

Gigerenzer, G. (2000). *Adaptive Thinking, Rationality in the Real World.* New York: Oxford University Press.

Gigerenzer, G. (2007). *Gut Feelings.* New York: Viking Press.

Gigerenzer, G., & Engel, C. (2006). *Heuristics and the Law.* Cambridge: MIT Press.

Gilmore, G. (1977). *The Ages of American Law.* New Haven: Yale University Press.

Gilmore, G. (1999). Some Reflections on Oliver Wendell Holmes Jr. *Green Bag, 12,* 374–379.

Glenn, A. L., & Raine, A. (2009). Psychopathy and Instrumental Aggression: Evolutionary, Neurobiological and Legal Perspectives. *International Journal of Law and Psychiatry, 32,* 253–258.

Glimcher, P. W. (2003). *Decision, Uncertainty and the Brain: The Science of Neuroeconomics.* Cambridge: MIT Press.

Goodenough, O. R. (2004). Responsibility and Punishment. *Philosophical Transactions of the Royal Society of London. Series B, 359,* 1805–1809.

Goodman, N. (1955/1973). *Fact, Fiction and Forecast.* New York: Bobbs-Merrill.

Goodman, R. B. (1990). *American Philosophy and the Romantic Tradition.* Cambridge: Cambridge University Press.

Gordley, J. (1997). When Paths Diverge: A Response to Albert Alschuler and Oliver Wendell Holmes Jr. *Florida Law Review, 49,* 441–462.

Gordon, C. (2015). *Romantic Outlaws: The Extraordinary Lives of Mary Wollstonecraft and Her Daughter Mary Shelley.* New York: Random House.

Gordon, R. W. (1982). Holmes's Common Law as Legal and Social Science. *Hofstra Law Review, 10,* 719–740.

Gordon, R. W. (Ed.). (1992). *The Legacy of Oliver Wendell Holmes Jr.* Palo Alto: Stanford University Press.

Gordon, R. W. (2017). *Taming the Past: Essays on Law in History and History in Law.* Cambridge: Cambridge University Press.

Gould, S. J. (1984). Carrie Buck's Daughter. *Natural History, 93,* 14–18.

Grant, S. M. (2016). *Oliver Wendell Holmes Jr. Civil War Soldier, Supreme Court Justice.* New York and London: Routledge.

Gray, J. C. (1909). *The Nature and Sources of Law.* New York: Columbia University Press.

Gray, J. G. (1970). *Understanding Violence and Other Essays.* New York: Harper & Row.

Gray, J. G. (1998). *The Warriors: Reflections on Men in Battle.* Lincoln: University of Nebraska Press (Bison Books).

Greely, H. T. (2008). Towards Responsible Use of Cognitive Enhancing Drugs by the Healthy. *Nature, 456,* 702–705.

Greely, H. T. (2009). Law and the Revolution in Neuroscience: An Early Look at the Field. *Akron Law Review, 42,* 687–715.

Greely, H. T. (2010). Enhancing Brains: What Are We Afraid Of? *Cerebrum, 2010, 14,* 8–18.

Greely, H. T. (2011). Reading Minds with Neuroscience—Possibilities for the Law. *Cortex, 47,* 1254–1255.

Greely, H. T. (2013). Mind Reading: Neuroscience and the Law. In S. J. Morse & A. L. Roskies (Eds.), *A Primer on Criminal Law and Neuroscience*. Oxford: Oxford University Press.

Greely, H. T., & Illes, J. (2007). Neuroscience-Based Lie Detection: The Urgent Need for Regulation. *American Journal of Law and Medicine, 33,* 377–431.

Green, J. M. (2008). *Pragmatism and Social Hope*. New York: Columbia University Press.

Green, St. J. (1870). Proximate and Remote Cause. *American Law Review, 201.*

Green, T. H. (1883). *Prolegomena to Ethics*. London: Thomas Cromwell.

Greene, J. D. (2014). Beyond Point and Shoot Morality: Why Cognitive Neuroscience Matters for Ethics. *Ethics, 124,* 695–726.

Greene, J. D., & Cohen, J. (2004). For the Law, Neuroscience Changes Nothing and Everything. *Philosophical Transactions of the Royal Society of London. Series B, 359,* 1775–1785.

Greene, J. D., Morelli, S. A., Lowenberg, K., Nystrom, L. E., & Cohen, J. D. (2008). Cognitive Load Selectively Interferes with Utilitarian Moral Judgment. *Cognition, 107,* 1144–1154.

Greene, J. D., Nystrom, L. E., Engell, A. D., Darley, J. M., & Cohen, J. D. (2004). The Neural Bases of Cognitive Conflict and Control in Moral Judgment. *Neuron, 44,* 389–400.

Greene, J. D., & Paxton, J. M. (2009). Patterns of Neural Activity Associated with Honest and Dishonest Moral Decision. *Proceedings of the National Academy of Sciences of the United States of America, 106,* 12506–12511.

Greene, J. D., Sommerville, R. B., Nystrom, L. E., Darley, J. M., & Cohen, J. D. (2001). An fMRI Investigation of Emotional Engagement in Moral Judgment. *Science, 293,* 2105–2108.

Gregg, P. L. (1942–1943). The Pragmatism of Mr Justice Holmes. *Georgetown Law Review, 252,* 263–295.

Grey, T. C. (1983). Langdell's Orthodoxy. *University of Pittsburgh Law Review, 45,* 1–53.

Grey, T. C. (1989). Holmes and Legal Pragmatism. *Stanford Law Review, 41,* 787–870.

Grey, T. C. (1990). Hear the Other Side: Wallace Stevens and Pragmatists Legal Theory. *Southern California Law Review, 63,* 1569–1595.

Grey, T. C. (1991a). *The Wallace Stevens Case: Law and the Practice of Poetry*. Cambridge: Harvard University Press.

Grey, T. C. (1991b). What Good Is Legal Pragmatism. In M. Brint & W. Weaver (Eds.), *Pragmatism in Law and Society*. Boulder, CO: Westview Press.

Grey, T. C. (1992). Holmes, Pragmatism and Democracy. *Oregon Law, 71*, 521–542.

Grey, T. C. (1995). Molecular Motions: The Holmesian Judge in Theory and Practice. *William and Mary Law Review, 37*, 19–45.

Grey, T. C. (2000). Holmes on the Logic of the Law. In S. J. Burton (Ed.), *The Path of the Law and Its Influence: The Legacy of Oliver Wendell Holmes Jr.* Cambridge: Cambridge University Press.

Grey, T. C. (2003). Judicial Review and Legal Pragmatism. *Wake Forest Law Review, 38*, 473–511.

Grey, T. C. (2014). *Formalism and Pragmatism in American Law*. Boston: Brill Press.

Gross, C. G. (1998). *Brain, Vision, Memory: Tales in the History of Neuroscience*. Cambridge: MIT Press.

Grotius, H. (1625/2009). *On the Law of War and Peace*. Alexandria: Library of Alexandria.

Grotius, H. (1625/2012). *The Rights of War and Peace: Including the Law of Nature and Nations*. Filiquarian Legacy Publishing.

Haack, S. (1998). *Manifesto of a Passionate Moderate*. Chicago: University of Chicago Press.

Haack, S. (2005). On Legal Pragmatism. *The American Journal of Jurisprudence, 50*, 71–105.

Haack, S. (2011). Pragmatism, Law, and Morality: The Lessons of Buck vs Bell. *European Journal of Pragmatism and American Philosophy, 2*, 67–87.

Hacking, I. (1965/1979). *Logic of Statistical Inference*. Cambridge: Cambridge University Press.

Hacking, I. (1990). *The Taming of Chance*. Cambridge: Cambridge University Press.

Haider, A. (2006). Roper vs Simmons: The Role of the Science Brief. *Ohio State Journal of Criminal Law, 3*, 369–377.

Haidt, J. (2007). The New Synthesis in Moral Psychology. *Science, 316*, 998–1002.

Hall, D. L. (1973). *The Civilization of Experience*. New York: Fordham University Press.

Hamilton, A., Jay, J., & Madison, J. (1818). *The Federalist Papers* (Gideon ed.). Indianapolis: Liberty Fund.

Hamilton, W. (1853). *Discussions in Philosophy and Literature*. Edinburgh: MacLachlan and Stewart.

Hamilton, W. (1859). *Metaphysics and Logic*. Boston: Gould and Lincoln.

Hand, L. (1944/1974). *The Spirit of Liberty*. New York: Knopf.

Hanson, N. R. (1958/1972). *Patterns of Discovery*. Cambridge, MA: Cambridge University Press.

Hanson, N. R. (1971). *Observation and Explanation*. New York, NY: Harper Press.

Harman, O. (2013). Uninformed Minds: Juveniles, Neuroscience and the Law. *Studies in History and Philosophy of Biological and Biochemical Sciences, 44,* 455–459.

Harrington, A. (1996). *Reenchanted Science*. Princeton: Princeton University Press.

Hart, H. L. A. (1958). Positivism and the Separation of Law and Morals. *Harvard Law Review, 71,* 593–629.

Hart, H. L. A. (1961). *The Concept of Law*. Oxford: Clarendon.

Hart, H. L. A. (1968). *Punishment and Responsibility*. Oxford: Oxford University Press.

Hart, H. L. A. (1977). American Jurisprudence Through English Eyes: The Nightmare and the Noble Dream. *Georgia Law Review, 11,* 1676–1677.

Hart, H. L. A. (1987). *Essays in Jurisprudence and Philosophy*. Oxford: Clarendon.

Hauser, N. (2003). Pragmatism and the Loss of Innocence. *Cognitio, 4,* 197–210.

Hawrylycz, M. J., Lein, E. S., Guillozet-Bongaarts, A. L., Shen, E. H., Ng, L., Miller, J. A., et al. (2012). An Anatomically Comprehensive Atlas of the Adult Human Brain Transcriptome. *Nature, 489,* 391–398.

Hawrylycz, M. J., Miller, J. A., Menon, V., Feng, D., Dolbeare, T., Guillozet-Bongaarts, A. L., et al. (2015). Canonical Genetic Signatures of the Adult Human Brain. *Nature Neuroscience, 18,* 1832–1842.

Hayek, F. A. (1960). *The Constitution of Liberty*. Chicago: University of Chicago Press.

Healy, T. (2013). *The Great Dissent: How Oliver Wendell Holmes Jr. Changes His Mind and Changes History of Free Speech in America*. New York: Henry Holt.

Heelan, P. A., & Schulkin, J. (1998). Hermeneutical Philosophy and Pragmatism: A Philosophy of Science. *Synthese, 115,* 269–302.

Herbert, J., & Schulkin, J. (2002). Neurochemical Coding of Adaptive Responses in the Limbic System. In D. Pfaff (Ed.), *Hormones, Brain and Behavior*. New York: Elsevier Press.

Hickman, L. A. (1992). *John Dewey's Pragmatic Technology.* Bloomington: Indiana University Press.

Hickman, M. B. (1952). Mr Justice Holmes: A Reapprasial. *The Western Political Quarterly, 5,* 66–83.

Hicks, F. C. (1946). Lincoln, Wright and Holmes at Fort Stevens. *Journal of the Illinois State Historical Society, 39,* 123–132.

Hignett, C. (1952/1975). *A History of the Athenian Constitution.* Oxford: Clarendon University Press.

Hirstein, W., Poland, J., & Radden, J. (2005). *Brain Fiction: Deception and the Riddle of Confabulation.* Cambridge: MIT Press.

Hirstein, W., Stifford, K. L., & Fagan, T. K. (2018). *Responsible Brains: Neuroscience, Law and Human Culpability.* Cambridge: MIT Press.

Hobbes, T. (1651/1958). *Leviathan.* New York: Bobbs-Merrill.

Hoffman, D. A. (2014). *The Punishers's Brain: The Evolution of Judge and Jury.* Cambridge: Cambridge University Press.

Hoffman, M. B. (2004). The Neuroeconomic Path of the Law. *Philosophical Transactions of the Royal Society of London. Series B, 359,* 1667–1676.

Hohwy, J. (2013). *The Predictive Mind.* Oxford: Oxford University Press.

Hollinger, D. A. (1968). Perry Miller and Philosophical History. *History and Theory, 7,* 189–202.

Hollinger, D. A. (1975). *Morris R. Cohen and the Scientific Ideal.* Cambridge: MIT Press.

Hollinger, D. A. (1977). Review: The Culture of Experience: Philosophical Essays in the American Grain. *Transactions of the C. S. Peirce Society, 13,* 312–315.

Hollinger, D. A. (1990). Free Enterprise and Free Inquiry: The Emergence If Laissez-Faire Communitarianism in the Ideology of Science in the United States. *New Literary History, 21,* 897–919.

Hollinger, D. A. (1992). The Tough Minded Justice Holmes, Jewish Intellectuals and the Making of an American Icon. In R.W. Gordon (Ed.), *The Legacy of O. W. Holmes.* Palo Alto: Stanford University Press.

Hollinger, D. A. (1996). *Science, Jews and Secular Culture.* Princeton: Princeton University Press.

Hollinger, D. A. (2005, Winter). The One Drop Rule and the One Hate Rule. *Deadalus, 18,* 28.

Holmes, O. W., Jr. (1864, May 16). Letter to Mr and Mrs Holmes.

Holmes, O. W., Jr. (1870). Codes and the Arrangement of the Law. *American Law Review.*

Holmes, O. W., Jr. (1873). The Gas-Stokers Strike. *American Law Review, 7,* 582–583.

Holmes, O. W., Jr. (1876). Primitive Notions of the Law. *American Law Review, 11,* 641–660.

Holmes, O. W., Jr. (1880a). Review of CC Langwell: A Selection of Cases of Contracts with a Summary of the Topics Covered by the Cases.

Holmes, O. W., Jr. (1880b). Trespass and Negligence. *American Law Review, 14.*

Holmes, O. W., Jr. (1881/1952). *The Common Law.* New York: Dover.

Holmes, O. W. Jr. (1884). *Sons of Harvard Who Fell in Battle.* Harvard Alumni Dinner Speech.

Holmes, O. W., Jr. (1891). Agency. *Harvard Law Review, 5.*

Holmes, O. W., Jr. (1894). Privilege, Malice and Intent. *Harvard Law Review, 8,* 1–14.

Holmes, O. W., Jr. (1895a). Scientific Proof and Relations of Law and Medicine: Learning and Science. *Boston University Law Review, 26.*

Holmes, O. W., Jr. (1895b). The Soldier's Faith. Graduating Class of Harvard University.

Holmes, O. W., Jr. (1896). *Vegelahn vs Gunther.*

Holmes, O. W., Jr. (1897a). The Path of the Law. *Harvard Law Review, 10*(8), 45.

Holmes, O. W., Jr. (1897b). *The Fraternity of Arms.* Remarks at a meeting of the 20th Regimental Association.

Holmes, O. W., Jr. (1899a). The Theory of Legal Interpretation. *Harvard Law Review, 12,* 417–420.

Holmes, O. W., Jr. (1899b). Law in Science and Science in the Law. *Harvard Law Review, 12,* 443–463.

Holmes, O. W., Jr. (1900). *Montesquieu: Critical and Biographical Introduction to the Spirit of the Laws.* New York: Appleton and Company.

Holmes, O. W., Jr. (1901). One Hundredth Anniversary of the day on which Marshall took his seat on the bench.

Holmes, O. W., Jr. (1902). *Twenty Years in Retrospect.* Speech at a Banauget of the Middlesex Bar Association.

Holmes, O. W., Jr. (1904). *Northern Securities Company vs United States.*

Holmes, O. W., Jr. (1905). *Lochner vs New York.*

Holmes, O. W., Jr. (1908). *Hudson Water co. v. McCarter.*

Holmes, O. W., Jr. (1913a). *Nash v United States.*

Holmes, O. W. Jr. (1913b). The Class of 61. Harvard University.

Holmes, O. W., Jr. (1913c). *Law and the Court*. Speech at a dinner at the Harvard Law School.

Holmes, O. W., Jr. (1914a). *Gompers v. United States*.

Holmes, O. W., Jr. (1914b). *International Harvester Co. vs Kentucky*.

Holmes, O. W., Jr. (1915). Ideals and Doubts. *Illinois Law Review, 10*, 14–22.

Holmes, O. W., Jr. (1917). *Southern Pacific Co. v. Jensen*.

Holmes, O. W., Jr. (1918). Natural Law. *Harvard Law Review, 32*, 40–44.

Holmes, O. W., Jr. (1919a). *Abrams vs United States*.

Holmes, O. W., Jr. (1919b). *Schneck vs United States*.

Holmes, O. W., Jr. (1920). *Collected Legal Papers*. London: Constable and Co.

Holmes, O. W., Jr. (1921). *American Banks and Trust Co. vs Federal Reserve Banks*.

Holmes, O. W., Jr. (1927a). *Buck vs Bell*.

Holmes, O. W., Jr. (1927b). *Baltimore and Ohio Railroad v. Goodman*.

Holmes, O. W., Jr. (1928). *Olmstead vs United States*.

Holmes, O. W., Jr. (1946). *Touched with Fire: Civil War Letters and Diary 1861–1864*. Cambridge: Harvard University Press.

Holmes, O. W., Jr. (1947). *Justice Holmes and Doctor Wu: An Intimate Correspondence*. New York: Central Books.

Holmes, O. W., Jr. (1994). *Collected Works of Justice Holmes* (R. Posner, Ed.). Chicago: University of Chicago Press.

Holmes, O. W., Jr. (1995). *The Collected Works of Justice Holmes* (S. M. Novick, Ed.). Chicago: University of Chicago Press.

Holmes, O. W., Jr., & Cohen, M. R. (1948). *Correspondence 1915–1934* (F. S. Cohen, Ed.). Yale Law School Legal Repository.

Holmes, O. W., Jr. & Einstein, L. (1964). *Letters 1903–1935* (B. Peabody, Ed.). London: Macmillan.

Holmes, O. W., Jr., & Frankfurter, F. (1996). *Correspondence 1912–1934* (R. M. Mennell & C. L. Compston, Eds.). Lebanon: University Press of New England.

Holmes, O. W., Jr., & Laski, H. J. (1941/1961). *Letters 1916–1935* (M. DeWolfe Howe, Ed.). Cambridge: Harvard University Press.

Holmes, O. W., Jr. & Pollock, F. (1942/2015). *Letters, Volumes 1 and 2* (M. DeWolfe Howe, Ed.). Cambridge: Cambridge University Press.

Holmes, O. W., Jr., & Sheehan, P. A. (1993). *Correspondence 1903–1913* (D. B. Burton, Ed.). New York: Fordham University Press.

Holmes, O. W., Sr. (1859). *The Autocrat at the Breakfast Table*. Boston: Phillips, Sampson and Company.

Holmes, O. W., Sr. (1861). *Currents and Counter Currents in Medical Science and Other Essays.*

Holmes, O. W., Sr. (1864). *Soundings from the Atlantic.* Boston: Ticknor and Fields.

Holmes, O. W., Sr. (1885/2004). *Ralph Waldo Emerson* (ebook).

Holzer, H. (2014). *Lincoln and the Power of the Press.* New York: Simon and Schuster.

Horn, S. R., Charney, D. S., & Feder, A. (2016). Understanding Resilience: New approaches for Preventing and Treating PTSD. *Experimental Neurology, 284,* 119–132.

Horwitz, M. J. (1992a). The Place of Holmes in American Legal Thought. In R. Gordon (Ed.), *Legacy of Oliver Wendell Holmes Jr.* Palo Alto: Stanford University Press.

Horwitz, M. J. (1992b). *The Transformation of American Law.* Oxford: Oxford University Press.

Howe, M. D. (1951). The Positivism of Mr. Justice Holmes. *Harvard Law Review, 64,* 529–546.

Howe, M. D. (1957/1963). *Justice Oliver Wendell Holmes: Volumes 1 and Volumes 2—The Proving Years.* Cambridge: Harvard University Press.

Hume, D. (1738). *A Treatise of Human Nature.* Start Publishing LLC.

Hurst, J. W. (1964). *Justice Holmes or Legal History.* New York: Macmillan.

Hutchinson, A. C. (2005). *Evolution and the Common Law.* Cambridge: Cambridge University Press.

Illes, J. (2006). *Neuroethics: Defining the Issues in Theory, Practice and Policy.* Oxford: Oxford University Press.

Illes, J., & Bird, S. J. (2006). Neuroethics: A Modern Context for Ethics in Neuroscience. *Trends in Neuroscience, 29,* 511–517.

Illes, J., Tairuan, K., Federico, C. A., Tabet, A., & Glover, G. H. (2010). Reducing Barriers to Ethics in Neuroscience. *Frontiers in Human Neuroscience, 4,* 167.

Israel, J. I. (2001). *Radical Enlightenment.* Oxford: Oxford University Press.

Jackson, J. H. (1884/1958). Evolution and Dissolution of the Nervous System. In *Selected Writings of John Hughlings Jackson.* London: Staples Press.

Jackson, P. L., & Decety, J. (2004). Motor Cognition: A New Paradigm to Self and Other Interactions. *Current Opinion in Neurobiology, 14,* 259–263.

James, W. (1868/1961). Letters O. W. Holmes Jr. In *Selected Letters.* New York: Farrar, Straus and Cudahy.

James, W. (1890/1952). *The Principles of Psychology.* New York: Dover Press.

James, W. (1896/1984). *Exceptional Mental States*. Amherst: University of Massachusetts Press.

James, W. (1897/1927). *The Will to Believe and Other Essays in Popular Philosophy*. New York: Longmans, Green and Co.

James, W. (1902/1961). *Varieties of Religious Experience: A Study of Human Nature*. New York: Collier Press.

James, W. (1907/1955). *Pragmatism*. New York: Meridian.

Janack, M. (2012). *What We Mean by Experience*. Palo Alto: Stanford University Press.

Jaspers, K. (1913/1997). *General Psychopathology* (Vols. I & II, J. Hoenig & M. W. Hamilton, Trans.) Baltimore: The Johns Hopkins University Press.

Jay, M. (2005). *Songs of Experience*. Berkeley: University of California Press.

Jeannerod, M. (1999). To Act or Not to Act: Perspectives on the Representation of Action. *Quarterly Journal of Experimental Psychology, 52,* 1–29.

Jefferson, T. (1787/1982). *Notes on the State of Virginia*. New York: Norton.

Joas, H. (1993). *Pragmatism and Social Theory*. Chicago: University of Chicago Press.

Johnson, M. (1993). *Moral Imagination*. Chicago: University of Chicago Press.

Johnson, M. (2002). Law Incarnate. *Brooklyn Law Review, 67,* 949–962.

Johnson, M. (2007). Mind, Metaphor, Law. *Mercer Law Review, 58,* 845–868.

Johnson, M. (2014). *Morality for Humans*. Chicago: University of Chicago Press.

Jones, O. D. (2004). Law, Evolution and the Brain. *Philosophical Transactions of the Royal Society of London. Series B: Biological Sciences, 29,* 1697–1707.

Jones, O. D., Bonnie, R. J., Casey, B. J., Davis, A., Faigman, D. L., Hoffman, M., et al. (2014). Law and Neuroscience: Recommendations Submitted to the President's Bioethics Commission. *Journal of Law and the Biosciences, 1,* 224–236.

Jones, O. D., & Goldsmith, T. H. (2005). Law and Behavioral Biology. *Columbia Law Review, 105,* 405–502.

Jones, O. D., Marois, R., Farah, M. J., & Greely, T. H. (2013). Law and Neuroscience. *The International Journal of Neuroscience, 33,* 17624–17630.

Jones, O. D., Schall, J. D., & Shen, F. S. (2014). *Law and Neuroscience*. Frederick, MD: Kluwer.

Joyce, R. (2000). *The Evolution of Morality*. Cambridge: MIT Press.

Justinian. *Corpus Juris Civilis*.

Kagan, J. (1984). *The Nature of the Child*. New York: Basic Books.

Kagan, J. (2002). *Surprise, Uncertainty and Mental Structure*. Cambridge: Harvard University Press.

Kahneman, D. (2011). *Thinking Fast and Slow*. New York: Farrar, Straux, and Giroux.

Kahneman, D., Slovic, P., & Tversky, A. (1982). *Judgment Under Uncertainty: Heuristics and Biases*. Cambridge, UK: Cambridge University Press.

Kamienski, L. (2016). *Shooting Up: A Short History of Drugs and War*. Oxford: Oxford University Press.

Kamm, F. M. (1993). *Morality, Morality*. Oxford: Oxford University Press.

Kandel, E. R., & Squire, L. R. (2000). Neuroscience: Breaking Down Scientific Barriers to the Study of Brain and Mind. *Science, 290,* 1113–1120.

Kant, I. (1788/1956). *Critique of Reason*. Indianapolis: Bobbs-Merrill.

Kant, I. (1789/1997). *Critique of Practical Reason*. Cambridge: Cambridge University Press.

Kant, I. (1792/1914). *Critique of Judgment*. London: Macmillan.

Kaplan, A. (2016). *Imbecility: The Supreme Court, American Eugenics and the Sterilization of Carrie Buck*. New York: Penguin.

Kelley, P. J. (1983). A Critical Analysis of Holmes Theory of Torts. *Washington University Law, 61,* 681–744.

Kelley, P. J. (1989–1990). Was Holmes a Pragmatist? Reflections on a New Twist to an Old Argument. *Southern Illinois University Law Review, 14,* 427–467.

Kelley, P. J. (1993). Holmes's Early Constitutional Law Theory and Its Applications in Taking Cases on the Massachusetts Supreme Court. *Southern Illinois University Law Review, 18,* 357–414.

Kelley, P. J. (2000). Critical Analysis of Holmes's Theory of Contract. *Notre Dame Law Review, 75,* 1681–1773.

Kelley, P. J. (2002). Holmes Langdell and Formalism. *Ration Juris, 1,* 26–51.

Kellogg, F. R. (1984). *Formative Essays of Oliver Wendell Holmes Jr.* London: Greenwood Press.

Kellogg, F. R. (1992). Who Owns Pragmatism? *Journal of Speculative Philosophy, 6,* 67–80.

Kellogg, F. R. (2003). Holmes, Common Law Theory, and Judicial Restraint. *The John Marshall Law Review, 36,* 457–502.

Kellogg, F. R. (2004). Holistic Pragmatism and Law: Morton White on Justice Oliver Wendell Holmes. *Transactions of the Charles S. Peirce Society, 40,* 559–567.

Kellogg, F. R. (2007). *Oliver Wendell Holmes: The Legal Theory as Judicial Restraint.* Cambridge: Cambridge University Press.

Kellogg, F. R. (2017). Take the Trolley Problem… Please! Pragmatism, Moral Particularism and the Continuum of Normative Inquiry. *Contemporary Pragmatism, 12,* 8–18.

Kellogg, F. R. (2018). *Oliver Wendell Holmes Jr. and Legal Logic.* Chicago: University of Chicago Press.

Kennedy, D. M. (2001). *Freedom from Fear.* Oxford: Oxford University Press.

Kent, J. (1873). *Commentaries* (O. W. Holmes Jr, Ed., 12th ed.). New York: Little, Brown.

Keynes, J. M. (1921/1957). *A Treatise on Probability.* New York: Harper.

Kiehl, K. A., & Sinnott-Armstrong, W. P. (2013). *Handbook on Psychopathology and Law.* Oxford: Oxford University Press.

Kitcher, P. (2012). *Preludes to Pragmatism.* Oxford: Oxford University Press.

Kitcher, P. (2014). *Life After Faith.* New Haven: Yale University Press.

Kloppenberg, J. T. (2010). James's Pragmatism and American Culture, 1907–2007. In J. Stuhr (Ed.), *One Hundred Years of Pragmatism* (pp. 7–41). Bloomington: Indiana University Press.

Kolber, A. J. (2014). Will There Be a Neurolaw Revolution. *Indiana law Journal, 89,* 808–845.

Koopman, C. (2009). *Pragmatism as Transition.* New York: Columbia University Press.

Koopman, C. (2011). Genealogical Pragmatism. *Journal of the Philosophy of History, 5,* 533–561.

Kosslyn, S. K. (1986). *Image and Mind.* Cambridge: Harvard University Press.

Kovecses, Z. (2005). *Metaphor in Culture.* Cambridge: Cambridge University Press.

Kozel, F. A., Johnson, K. A., Mu, Q., Grenesko, E. L., Laken, S. J., & George, M. S. (2005). Detecting Deception Using fMRI. *Biological Psychiatry, 58,* 605–613.

Kuhn, T. S. (1962). *The Structure of Scientific Revolution.* Chicago, IL: University of Chicago Press.

Kuklick, B. (2000). *A History of Philosophy in America.* Oxford: Oxford University Press.

Lacey, N. (2004). *A Life of H. L. A. Hart.* Oxford: Oxford University Press.

Lakoff, G., & Johnson, M. (1980/2003). *Metaphors We Live By.* Chicago: University of Chicago Press.

Lakoff, G., & Johnson, M. (1999). *Philosophy in the Flesh.* New York: Basic Books.

Lakoff, G., & Nunez, R. E. (2000). *Where Mathematics Comes From*. New York: Basic Books.

Langdell, C. C. (1880). *Summary of the Law of Contracts* (2nd ed.). Boston: Little, Brown.

LaPiana, W. P. (1990). Victorian from Beacon Hill: Oliver Wendell Holmes's Early Legal Scholarship. *Columbia Law Review, 90*, 809–833.

LaPiana, W. P. (1994). *Logic and Experience*. Oxford: Oxford University Press.

Laski, H. J. (1916). The Basis of Vicarious Liability. *Yale Law Review, 26*, 105.

Laski, H. J. (1925). *Socialism and Freedom*. London: Fabian Society.

Laski, H. J. (1931). The Political Philosophy of Mr. Justice Holmes. *Yale Law Review, 40*, 683–695.

LeDoux, J. (2015). *Anxious: Using the Brain to Understand and Treat Fear and Anxiety*. New York: Viking Press.

Leibniz, G. W. (1671/1951). Elements of Law and Justice. In P. P. Wiener (Ed.), *Leibniz Selections*. New York: Scribner.

Leiter, B. (2000). Holmes, Economics and Classical Realism. In S. J. Burton (Ed.), *The Path of the Law and Its Influence: The Legacy of Oliver Wendell Holmes Jr*. Cambridge: Cambridge University Press.

Leiter, B. (2007). *Naturalizing Jurisprudence*. Oxford: Oxford University Press.

Leonard, G. (2006). Holmes and the Lochner Court. *Boston University School of Law*. No. 06-38.

Leopold, A. (1949/1987). *A Sand County Almanac*. Oxford: Oxford University Press.

Lerner, M. (1943). *The Mind and Faith of Justice Holmes*. New York: Random House.

Levin, J. (1999). *The Poetics of Transition: Emerson, Pragmatism and American Literary Modernism*. Durham: Duke University Press.

Levinson, S. (2000). Emerson and Holmes: Serene Skeptics. In S. J. Burton (Ed.), *The Path of the Law and Its Influence: The Legacy of Oliver Wendell Holmes Jr*. Cambridge: Cambridge University Press.

Levy, N. (2007). *Neuoethics*. Cambridge: Cambridge University Press.

Levy, N. (2011). *Hard Luck: How Luck Undermines Free Will and Responsibility*. Oxford: Oxford University Press.

Lewis, D. L. (2000). *W. E. B. Dubois: The Fight for Equality and the American Century*. New York: Holt.

Libet, B. (2004). *Mind Time*. Cambridge: Harvard University Press.

Lieberman, J. K. (1992). *The Evolving Constitution*. New York: Random House.

Liet, C. A. (1961). *The Dissenting Opinions of Justice Holmes*. Littleton, CO: Rottham and Co.

Liewellyn, K. N. (1962). *Jurisprudence: Realism in Theory and Practice*. Chicago: University of Chicago Press.

Locke, J. (1689). *A Letter Concerning Toleration*. Indianapolis: Liberty Fund.

Locke, J. (1690). *Two Treatises on Government*. London: Awnsham Churchill.

Lombardo, P. A. (2008). *Three Generations, No Imbeciles: Eugenics and the Supreme Court and Buck vs Bell*. Baltimore: Johns Hopkins University Press.

Lovejoy, A. O. (1936/1954). *The Great Chain of Being*. Cambridge: Harvard University Press.

Luban, D. (1983). Dark Times: Hannah Arendt's Theory of Theory. *Social Research, 50,* 215–248.

Luban, D. (1992). Justice Holmes and Judicial Virtue. *Nomos, 34,* 235–264.

Luban, D. (1994). Justice Holmes and the Metaphysics of Judicial Restraint. *Duke Law Journal, 44,* 449–523.

Luban, D. (1998). What's Pragmatic About Legal Pragmatism. In M. Dickstein (Ed.), *The Revival of Pragmatism*. Durham: Duke University Press.

Luban, D. (2010). The Rule of Law and Human Dignity: Reexamining Fuller's Canons. *Hague Journal on the Rule of Law, 29,* 1–16.

Lucey, F. E. (1941). Natural Law and American Legal Realism. *Georgetown Law Review, 30,* 493–533.

Lyons, D. (1993). *Moral Aspects of Legal Theory.* Cambridge: Cambridge University Press.

Lyons, D. (2010). *Ethics and the Rule of Law.* Cambridge: Cambridge University Press.

Machiavelli, N. (1513/1992). *The Prince*. New York: Knopf.

Machiavelli, N. (1520/1994). *The Art of War*. Chicago: Chicago University Press.

MacIntyre, A. (1990). *Three Rival Versions of Moral Inquiry*. Notre Dame, IN: University of Notre Dame Press.

Madden, E. H. (1963). *Chauncey Wright and the Foundations of Pragmatism*. Seattle: University of Washington Press.

Maier, P. (2011). *Ratification: The People Debate the Constitution 1877–1878*. New York: Simon and Shuster.

Maier, P. (2012). *American Scripture*. New York: Vintage Books.

Maine, H. S. (1861/1986). *Ancient Law*. New York: Dorset Press.

Makino, S., Gold, P. W., & Schulkin, J. (1994). Corticosterone Effects on Corticotropin-Releasing Hormone mRNA in the Central Nucleus of the

Amygdala and the Parvocellular Region of the Paraventricular Nucleus of the Hypothalamus. *Brain Research, 640,* 105–112.

Malthus, R. B. (1839/1970). *An Essay on the Principle of Population.* New York: Pelican.

Margolis, J. (2002). *Reinventing Pragmatism.* Ithaca: Cornell University Press.

Markowitsch, H. J., & Staniloiu, A. (2011). Neuroscience, Neuroimaging and the Law. *Corte, 47,* 1248–1251.

Marshall, J. (2000). *The Life of George Washington.* Indianapolis: Liberty Fund.

Marsur, L. (2015). *Lincoln's Last Speech.* Oxford: Oxford University Press.

Martin, A. (2009). Circuits in Mind: The Neural Foundations for Object Concepts. In *The Cognitive Neurosciences* (4th ed., pp. 1031–1045). Cambridge: MIT Press.

Martin, J. (2002). *The Education of John Dewey.* New York: Columbia University Press.

McDermott, J. J. (2007). *The Drama of Possibility: Experience as Philosophy of Culture.* Bronx: Fordham University Press.

McDowell, G. L. (2010). *The Language of Law and the Foundations of American Constitutionalism.* Cambridge: Cambridge University Press.

McEwen, B. S. (2007). Physiology and Neurobiology of Stress and Adaptation: Central Role of the Brain. *Physiological Reviews, 87,* 873–904.

McEwen, B. S., Gray, J. D., & Nasca, C. (2014). Recognizing Resilience: Learning from the Effects of Stress on the Brain. *Neurobiology of Stress, 1,* 1–11.

McGaugh, J. L. (2003). *Memory and Emotion: The Making of Lasting Memories.* New York: Columbia University Press.

McHugh, P. R., & Slavney, P. R. (1983/1989). *The Perspectives of Psychiatry.* Baltimore: Johns Hopkins University Press.

McPherson, J. M. (1975). *The Abolitionist Legacy.* Princeton: Princeton University Press.

McPherson, J. M. (1988). *Battle Cry of Freedom.* New York: Ballantine Books.

McPherson, J. M. (1991). *Abraham Lincoln and the Second American Revolution.* Oxford: Oxford University Press.

McPherson, J. M. (2008). *Tried by War: Abraham Lincoln as Commander in Chief.* Princeton: Princeton University Press.

Mead, G. H. (1934). *Mind, Self, and Society.* Chicago: University of Chicago Press.

Mead, M. (1964). *Continuities of Cultural Evolution.* New Haven: Yale University Press.

Mellars, P. (2006). Why Did Modern Human Populations Disperse from Africa 60,000 Years Ago? *Proceedings of the National Academy of Sciences of the United States of America, 103,* 9381–9386.

Menand, L. (2001). *The Metaphysical Club.* New York: Farrar, Straus, Giroux.

Menand, L. (2002). *American Studies.* New York: Farrar, Straus and Giroux.

Mencken, H. L. (1930). Review of the Dissenting Opinions of Mr. Justice Holmes. *The American Mercury,* 122–124.

Mencken, H. L. (1932). The Great Holmes Mystery. *The American Mercury.*

Mendenhall, A. P. (2009). Oliver Wendell Holmes Jr. and the Darwinian Common-Law Paradigm. *European Journal of Pragmatism and American Philosophy., 2,* 129–148.

Mendenhall, A. P. (2011). *The Emersonian Oliver Wendell Holmes.* The Literary Lawyer.

Mendenhall, A. P. (2013). Justice Holmes and Conservatism. *Texas Review of Law and Politics, 2,* 305–314.

Mendenhall, A. P. (2015a). Pragmatism on the Shoulders of Emerson: Oliver Wendell Holmes Jr.'s Jurisprudence as a Synthesis of Emerson, James and Dewey. *The South Carolina Review, 48,* 93–109.

Mendenhall, A. P. (2015b). *Oliver Wendell Holmes Jr., Pragmatism and the Jurisprudence of Agon.* Lewisburg: Bucknell University Press.

Mettrie, J. (1912). *Man a Machine.* Chicago: Open Court.

Meyen, G. (2014). Neurolaw. Ethics Theory Moral. *Practice, 17,* 8119–8129.

Midgley, M. (1989). *Wisdom, Information and Wonder.* New York: Routledge.

Milgram, S. (1992/1997). *The Individual in a Social World.* New York: McGraw Hill.

Milgram, S. (2009). *Obedience to Authority.* New York: Harper Perennial Classics.

Mill, J. S. (1843/1872). *A System of Logic* (2 Vols.). London: Longmans, Green and Co.

Mill, J. S. (1859/1974). *On Liberty.* New York: Pelican.

Mill, J. S. (1861/1987). *Utilitarianism.* New York: Penguin Press.

Mill, J. S. (1869). *1970 the Subjection of Women.* Cambridge: MIT Press.

Miller, P. (1939/1982). *The New England Mind of the Seventeenth Century.* Cambridge: Harvard University Press.

Miller, R. F. (2005). *Harvard's Civil War.* New Hampshire: University of New England Press.

Mirandola, G. P. D. (1486/1956). *Oration of the Dignity of Man.* Chicago: Gateway Edition.

Mithen, S. (1996). *The Prehistory of the Mind.* London: Thames and Hudson.

Mithen, S. (2006). *The Signing Neanderthal.* Cambridge: Harvard University Press.

Moll, H., & Tomasello, M. (2007). Cooperation and Human Cognition: The Vygotskian Intelligence Hypothesis. *Philosophical Transactions of the Royal Society of London. Series B: Biological Sciences, 362,* 639–648.

Moll, J., de Oliveira-Souza, R., Moll, F. T., Ignácio, F. A., Bramati, I. E., Caparelli-Dáquer, E. M., et al. (2005). The Moral Affiliations of Disgust: A Functional MRI Study. *Cognitive and Behavioral Neurology, 18,* 68–78.

Moll, J., Krueger, F., Zahn, R., Pardini, M., de Oliveira-Souza, R., & Grafman, J. (2006). Human Fronto-Mesolimbic Networks Guide Decisions About Charitable Donation. *Proceedings of the National Academy of Sciences of the United States of America, 103,* 15623–15628.

Moll, J., & Schulkin, J. (2009). Social Attachment and Aversion: On the Humble Origins of Human Morality. *Neuroscience and Biobehavioral Reviews, 33,* 456–465.

Monagan, J. S. (1988). *The Grand Panjandrum.* Lanham: University Press of America.

Moore, G. E. (1903/1968). *Principia Ethica.* Cambridge: Cambridge University Press.

Moreno, J. D. (1995). *Deciding Together.* Oxford: Oxford University Press.

Moreno, J. D. (1999/2003). Bioethics Is a Naturalism. In G. McGee (Ed.), *Pragmatic Bioethics.* Cambridge: MIT Press.

Moreno, J. D. (2003). Neuroethics: An Agenda for Neuroscience and Society. *Nature Reviews, 4,* 149–153.

Moreno, J. D. (2012). *The Body Politic.* New York: Bellevue Literary Press.

Moreno, J. D., & Farah, M. J. (2012). Neuroethics. *Science Progress, 3,* 13–18.

Moreno, J.D., & Schulkin, J. (2020, in press). *The Brain in Context: A Pragmatic Guide to Neuroscience.* New York: Columbia University Press.

Morse, S. J. (2008). Determinism and the Death of Folk Psychology: Two Challenges to Responsibility from Neuroscience. *Minnesota Journal of Law, Science & Technology, 9*(1), 1–36.

Morse, S. J. (2010). Lost in Translation: An Essay on Law and Neuroscience. In *Law and Neuroscience.* Oxford: Oxford University Press.

Morse, S. J. (2011). *Avoiding Irrational NeuroLaw Exuberance: A Plea of Neuromodesty.* Neuroethics Publications.

Morse, S. J. (2012). New Therapies, Old Problems, or a Plea for Neuromodesty. *American Journal of Bioethics, 3,* 60.

Morse, S. J. (2013a). Common Criminal Law Compatibilism. In N. A. Vincent (Ed.), *Neuroscience and Legal Responsibility*. Oxford: Oxford University Press.

Morse, S. J. (2013b). Brain Overclaim Redux. In *Law and Equality*. Minneapolis: University of Minnesota Law School.

Morse, S. J., & Roskies, A. L. (Eds.). (2013). *A Primer on Criminal Law and Neuroscience*. Oxford: Oxford University Press.

Mounce, H. O. (1997). *The Two Pragmatisms*. London: Routledge.

Murphy, E., & Greely, H. T. (2011). What Will Be the Limits of Neuroscience-Based Mind-Reading and the Law. In *Oxford Handbook of Neuroethics*. Oxford: Oxford University Press.

Nadel, L., & Sinnott-Armstrong, W. P. (2012). *Memory and Law*. Oxford: Oxford University Press.

Nagan, W. P. (1997). Not Just a Descending Trail: Traversing Holmes's Many Paths of the Law. *Florida Law Review, 49,* 463–482.

Nagel, T. (1970). *The Possibility of Altruism*. Princeton: Princeton University Press.

Nagel, T. (1991). *Equality and Partiality*. Oxford: Oxford University Press.

Neville, R. C. (1974). *The Cosmology of Freedom*. New Haven: Yale University Press.

Neville, R. C. (1978). *Soldier, Sage, Saint*. New York: Fordham University Press.

Newmyer, R. K. (2001). *John Marshall and the Heroic Age of the Supreme Court*. Baton Rouge: Louisiana State University Press.

Niebuhr, R. (1932/1960). *Moral Man and Immoral Society*. New York: Charles Scribner.

Nietzsche, F. (1878/1996). *Human All Too Human*. Cambridge: Cambridge University Press.

Nietzsche, F. (1886/1972). *Beyond Good and Evil*. Mineola: Dover.

Nietzsche, F. (1901/2017). *Will to Power*. New York: Penguin.

Norgren, R. (1995). Gustatory System. In *The Nervous System*. San Diego: Academic Press.

Novak, B. (2007). *Voyages of the Self*. Oxford: Oxford University Press.

Novick, S. M. (1989). *Honorable Justice: The Life of Oliver Wendell Holmes*. Boston: Little, Brown.

Novick, S. M. (1992a). Justice Holmes and the Art of Biography. *William and Mary Law Review, 33,* 1219–1248.

Novick, S. M. (1992b). Justice Holmes's Philosophy. *Washington University Law Review, 70,* 703–753.

Nussbaum, M. C. (1997). *Cultivating Humanity*. Princeton: Princeton University Press.

Nussbaum, M. C. (2000). Why Practice Needs Ethical Theory. In S. J. Burton (Ed.), *The Path of the Law and Its Influence: The Legacy of Oliver Wendell Holmes Jr.* Cambridge: Cambridge University Press.

Nussbaum, M. C. (2004). *Hiding from Humanity: Disgust, Shame and the Law*. Princeton: Princeton Law Review.

O'Hara, E. A. (2004). How Neuroscience Might Advance the Law. *Philosophical Transactions of the Royal Society of London. Series B, 359,* 1677–1684.

Oliver, K. M. (2015). *Earth and World*. New York: Columbia University Press.

Olson, A., & Phelps, E. A. (2007). Social Learning of Fear. *Nature Neuroscience, 9,* 1095–1101.

Paine, T. (1776/1996). *Common Sense and the Crisis*. New York: Anchor Books.

Papineau, D. (1993). *Philosophical Naturalism*. Oxford: Blackwell.

Pappas, G. P. (2008). *John Dewey's Ethics: Democracy as Experience*. Bloomington: Indiana University Press.

Pardo, M. S., & Patterson, D. (2013). *Minds, Brain and Law: The Conceptual Foundations of Law and Neuroscience*. Oxford: Oxford University Press.

Parker, K. (2003). The History of Experience: On the Historical Imagination of Oliver Wendell Holmes Jr. *Political and Legal Anthropology Review, 26,* 60–83.

Parrott, W. G., & Schulkin, J. (1993). Neuropsychology and the Cognitive Nature of Emotions. *Cognition and Emotion, 7,* 43–59.

Passingham, R. (1993). *The Frontal Lobes and Voluntary Action*. Oxford: Oxford University Press.

Patterson, D. M. (1990). Law's Pragmatism: Law as Practice and Narrative. *Virginia Law Review, 76,* 937–991.

Pearcey, N. R. (2001). Darwin's New Bulldogs: Scopes and American Legal Philosophy. *UL Review, 13,* 483–514.

Peirce, C. S. (1868). Questions Concerning Certain Faculties Claimed for Man. *Journal of Speculative Philosophy, 2,* 103–114.

Peirce, C. S. (1873/1959). Logic. In A. W. Burns (Ed.), *Collected Papers* (Vols. 7 & 8). Cambridge: Harvard University Press.

Peirce, C. S. (1877/1992). The Fixation of Belief. In C. Kloesel & N. Houser (Eds.), *The Essential Peirce: Selected Philosophical Writings Volume 1* (p. 115). Bloomington: Indiana University Press.

Peirce, C. S. (1878a). Deduction, Induction and Hypothesis. *Popular Science Monthly, 13,* 470–482.

Peirce, C. S. (1878b). Doctrine of Chances. *Popular Scientific Monthly, 12,* 604–615.

Peirce, C. S. (1898/1992). *Reasoning and the Logic of Things: The Cambridge Conferences Lectures of 1898 (Harvard Historical Studies)* (K. L. Ketner & H. Putnam, Eds.). Cambridge, MA: Harvard University Press.

Peirce, C. S. (1958). *Collected Papers.* Cambridge: Harvard University Press.

Perry, R. B. (1935). *The Thought and Character of William James* (Vol. 1 & 2). Boston: Little, Brown.

Peters, A., McEwen, B. S., & Friston, K. (2017). Uncertainty and Stress: Why It Causes Diseases and How It Is Mastered by the Brain. *Progress in Neurobiology, 156,* 164–188.

Peters, J. D. (2005). *Courting the Abyss: Free Speech and the Liberal Tradition.* Chicago: University of Chicago Press.

Petoff, A. (2015). Neurolaw: A Brief Introduction. *Iranian Journal of Neurology, 5,* 53–58.

Phelps, E. A., Delgado, M. R., Nearing, K. I., & LeDoux, J. E. (2004). Extinction Learning in Humans: Role of the Amygdala and vmPFC. *Neuron, 43,* 897–905.

Phelps, E. A., O'Connor, K. J., Cunningham, W. A., Funayama, E. S., Gatenby, J. C., Gore, J. C., et al. (2000). Performance on Indirect Measures of Race Evaluation Predicts Amygdala Activation. *Journal of Cognitive Neuroscience, 12,* 729–738.

Phillips, M. L., Young, A. W., Senior, C., Brammer, M., Andrew, C., Calder, A. J., et al. (1997). A Specific Neural Substrate for Perceiving Facial Expression of Disgust. *Nature, 389,* 495–498.

Pinker, S.J. (2011). *The Better Angels of our nature.* New York: Viking Press.

Plutarch, R. W. (1971). *Moral Essays.* New York: Penguin.

Pohlman, H. L. (1984). *Justice Oliver Wendell Holmes Jr.* Cambridge: Harvard University Press.

Poirier, R. (1992). *Poetry and Pragmatism.* Cambridge: Harvard University Press.

Pollock, F. (1912). *The Genius of the of Common Law.* New York: Columbia University Press.

Posner, R. A. (1985). Book Review of Kellogg and Pohlman Books on Holmes. *University of Chicago Law School, 53,* 870–875.

Posner R. A. (1988/2009). *Law and Literature.* Cambridge: Harvard University Press.

Posner, R. A. (1990a). *Cardozo: A Study in Reputation.* Chicago: University of Chicago Press.

Posner, R. A. (1990b). What Pragmatism Has to Offer the Law. *California Law Review, 63,* 1653–1670.

Posner, R. A. (1992). *The Essential Holmes.* Chicago: University of Chicago Press.

Posner, R. A. (1995). *Overcoming Law.* Cambridge: Harvard University Press.

Posner, R. A. (2000). Savigny, Holmes and the Law and Economics of Possession. *Virginia Law Review, 86,* 535–567.

Posner, R. A. (2002). Is Pragmatic Adjudication Inescapable? In *How Judges Think.* Cambridge: Harvard University Press.

Posner, R. A. (2003). *Law, Pragmatism and Democracy.* Cambridge: Harvard University Press.

Posner, R. A. (2007). *Economic Analysis of the Law.* Alphen aan den Rijn: Wolters Kluwer.

Posner, R. A. (2008). *How Judges Think.* Cambridge: Harvard University Press.

Postrema, G. J. (2011). Justice Holmes: A New Path for American Jurisprudence. In G. J. Postrema & E. Pattaro (Eds.), *Treatise on Legal Philosophy and General Jurisprudence. Volume 11 Legal Philosophy in the 20th Century: The Common Law World.* Dordrecht: Springer.

Pound, R. (1908). Mechanical Jurisprudence. *Columbia Law Review, 8,* 609–610.

Pound, R. (1921a). A Theory of Social Interests. *American Sociological Society, 15,* 16–45.

Pound, R. (1921b). Judge Holmes's Contributions to the Science of the Law. *Harvard Law Review, 34,* 449–453.

Power, L. A. (2009). *The Supreme Court and the American Elite.* Cambridge: Harvard University Press.

Power, M. L., & Schulkin, J. (2009). *The Evolution of Obesity.* Baltimore: Johns Hopkins University Press.

Pratt, S. L. (2002). *Native Pragmatism.* Bloomington: Indiana University Press.

Premack, D., & Premack, A. J. (1995). Origins of Human Social Competence. In M. S. Gazzaniga (Ed.), *The Cognitive Neurosciences* (pp. 205–218). Cambridge: MIT Press.

Previc, F. H. (2009). *The Dopaminergic Mind in Human Evolution and History.* Cambridge: Cambridge University Press.

Putnam, H. (1990). *Realism with a Human Face.* Cambridge: Harvard University Press.

Putnam, H. (2014). *Philosophy in the Age of Science.* Cambridge: Harvard University Press.

Quine, W. V. O. (1953/1961). *From a Logical Point of View*. New York: Harper & Row.

Quine, W. V. O. (1969). Epistemology Naturalized. In *Ontological Relativity and Other Essays*. New York, NY: Columbia University Press.

Quine, W. V. O., & Ullian, J. S. (1970). *The Web of Belief*. New York: Random House.

Rabban, D. M. (2013). *Law's History: American Legal Thought and the Transatlantic Turn of History*. Cambridge: Cambridge University Press.

Racine, E. (2010). *Pragmatic Neuroethics: Improving Treatment and Understanding of the Mind/Brain*. Cambridge: MIT Press.

Raine, A., et al. (2003). Corpus Callosum Abnormalities in Psychopathic Antisocial Individuals. *Archives of General Psychiatry, 60,* 1134–1142.

Rakic, P. (2000). Setting the Stage for Cognition: Genesis of the Primate Cerebral Cortex. In M. S. Gazzaniga (Ed.), *The New Cognitive Neurosciences*. Cambridge: MIT Press.

Rakoff, J. S. (2011). Science and the Law: Uncomfortable Bedfellows. *Seton Hall Law Review, 38,* 1379–1393.

Rakoff, J. S. (2016, May). Neuroscience and the Law: Don't Rush In. *The New York Review of Books*.

Rakove, J. N. (1990). *James Madison and the Creation of the American Republic*. New York: HarperCollins.

Rakove, J. N. (1996). *Original Meanings*. New York: Knopf.

Rakove, J. N. (2010). *The Revolutionaries*. New York: Houghton.

Randall, J. H. (1958). *Nature and Historical Experience*. New York: Columbia University Press.

Rawls, J. (1971). *A Theory of Justice*. Cambridge: Harvard University Press.

Raz, J. (1970/1980). *The Concept of a Legal System*. Oxford: Clarendon Press.

Reece, H. (1998). *Law and Science*. Oxford: Oxford University Press.

Reid, T. (1764/1997). *An Inquiry into the Human Mind*. Edinburgh: Edinburgh University Press.

Reimann, M. (1990). Nineteenth Century German Legal Science. *Boston College Law Review, 31,* 837–897.

Reimann, M. (2002). Lives in the Law: Horrible Holmes. *Michigan Law Review, 100,* 1676–1689.

Reisman, W. M., Wiessner, S., & Willard, A. R. (2007). The New Haven School: A Brief Introduction. *Yale Journal of International Law, 32,* 575–582.

Reyna, V. F., & Zayas, V. (2016). *Neuroscience of Risky Decision Making*. Washington, DC: American Psychological Association.

Reynolds, D. S. (1995). *Walt Whitman's America: A Cultural Biography*. New York: Knopf.

Ricardo, D. (1817/1973). *The Principles of Political Economy and Taxation*. London: Guernsey.

Richards, R. J. (2002). *The Romantic Conception of Life: Science and Philosophy in the Age of Goethe*. Chicago: University of Chicago Press.

Richardson, J. (2007). *A Natural History of Pragmatism*. Cambridge: Cambridge University Press.

Richardson, J. (2014). *Pragmatism and American Experience*. Cambridge: Cambridge University Press.

Richardson, R. D. (1995). *Emerson: The Mind on Fire*. Berkeley: University of California Press.

Richerson, P. J., & Boyd, R. (2005). *Not by Genes Alone*. Chicago: University of Chicago press.

Richter, C. P. (1942–1943). Total Self-Regulatory Functions in Animals and Human Beings. *Harvey Lectures, 38*, 367–371.

Rogat, Y. (1962). Mr Justice Holmes: A Dissenting Opinion. *Stanford Law Review, 3*, 254–276.

Rogat, Y. (1964). The Judge as Spectator. *University of Chicago Law Review, 31*, 213–256.

Rogat, Y., & O'Fallon, J. M. (1984). Mr Justice Holmes: A Dissenting Opinion. The Free Speech Cases. *Stanford Law Review, 36*, 1349–1406.

Rolls, E. T. (1999). *The Brain and Emotion*. Oxford: Oxford University Press.

Roosevelt, T. (1901/2010). *A History of the United States*. New York: HarperCollins.

Rorty, R. (1979). *Philosophy and the Mirror of Nature*. Princeton: Princeton University Press.

Rorty, R. (1982). *Consequences of Pragmatism*. Minneapolis: University of Minnesota Press.

Rorty, R. (1990a). *Philosophy and Social Hope*. New York: Penguin.

Rorty, R. (1990b). The Banality of Pragmatism and the Poetry of Justice. *Southern California Law Review, 63*, 1811–1819.

Rose, N., & Abi-Rached, J. M. (2013). *Neuro: The New Brain Sciences and the Management of the Mind*. Princeton: Princeton University Press.

Rosen, J. (2007). *The Supreme Court*. New York: Henry Holt.

Rosen, J. B., & Schulkin, J. (1998). From Normal Fear to Pathalogical Anxiety. *Psychological Review, 105*, 325–350.

Rosenberg, C. E. (2006). Contested Boundaries: Psychiatry, Disease and Diagnosis. *Perspectives in Biology and Medicine, 49*, 407–424.

Rosenberg, D. (1995). *The Hidden Holmes.* Cambridge: Harvard University Press.

Rosenblatt, R. (1975). Holmes, Peirce and Legal Pragmatism. *Yale Law Review, 84,* 1123–1140.

Rosenfeld, M. (1996). Pragmatism, Pluralism and Legal Interpretation: Posner's and Rorty's Justice Without Metaphysics Meets Hate Speech. *Cardozo Law Review, 18,* 97–152.

Roskies, A. L. (2007). Are Neuroimages Like Photographs of the Brain? *Philosophy of Science, 74,* 660–672.

Roskies, A. L. (2010). How Does Neuroscience Affect Our Conception of Volition. *Annual Review of Neuroscience, 33,* 109–130.

Roskies, A. L., & Morse, S. J. (2013). Neuroscience and the Law: Looking Forward. In S. J. Morse & A. L. Roskies (Eds.), *A Primer on Criminal Law and Neuroscience.* Oxford: Oxford University Press.

Rousseau, J. (1755/1984). *A Discourse on Inequality.* New York: Penguin.

Rousseau, J. (1769). *The Confessions.* Hertfordshire: Wordsworth Editions.

Rousseau, J. (1776/1980). *Reveries of a Solitary Walker.* New York: Penguin.

Rozin, P. (1976). The Evolution of Intelligence and Access to the Cognitive Unconscious. In J. Sprague & A. N. Epstein (Eds.), *Progress in Psychobiology and Physiological Psychology.* New York: Academic Press.

Rozin, P. (1998). Evolution and Development of Brains and Cultures: Some Basic Principles and Interactions. In M. S. Gazzaniga & J. S. Altman (Eds.), *Brain and Mind: Evolutionary Perspectives.* Strasbourg: Human Frontiers Science Program.

Sabini, J., & Schulkin, J. (1994a). Reconciling Social Constructions and Biological Realism. *Journal for the Theory of Social Behavior, 24,* 207–217.

Sabini, J., & Schulkin, J. (1994b). Biological Realism and Social Constructivism. *Journal for the Theory of Social Behavior, 224,* 207–217.

Sabini, J., & Silver, M. (1982). *Moralities of Everyday Life.* Oxford: Oxford University Press.

Salvato, G., Dings, R., & Reuter, L. (2014). Culture, Neuroscience and Law. *Frontiers in Psychology, 5,* 1–3.

Sandel, M. J. (2009). *Justice: What the Right to Do?.* New York: Farrar, Straus, Giroux.

Santayana, J. (1967). *Animal Faith and Spiritual Life.* New York: Appleton-Century-Crofts.

Sapolsky, R. M. (1992). *Stress: The Aging Brain and the Mechanisms of Neuron Death.* Cambridge: MIT Press.

Sapolsky, R. M. (2004). The Frontal Cortex and the Criminal Justice System. *Philosophical Transactions of the Royal Society of London. Series B, 359,* 1787–1796.

Sapolsky, R. M. (2017). *Behave*. New York: Penguin.

Sarokin, D. J., & Schulkin, J. (2016). *Missed Information*. Cambridge: MIT Press.

Savigny, R. C. (1831). *The Vocation of Our Age for Legislation and Jurisprudence*. Clark: The Lawbook Exchange.

Scanlon, T. M. (1998). *What We Owe Each Other*. Cambridge: Harvard University Press.

Scanlon, T. M. (2014). *Being Realistic About Reasons*. Oxford: Oxford University Press.

Schacter, D. L. (1996). *Searching for Memory*. New York: Basic Books.

Schlag, P. (2002). The Aesthetics of Common Law. *Harvard Law Review, 115,* 1005–1118.

Schneider, H. W. (1946/1963). *A History of American Philosophy*. New York: Columbia University Press.

Schoenbach, L. (2012). *Pragmatic Modernism*. Oxford: Oxford University Press.

Schulkin, J. (1982). *The Pursuit of Inquiry*. New York: SUNY Press.

Schulkin, J. (1991). Science and Human Rights. *The Journal of General Evolution, 32,* 243–253.

Schulkin J. (1992). *The Puruit of Inquiry*. New York: SUNY Press

Schulkin, J. (2000). *Roots of Social Sensibility*. Cambridge: MIT Press.

Schulkin, J. (2004). *Bodily Sensibility: Intelligent Action*. Oxford: Oxford University Press.

Schulkin, J. (2007). *Effort: A Behavioral Neuroscience Perspective on the Will*. Mahway: Erlbaum Press.

Schulkin, J. (2011). *Adaptation and Well Being*. Cambridge: Cambridge University Press.

Schulkin, J. (2012). *Naturalism and Pragmatism*. London: Palgrave Macmillan.

Schulkin, J. (2013). *Reflections on the Musical Mind*. Princeton: Princeton University Press.

Schulkin, J. (2015). *Pragmatism and the Search for Coherence in Neuroscience*. London: Palgrave Macmillan.

Schulkin, J. (2017). *The CRF Signal: Uncovering and Information Molecule*. Oxford: Oxford University Press.

Schulkin, J., McEwen, B. S., & Gold, P. W. (1994). Allostasis, Amygdala, and Anticipatory Angst. *Neuroscience & Biobehavioral Reviews, 18,* 385–396.

Schultz, A., & Luckman, T. (1973). *The Structures of the Life World*. Evanston: Northwestern University Press.

Schultz, W. (2002). Getting Formal with Dopamine and Reward. *Neuron, 36,* 241–263.

Schultz, W. (2004). Neural Coding of Basic Reward Terms of Animal Learning, Game Theory, Microeconomics and Behavioral Ecology. *Current Opinion in Neurobiology, 14,* 139–147.

Schutz, A. (1932/1967). *The Phenomenology of the Social World* (G. Walsh & F. Lehnert, Trans.). Chicago, IL: Northwestern University Press.

Schwartz, B. (1994). Holmes vs Hand: Clear and Present Danger or Advocacy of Unlawful Action? *The Supreme Court Review, 1994,* 209–245.

Schwarz, J. (1985). Oliver Wendell Holmes's "The Path of the Law". *American Journal of Legal History, 29,* 235–250.

Scott, T. R. (2012). Neuroscience May Supersede Ethics and Law. *Science and Engineering Ethics, 18,* 433–437.

Scull, A. (2015). *Madness in Civilization.* Princeton: Princeton University Press.

Seigfried, C. H. (1996). *Pragmatism and Feminism.* Chicago: University of Chicago Press.

Sellars, W. (1956/1997). *Empiricism and the Philosophy of Mind.* Cambridge: Harvard University Press.

Shappin, S., & Schaffer, S. (1985). *Leviathan and the Air Pump.* Princeton: Princeton University Press.

Shariff, A. F., Greene, J. D., Karremans, J. C., Luguri, J. B., Clark, C. J., Schooler, J., et al. (2014). Free Will and Punishment: A Mechanistic View of Human Nature Reduces Retribution. *American Psychological Science., 20,* 1–8.

Shen, F. X., Hoffman, M. B., Jones, O. D., Greeme, J. D., & Marois, R. (2011). Sorting Guilty Mind. *New York University Law Review, 86,* 1306.

Shenhav, A., & Greene, J. D. (2014). Integrative Moral Judgment: Dissociating the Roles of the Amygdala and Ventromedial Prefrontal Cortex. *Journal of Neuroscience, 34,* 4741–4749.

Sherman, N. (2005). *Stoic Warriors.* Oxford: Oxford University Press.

Sherratt, Y. (2013). *Hiltler's Philosophers.* New Haven: Yale University Press.

Shook, J. R. (2000). *Dewey's Empirical Theory of Knowledge and Reality.* Nashville: Vanderbilt University.

Shook, J. R. (2010). Dewey's Naturalized Philosophy of Spirit and Religion. In *John's Dewey's Philosophy of Spirit. With the 1897 Lecture on Hegel.* New York: Fordham University Press.

Shusterman, R. (1992). *Pragmatist Aesthetics.* Oxford: Blackwell.

Sidgwick, H. (1902/1988). *Outlines of the History of Ethics*. Indianapolis: Hackett.

Silk, J. B. (2007). The Adaptive Value of Sociality in Mammalian Groups. *Philosophical Transactions of the Royal Society, 362*, 539–559.

Simon, H. A. (1957). *Models of Man, Social and Rational: Mathematical Essays on Rational Human Behavior in a Social Setting*. New York: Wiley.

Simon, H. A. (1962). The Architecture of Complexity. *Proceedings of the American Philosophical Society, 106*, 470–473.

Simon, H. A. (1982). *Models of Bounded Rationality*. Cambridge: MIT Press.

Simon, H. A. (1996). *The Sciences of the Artificial*. Cambridge: MIT Press.

Simpson, G. G. (1949). *The Meaning of Evolution*. New Haven: Yale University Press.

Singer, P. (2015). *The Most Good You Can Do*. New Haven: Yale University Press.

Sinnott-Armstrong, W. (2008). *Moral Psychology: The Neuroscience of Morality: Emotion, Brain Disorders and Development*. Cambridge: MIT Press.

Sloane, C., & McKean, D. (2009). *The Great Decision*. New York: Public Affairs.

Smith, A. (1759/1974). *The Theory of Moral Sentiments*. Indianapolis: Liberty Classics.

Smith, J. E. (1970). *Themes in American Philosophy: Purpose, Experience and Community*. New York: Harper & Row.

Smith, J. E. (1978). *Purpose and Thought*. New Haven: Yale University Press.

Smith, J. E. (1992). *America's Philosophical Vision*. Chicago: University of Chicago Press.

Smith, J. E. H. (2011). *Divine Machines: Leibniz and the Sciences of Life*. Princeton, NJ: Princeton University Press.

Smith, K. S., & Graybiel, A. M. (2013). A Dual Operator View of Habitual Behaviors Reflecting Cortical and Striatal Dynamics. *Neuron, 2*, 361–374.

Smith, S. D. (1990). The Pursuit of Pragmatism. *Yale Law Review, 100*, 409–449.

Smith, S. R. (2012). Neuroscience, Ethics and Legal Responsibility. *Science and Engineering Ethics, 18*, 475–481.

Solymosi, T., & Shook, J. R. (2014). *Neuroscience, Neurophilosophy and Pragmatism: Brains at Work with the World*. London: Palgrave Macmillan.

Spencer, H. (1855). *Principles of Psychology*. London: Williams and Norgate.

Spinoza, B. (1668/1955). *On the Improvement of the Understanding*. New York: Dover Press.

Squire, L. R. (2004). Memory Systems of the Brain: A Brief History and Current Perspective. *Neurobiology of Learning and Memory, 82,* 171–177.

Stein, P. (1980). *Legal Evolution: The Story of an Idea.* Cambridge: Cambridge University Press.

Sterling, P. (2004). Principles of Allostasis: Optimal Design, Predictive Regulation, Psychopathology and Rational Therapeutics. In J. Schulkin (Ed.), *Allostasis, Homeostasis and the Costs of Physiological Adaptation.* Cambridge: Cambridge University Press.

Sterling, P., & Eyer, J. (1981). Biological Basis of Stress-Related Mortality. *Social Science & Medicine, 15,* 3–42.

Sterling, P., & Laughlin, S. (2015). *Principles of Neural Design.* Cambridge: MIT Press.

Stevens, W. (1997). *Collected Poetry and Prose* (F. Kermode & J. Richardson, Eds.). New York: The Library of America.

Strauss, L. W. (1970). *Natural Right and History.* Chicago: University of Chicago Press.

Stone, G. R. (2004). *Perilous Times.* New York: Norton.

Stone, M. (2002). Formalism. In J. Coleman & S. Shapiro (Eds.), *Jurisprudence and the Philosophy of Law.* Oxford: Oxford University Press.

Stuhr, J. (1997). *Geneological Pragmatism.* Albany, NY: SUNY Press.

Stuhr, J. (2016). *Pragmatic Fashions: Pluralism, Relativism and the Absurd.* Bloomington: Indiana University Press.

Sullivan, M. (2007). *Legal Pragmatism.* Bloomington: Indiana University Press.

Sullivan, M., & Solove, D. J. (2003). Can Pragmatism Be Radical? Richard Posner and Legal Pragmatism. *Yale Law Journal, 113,* 687–739.

Sullivan, M., & Solove, D. J. (2013). Radical Pragmatism. In A. Malachowski (Ed.), *The Cambridge Companion to Pragmatism.* Cambridge: Cambridge University Press.

Sunstein, C. R. (2003). *Why Societies Need Dissent.* Cambridge: Harvard University Press.

Sunstein, C. R. (2004). *The Second Bill of Rights.* New York: Basic Books.

Sunstein, C. R. (2005). *Laws of Fear: Beyond the Precautionary Principle.* Cambridge: Cambridge University Press.

Swanson, L. W. (2000a). What Is the Brain? *Trends Neuroscience, 23,* 519–527.

Swanson, L. W. (2000b). Cerebral Hemisphere Regulation of Motivated Behavior. *Brain Research, 886,* 113–164.

Talisse, R. (2014). Deweyan Pragmatism. In G. Hubbs & D. Lind (Eds.), *Pragmatism, Law and Language.* London: Routledge.

Tattersall, I., & Schwartz, J. (2001). *Extinct Humans.* Boulder, CO: Westview Press.

Taylor, J. S., Harp, J. A., & Elliott, T. (1991). Natural Psychologists and Neurolawyers. *Neuropsychology, 5*, 293–305.

Tenenbaum, J., Kemp, C., Griffiths, T., & Goodman, N. (2011). How to Grow a Mind: Statistics, Structure, and Abstraction. *Science, 331*, 1279–1285.

Tennison, M. N., & Moreno, J. D. (2012). Neuroscience, Ethics, and National Security: The State of the Art. *PLoS Biology, 10*, 1–4.

Thomas, B. (1997). *American Literary Realism and the Failed Promise of Contract*. Berkeley: University of California Press.

Thompson, E. (2015). *Waking, Dreaming and Being*. New York: Columbia University Press.

Thoreau, H. D. (1849). *Civil Disobedience*. Mineola: Dover Thrift Editions.

Thoreau, H. D. (1971). *Great Short Works*. New York: Harper & Row.

Tinbergen, N. (1951/1969). *The Study of Instinct*. New York: Oxford University Press.

Tomasello, T. (2014). *A Natural History of Human Thinking*. Cambridge: Harvard University Press.

Toulmin, S. (1950). *Reason in Ethics*. Cambridge: Cambridge University Press.

Toulmin, S. (1958). *The Uses of Argument*. Cambridge: Cambridge University Press.

Toulmin, S. (1972). *Human Understanding: The Collective Use and Evolution of Concepts*. Princeton: Princeton University Press.

Touster, S. (1982). Holmes a hundred Years Ago: The Common Law and Legal Theory. *Hofstra Law Review, 10*, 673–708.

Treanor, W. M. (1998). Jam for Justice Holmes. *Georgetown Law Review, 86*, 813–874.

Tsomo, K. L. (2012). Compassion, Ethics and Neuroscience: Neuroethics Through Buddhist Eyes. *Science and Engineering Ethics, 18*, 529–537.

Tushnet, M. (1977). The Logic of Experience: Oliver Wendell Holmes on the Supreme Court. *Virginia Law Review, 63*, 975–1052.

Tze, L. (1906/1973). *Treatise on Response and Retribution*. LaSalle: Open Court Press.

Urofsky, M. I. (2009). *Louis D. Brandeis*. New York: Pantheon Books.

Urofsky, M. I. (2015). *Dissent and the Supreme Court*. New York: Pantheon Books.

Vannatta, S. (2014). Pragmatism Without the Fighting Tag: Functional Realism in Holmes's Jurisprudence and Moral Philosophy. In G. Hubbs & D. Lind (Eds.), *Pragmatism, Law and Language*. London: Routledge.

Vannatta, S., & Mendenhall, A. (2016). The American Nietzsche? Fate and Power in the Pragmatism of Justice Holmes. *UKMC Law Review, 85,* 187–205.

Van Woodard, C. (1954/1966). *The Strange Career of Jim Crow.* Oxford: Oxford University Press.

Varela, F. J., Thompson, E., & Rosch, E. (1990). *The Embodied Mind.* Cambridge: MIT Press.

Vetter, J. (1984). The Evolution of Holmes, Holmes and Evolution. *California Law Review, 72,* 343–368.

Vico, G. (1699–1707/1993). *On Humanistic Education.* Ithaca: Cornell University Press.

Vico, G. (1709/1990). *On the Study Methods of Our Time.* Ithaca: Cornell University Press.

Vilares, I., Wesley, M. J., Ahn, W.-Y., Bonnie, R. J., Hoffman, M., Jones, O. D., et al. (2017). Predicting the Knowledge-Reckless Distinction in the Human Brain. *Proceedings of the National Academy of Sciences of the United States of America, 114,* 3222–3327.

Vincent, N. A. (2013a). Enhancing Responsibility. In N. A. Vincent (Ed.), *Neuroscience and Legal Responsibility.* Oxford: Oxford University Press.

Vincent, N. A. (Ed.). (2013b). *Neuroscience and Legal Responsibility.* Oxford: Oxford University Press.

Von Ihering, R. (1877/1914). *Law as Means to an End.* Boston: Boston Book Co.

Vul, E., Harris, C., Winkielman, P., & Pashler, H. (2009). Puzzling High Correlations of fMRI Studies of Emotion, Personality and Social Cognition. *Perspectives in Psychological Science, 4,* 274–290.

Vygotsky, L. (1926/1962). *Thought and Language.* Cambridge: MIT Press.

Wagner, R. H. (2002). The Falling Out: The Relationship Between Oliver Wendell Holmes Jr. and Theodore Roosevelt. *Journal of Supreme Court History, 27,* 114–137.

Wahman, J. (2015). *Narrative Pragmatism: An Alternative Framework for Philosophy of Mind.* London: Rowman and Littlefield.

Waldbauer, J. R., & Gazzaniga, M. S. (2001). The Divergence of Neuroscience and the Law. *American Bar Association, 41,* 357–364.

Warner, R. (1993). Why Pragmatism? The Puzzling Place of Pragmatism in Critical Theory. *University of Illinois Law Review, 3,* 535–550.

Warner, R. (2010). *Legal Pragmatism: A Companion to Philosophy of Law and Legal Theory.* New York: Wiley.

Weaver, W. G. (2003). The Democracy of Self Devotion: Oliver Wendell Holmes Jr. and Pragmatism. In A. Morales (Ed.), *Renascent Pragmatism: Studies in Law and Social Science.* Burlington, VA: Ashgate.

Weber, M. (1903/1917/1949). *The Methodology of the Social Sciences*. New York: Free Press.

Weber, M. (1920/1958). *The Protestant Ethic*. New York: Scribner.

Wegner, D. M. (2002). *The Illusion of Conscious Will*. Cambridge: Harvard University Press.

Weinberg, L. (1997). Holmes's Failure. *Michigan Law Review, 96,* 691–723.

Weissman, D. (1993). *Truth's Debt to Value*. New Haven: Yale University Press.

Weissman, D. (2000). *A Social Ontology*. New Haven, CT: Yale University Press.

Weissman, D. (2014). *Zone Morality*. Berlin: Walter de Gruyter.

Wellington, H. H. (1990). *The Supreme Court and the Process of Adjudication: Interpreting the Constitution*. New Haven: Yale University Press.

Wells, C. P. (1997). Old Fashioned Postmodernism and the Legal Theories of Oliver Wendell Holmes Jr. *Brooklyn Law Review, 63,* 59–84.

Wells, C. P. (2000). Holmes and William James. In S. J. Burton (Ed.), *The Path of the Law and Its Influence: The Legacy of Oliver Wendell Holmes Jr*. Cambridge: Cambridge University Press.

Wells, C. P. (2015). Oliver Wendell Holmes Jr. and the American Civil War. *Journal of Supreme Court History, 40,* 282–313.

Wells-Hantzis, C. (1988). Legal Innovation Within the Wider Intellectual Tradition: The Pragmatism of Oliver Wendell Holmes Jr. *Northwestern Law Review, 82,* 541–595.

West, C. (1989). *American Evasion of Philosophy*. Madison: University of Wisconsin Press.

West, R. L. (1985a). Jurisprudence as Narrative: An Aesthetic Analysis of Modern Legal Theory. *New York University Law Review, 60,* 145–211.

West, R. L. (1985b). Liberalism Rediscovered: A Pragmatic Definition of the Liberal Vision. *Pittsburgh Law Review, 46,* 673–738.

Westbrook, R. B. (1993). *John Dewey and American Democracy*. Ithaca: Cornell University Press.

Whewell, W. (1989). *Theory of Scientific Method*. Indianapolis: Hackett Publishing.

White, G. E. (1971). The Rise and Fall of Justice Holmes. *University of Chicago Law Review, 39,* 51–77.

White, G. E. (1986). Looking at Holmes in the Mirror. *Law and History Review, 4,* 439–465.

White, G. E. (1988–1989). Chief Justice Marshall, Justice Holmes and the Discourse of Constitutional Adjudication. *William and Mary Law Review, 30,* 131–148.

White, G. E. (1990). Holmes as Correspondent. *Vanderbilt Law Review, 43,* 1707–1760.

White, G. E. (1990). Holmes's Life Plan: Confronting Ambition, Passion and Powerlessness. *New York University Law Review, 65,* 1409–1479.

White, G. E. (1993). *Justice Oliver Wendell Holmes: Law and the Inner Self.* Oxford: Oxford University Press.

White, G. E. (1995). The Canonization of Holmes and Brandeis: Epistemology and Judicial Reputations. *New York Law Review, 70,* 576–621.

White, G. E. (1997). Investing in Holmes at the Millennium. *Harvard Law Review, 110,* 1049–1054.

White, G. E. (2000). *Oliver Wendell Holmes Sage of the Supreme Court.* Oxford: Oxford University Press.

White, G. E. (2016). *Law in American History.* Oxford: Oxford University Press.

White, J. W. (2015). *Lincoln on Law, Leadership and Life.* Naperville, IL: Source Books.

White, M. (1947). *Social Thought in America: The Revolt Against Formalism.* Boston: Beacon Press.

White, M. (2002). *A Philosophy of Culture: The Scope of Holistic Pragmatism.* Princeton: Princeton University Press.

White, R. (2015). *The Hidden God: Pragmatism and Posthumanism in American Thought.* New York: Columbia University Press.

White, S. E. (2008). Brave New World: Neurowarfare and the Limits of Humanitarian Law. *Cornell International Law Journal, 41,* 177–210.

Whitehead, A. N. (1925/1997). *Science and the Modern World.* New York: Free Press.

Whitehead, A. N. (1933). *Adventures of Ideas.* New York: Free Press.

Whitehead, A. N. (1938). *Modes of Thought.* New York: Free Press.

Whitehead, A. N., & Russell, B. (1910–1913). *Principia Mathematica.* Cambridge: Cambridge University Press.

Whitman, W. (1855/1965). *Leaves of Grass.* New York: Norton.

Wiener, P. P. (1948). The Pragmatic Legal Philosophy of N. St. John Green. *Journal of the History of Ideas, 9,* 70–92.

Wiener, P. P. (1949). *Evolution and the Foundations of Pragmatism.* Cambridge: Harvard University Press.

Wilshire, B. (2000). *The Primal Roots of American Philosophy.* University Park: Penn State University Press.

Wilson, E. (1994). *Patriotic Gore.* New York: W. W. Norton.

Wilson, J. (1993). *The Moral Sense*. New York: Free Press.

Winter, S. L. (2001). *A Clearing in the Forest: Law, Life and Mind*. Chicago: University of Chicago Press.

Wittgenstein, L. (1922). *Tractatus Logico-Philosophicus*. London: Routledge and Kegan Paul.

Wittgenstein, L. (1953/1958). *Philosophical Investigations*. New York: Macmillan.

Wollstonecraft, M. (1792/1982). *A Vindication of the Rights of Women*. Buffalo: Prometheus Press.

Wood, G. S. (1993). *The Radicalism of the American Revolution*. New York: Vintage.

Woodward, C. (1955/1966). *The Strange Career of Jim Crow*. Oxford: Oxford University Press.

Wright, C. (1877/1971). *Philosophical Discussions*. New York: Burt Franklin.

Wright, C. (1878). *Letters of Chauncey Wright with Some Account of His Life* (J. Thayer, Ed.). Cambridge: Press of John Wilson and Son.

Yang, Y., & Raine, A. (2009). Prefrontal Structural and Functional Brain Imaging Findings in Antisocial, Violent, and Psychopathic Individuals. *Psychiatry Review, 174*, 81–88.

Zak, P. J., Kurzban, R., & Matzner, W. T. (2005). Oxytocin Is Associated with Human Trustworthiness. *Hormones and Behavior, 48*, 522–527.

Index

The manufacturer's authorised representative in the EU is Springer
Nature Customer Service Centre GmbH, Europaplatz 3, 69115 Heidelberg,
Germany. If you have any concerns regarding our products, please
contact ProductSafety@springernature.com

Printed and bound by CPI Group (UK) Ltd, Croydon, CR0 4YY
29/04/2026
02099478-0013